MICROFLUIDICS
AND
MICROSYSTEMS

MICROFLUIDICS AND MICROSYSTEMS

Editor

Kenneth Charles

DPH

DISCOVERY PUBLISHING HOUSE
INDIA

Published by:

DISCOVERY PUBLISHING HOUSE
4383/4B, Ansari Road, Darya Ganj
New Delhi-110 002 (India)
Phone : +91-11-23279245; 23253475; 43596065
Mobile : +91 9811179893 / +91 9871656464
E-mail : discoverybooksindia@gmail.com
orderdphbooks@gmail.com
namitwasan9@gmail.com
web : www.discoverypublishinggroup.com

Microfluidics and Microsystems
Edited by: **Kenneth Charles**

International Standard Book Number: **978-81-19205-82-0** (Hardback)

PREFACE

Microfluidics has gained significant attention in recent years due to its potential in various fields, particularly in diagnostics and bioanalysis. This emerging field involves the precise control and manipulation of fluids at the microscale, offering advantages such as high-throughput processing, reduced sample volumes, and integration with other components in compact systems. It has proven to be well-suited for applications in clinical settings as well as point-of-care testing, where rapid and accurate results are crucial. One of the key contributions of microfluidics technology is its impact on the advancement of biology. By enabling researchers to manipulate and analyze fluids at the microscale, microfluidic platforms have facilitated breakthroughs in areas such as genomics, proteomics, cell biology, and drug discovery. These platforms allow for the miniaturization and parallelization of experiments, leading to improved efficiency, cost-effectiveness, and the ability to work with limited sample volumes.

In the fields of chemistry and biochemistry, microfluidics has opened up new avenues for research and development. The development of novel components and processes for fluid injection, mixing, pumping, and storage in microchannels has expanded the capabilities of microfluidic systems. Researchers can now perform precise chemical reactions, monitor reaction kinetics in real-time, and explore the behavior of fluids in confined spaces. This has led to advancements in areas such as chemical synthesis, drug formulation, and environmental analysis.

Despite these advancements, challenges still exist in the field of microfluidics. One such challenge is the introduction of samples into microchannels and the processing of a wide range of sample volumes. Ensuring reliable and efficient sample loading and handling is essential for accurate and reproducible results. Additionally, issues such as clogging, bubble formation, and sample evaporation can impact the performance of microfluidic devices. Overcoming these challenges requires innovative solutions and further advancements in nanotechnologies, which have the potential to enhance the functionality and reliability of microfluidic systems.

The development of microfluidics has also led to the emergence of microsystems, which are highly sensitive bioanalytical platforms capable of analyzing complex samples. These microsystems combine microfluidic components with other technologies, such as biosensors, microarrays, and imaging systems, to enable a wide range of applications. They have found applications in pharmaceutical research, biomedical diagnostics, environmental monitoring, and chemical engineering, among others. The integration of multiple functionalities within a compact and portable device has the potential to revolutionize the way we perform diagnostics and monitor health conditions.

The book "Microfluidics and Microsystems" serves as an introduction to the emerging field of microfluidics, offering insights into the latest research and developments in this area. It is specifically designed to benefit scientists and students who are working in the fields of MEMS (Microelectromechanical Systems) and microfluidics.

The book provides readers with a comprehensive overview of the discipline, highlighting its importance and potential applications. It covers various topics related to microfluidics, including the precise control and manipulation of fluids at the micro scale, the integration of microfluidic components with other technologies, and the advancements in nanotechnologies that have enhanced microfluidic techniques.

By delving into the research and developments in microfluidics, the book aims to equip scientists and students with a solid understanding of the field's fundamental principles and practical applications. It explores the advancements in fluid injection, mixing, pumping, and storage within micro channels, as well as the challenges that researchers face in sample introduction and processing.

With its focus on MEMS and microfluidics, the book offers valuable insights into the design, fabrication, and characterization of microsystems that incorporate microfluidic components. It emphasizes the significance of highly sensitive bioanalytical systems and their ability to analyze complex samples. The book also underscores the potential impact of microfluidics in various fields, including pharmaceutical research, biomedical diagnostics, and chemical engineering.

The editor has collected reliable and scientifically proven information from various credible sources to provide readers with the latest and most accurate information available. The book is a compendium of contributions from various experts in the field, and the editor expresses gratitude to these contributors for sharing their knowledge and expertise.

Editor

CONTENTS

1 CHAPTER

Overview of Materials for Microfluidic Applications

Emmanuel Roy, Antoine Pallandre, Bacem Zribi, Marie-Charlotte Horny, François Damien Delapierre, Andrea Cattoni, Jean Gamby and Anne-Marie Haghiri-Gosnet

Centre de Nanosciences et de Nanotechnologies, CNRS, Université Paris-Saclay, C2N Marcoussis, France

ABSTRACT

For each material dedicated to microfluidic applications, inherent microfabrication and specific physico-chemical properties are key concerns and play a dominating role in further microfluidic operability. From the first generation of inorganic glass, silicon and ceramics microfluidic devices materials, to diversely competitive polymers alternatives such as soft and rigid thermoset and thermoplastics materials, to finally various paper, biodegradable and hydrogel materials; this chapter will review their advantages and drawbacks regarding their microfabrication perspectives at both research and industrial scale. The chapter will also address, the evolution of the materials used for fabricating microfluidic chips, and will discuss the application-oriented pros and cons regarding especially their critical strategies and properties for devices assembly and biocompatibility, as well their potential for downstream biochemical surface modification are presented.

Keywords: *microfabrication; prototyping; manufacturing; thermoplastics; thermoplastics elastomers; assembly; bonding*

1. INTRODUCTION

Following Pasteur microbiology and Curie radiotherapy, the next coming and significant medical revolution is geared toward the foreseen convergence and integration of the following innovative science and technology components for the emergence of a new predictive, personalized, and preemptive medicine and practice. Bioengineering, microbiology, microfluidics, optics, and electrochemistry are today the most significant key enabler disciplines. Their complementary features lead to complex systems and democratization and wide public access to personalized and new medical products that will promote both wealth and innovation at the same time, as the IT industry did in the last decades. This technological trend is of tremendous impacts in both sociological and economical applications. For example, the needs of *in situ* personal monitoring are crucial for the comfort and the health surveillance for our specific aging modern society, and it has become a major concern for the health care system. It offers additional opportunities to lower the hospital workloads and expense. It also promotes enhanced therapy efficiency through real-time and accurate monitoring for proper therapy delivery. The advent of rapid molecular diagnostic systems shortening pathogenic genomic identification down to 1–2 h in an integrated manner, beyond the standard cellular culture identification that may extend from several days up to few weeks is one of the most promising paths for decentralized and personalized medicine. In addition, it is also a powerful asset to address major problems in bacterial antibiotic resistance. Beyond human health, rapid molecular diagnostic approaches deserve multiple segments from water quality control to veterinary and agronomy fields for widely accessible, deployable, and low-cost analytical systems. The others aspects of the emergence of microfluidic polymer systems are to act as template for dedicated and addressable microenvironments, thus promoting the organogenesis of organs reconstruction and substitution and sustaining the development of novel drugs that are cost effective.

Moreover, the microfluidic science and applications are by essence multidisciplinary. For a targeted application, a seamless and holistic vision is of tremendous importance. In this chapter, we aim to provide an overview of materials used for microfluidic applications. Their selection, evaluation, and integration at both prototyping scale and toward mass production are discussed. Due to the fact that some materials and strategy might be suited for specific goals yet not fit for others, the specific drawbacks and advantages regarding their status and potentials toward a path to products are analyzed.

2. MATERIALS FOR MICROFLUIDIC APPLICATIONS

From research development to microfluidic platform applications and product transfer, a holistic and seamless strategy has to consider from the start point to the end. The right selection of the material interface, which might be suited for certain targeted applications, can be detrimental for another purpose. Therefore, navigating material sciences for microfluidic technology is somehow a nebulous nightmare. These difficulties are also enhanced according to the fact that these microfluidic techniques were developed by the physics and engineering landscapes involved in information technology (IT) research and development (R&D), and therefore they exploited silicon-based clean room technologies. The lack of an adequate material culture related to biomaterials and polymer-based microstructure induced obvious limitations, which is a significant inhibiting factor for further industrial transfers. Besides the complexity and the interdisciplines involved in microfluidic applications, where bioengineering, microbiology, microfluidics, optics, materials science, and electrochemistry are concerned, proper and deep understanding and paradigms of each community are a must for any achievement of a significant system. Still at its early age, microfluidic science and technology have already made successful demonstrations addressing the issues; however, this knowledge is still mostly limited to chemistry, biology, and physics researchers; and unfortunately, it has not sufficiently penetrated yet the biomedical research and clinical environments. The goal of this chapter is thus to fill the gap and promote pertinent highlights on materials performance for microfluidic applications.

Legacies from microelectronics industry and the glassware history in biomedical and chemistry areas, both silicon and glass materials, constitute the initial materials for microfluidic device fabrication. Currently, materials dedicated to microfluidics can be categorized into three broad groups: inorganic, polymers, and paper. Beyond silicon and glass, inorganic materials extend over co-fired ceramics and vitroceramics. The second polymer-based category can be divided into two subcategories (i) thermoset materials, which are thermal or UV curable materials from a low viscosity precompound dispensed over a mold and (ii) thermoplastic materials, which are thermoformable materials amendable for rapid prototyping and manufacturing. Both polymer subcategories display rigid to elastomer mechanical properties, and through adaptable formulation and enriched chemical modification, offer a broad range of physicochemical surface properties. Finally, paper microfluidics is an emerging technology based on a patterning approach, where devices drive liquid through capillary actions via wicking in a cellulose matrix.

2.1. Inorganic materials: silicon, glass and ceramics

Silicon was the first material used for microfluidics [1]. Indeed, in the mid-80s, the microelectromechanical systems (MEMS), from which microfluidic technology is one of the branches that handle fluid, were developed by microelectronic technology. In MEMS, silicon material was the standard material interface. Fabrication of silicon and glass devices involves either subtractive methods (e.g., wet and dry etching) or additive methods (such as metal/dielectric/insulate deposition). Silicon surface chemistry based on the silanol group (-Si-OH) is also well developed and large panels of surface biochemical modification accomplished via silanes chemistry are accessible. Silicon displays a high elastic modulus (~150 GPa) and so, in such a way, active pumping and valving integration, as well as Si brittle characteristics induced overall fragility. Therefore the narrow windows of mechanical properties are limiting factors. Transparent to IR light, but not in visible range, can cause serious issues and limitations that are obvious for biological fluorescence-based optical detection methods and for direct fluid imaging. Those issues can be partially solved via several hybrid system approaches, where Si channels can be sealed with transparent materials, such as glass or polymers, leading to a renaissance for Si-based detectors for microfluidic systems. However, Si microfluidic systems, where reagent storage and other embedded biochemical surface functionalization (e.g., DNA, proteins and cells patternings) are required, feature limitations when considering final assembly and package steps. Indeed, high pressure, temperature, and voltage constraints induced complex strategies for overall device integration concerns. Current approaches such as reactive ion etching (RIE) for plasma exposure in order to activate surface for sealing purposes or high temperature, high pressure parameters for anodic bonding methods are definitively not suitable when considering reagent integration, such as proteins, cells, and nucleic acids species. Even recent efforts developed by Ruchi et al. [2], who reported a low-temperature bonding strategy and Si-Si interface bonding at low-voltage direct current of about 80 V, the procedure was still performed at a processing temperature of 365°C. For the aforementioned strategy, where a hybrid system is implemented, the definite air-tight and water-tight sealing achievements, which are also of priority for biomedical applications considering

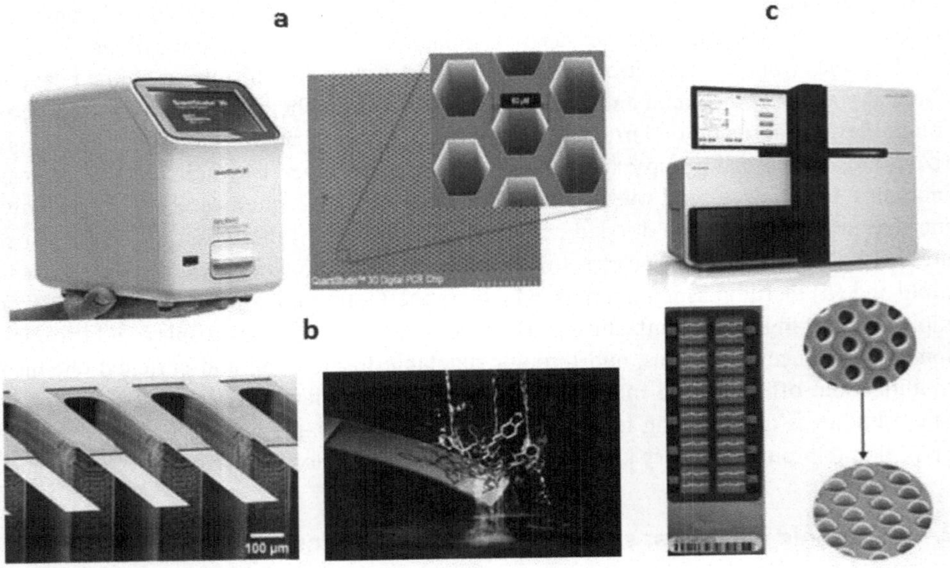

Figure 1. *(a) QuantStudionÔÔ3D Digital PCR system from Life Sciences and its silicon chip, 60 μm hexagonal wells organized in a honeycomb structure. (b) Array of Si cantilevers for label-free biodetection of illicit drugs in water. (c) Next generation sequencing HiSeq 2000v3 system developed by Illumina and the silicon chip for microbeads immobilization.*

contamination issues, suffer from the rather similar limitations. Indeed, hybrid approaches in most cases currently consist of realizing a pressure-free soft contact with the rigid silicon part that needs either oxygen plasma exposure or thermal treatment for sealing and device completion. However, at the research level, due to the high resolution of Si nanofabrication capabilities realized by either electronic beam or nanoimprint lithography among others, and its enriched surface chemistry means for biomolecular grafting, the following highly representative examples and significant achievements for the microfluidic community are highlighted. Applications related to the high-resolution capabilities of silicon extend over plasmonic, resonators, and microcantilevers systems. Indeed, high-quality nano-optomechanical resonators exhibiting mass sensitivity at the attogram level in the flow through operating environment have already been reported [3]. Si plasmonic microarrays have been interfaced for real-time and label-free monitoring of biomolecule interactions of A/G with immunoglobulin G (IgG) antibody [4]. Finally, as a tremendous achievement for the Si-based microfluidic approach, beyond the aforementioned limitations, we wish to underline the obvious advantages of the stationary Si-based system for complex digital qPCR platforms for genomic applications. Quantitative determination of pathogenic loads with such Si platforms definitely opens an area for medical research that benefits patients. In comparison to low-cost and single-use POC (Point-of-care) devices for decentralized purposes, where challenges involved with silicon materials imposed challenging technical needs, Si interface and glass are strategic materials of interests for stationary and highly complex systems. Figure 1(a) presents the QuantStudio® 3D Digital PCR system and the related silicon chip developed by Life Sciences Technologies Inc. (St. Petersburg, FL, US).

We also refer the readers to the silicon EWOD qPCR platform developed by Gidrol et al. [5] and for further interests to a recent review by V. Marx [6]. Finally, the silicon interface is the material of reference for the vast majority of nanomechanical biosensor [7] systems based on the cantilever approach as illustrated in Figure 1(b). Recently, an array of Si cantilevers for label-free biodetection of illicit drugs in water has been reported [8]. For next generation sequencing (NGS) purposes, Figure 1(c) displays the HiSeq 2000v3 system developed by Illumina Inc. (San Diego, CA, US) and the silicon chip realized for microbeads immobilization.

Besides Si, glass, due to its wide applications in chemistry has been adopted as a key material for lab-on-a-chip fabrication. Optically transparent with excellent and low-fluorescence background, glass also displays highly suitable chemical resistance. Thus it has emerged as a material of choice. Its transfer to microfluidics has been facilitated due to the traditional "off-chip" chemistry developed over several decades. Glass microfabrication involves well-established processes such as UV photolithography and chemical etching; however, such technologies rank relatively low when considering further manufacturing perspectives, compared to rapid thermal molding technology for polymer-based microfluidic system. It is for this reason the glass is generally viewed as an application material. Conversely, it may be easier and potentiality less cost effective to fabricate small number of microfluidic devices in glass than making molds for replication purposes using polymer-based systems. The chemical and thermal properties of glass enable a wide range of functions to be undertaken on the device, including the requiring aggressive and corrosive solvents, chemical agents, and others extreme temperature applications. Finally, assembly, conditioning, and on-chip reagent storage steps suffer similar limitations to the Si material. In particular, the glass fusion bonding is extremely slow and requires very high temperature. However, for high-pressure demanding applications associated with its strength, maintaining high level of channel integrity over operation, glass is definitively positioned at the preferred place. Compatible with electrophoretic-based separation techniques, microfluidic glass chips have been used for numerous demonstrators and coupled to either mass spectroscopy, electrochemical, or chemiluminescence detection means among others. From peptides to vascular biomarkers and DNA identification, we refer the reader to recent comprehensive reviews on advances for glass μCE chips [9, 10]. Glass mostly used the microfabrication approach that involves standard UV photolithography in combination with either dried or wet etching protocols. Frequently used wet etching chemical solutions feature a significant isotropic etchant property, therefore, the achievement of microchannels with straight and high aspect ratio structures are challenging. Alternative etching profiles such as anisotropic or straight-walled patterns can be done with dried RIE processes. Micromilling of glass is the second-most used approach. It is a subtractive process, where a rotating cutting tool removes material from a workpiece. Compared to micromilling of polymers, more attention for glass is required. This is primarily due to its high hardness and low tensile strength. Most milling is done using computer numerical control system (CNC micromachining), which represents a great deal of opportunity and flexibility in terms of pattern generation. However during processing, overheating of the diamond-coated drilling tool has to be considered, and cooling strategies therefore are implemented. Definitively, CNC micromachining offers a high level of interest for rapid prototyping of glass microchip, we refer the readers to the recently published micromilling tutorial review [11, 12].

Illustrating the high-performance capabilities of the glass interface for highly demanding applications, Figure 2(a) presents the microchip developed by Chan et al. [13] for the high-temperature synthesis of CdSe nanoparticles in nanoliter-volume droplets in a perfluorinated fluid carrier. Figures 2(b) and (c) highlight the recent efforts by Klearia (Marcoussis, France) and Dolomite (Royston, UK) companies for the development of low-temperature bonding protocols for embedded technologies. Finally, related to the development of NGS technologies, Figures 2(d) and (e) depict the glass chips made by Pacific Biosciences (Palo Alto, CA, USA) and Roche Diagnostics (Indianapolis, IN, USA) for their single molecule real-time sequencing technology (SMRT) and their 454 sequencing systems, respectively. Distributed by Weill Cornell Medical Researchers (Ithaca, NY, USA), each SMRT cell is provided at a cost of $ 400.

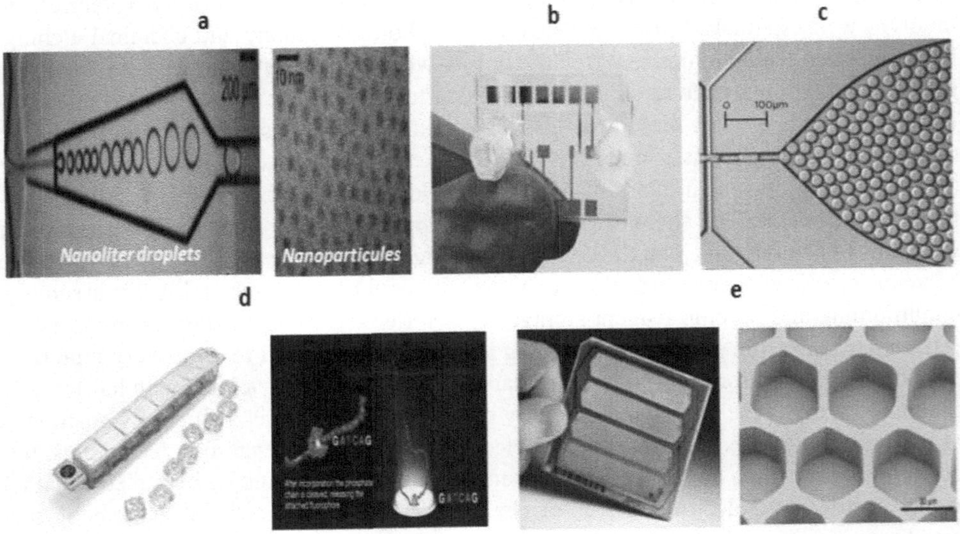

Figure 2. *(a) Glass chip for high-temperature (300°C) synthesis of CdSe nanoparticles from nanoliter-volume droplets. (b) Array of Si cantilevers for label-free biodetection of illicit drugs in water. (c) High demanding (500°C and 300 bar) glass chip for droplet generation form dolomite. (d) Pacific Biosciences single molecule Real-Time Sequencing glass chip. (e) Glass PicoTiter plate for DNA sequencing systems from 454 DNA sequencing system.*

Ceramic microfluidic platforms can be fabricated using low temperature co-fired ceramic (LTCC) technology for the achievement of hermetically sealed monolithic microfluidic platforms with homogeneous surface chemistry and physical properties through a pertinent cost-effective manner compared to Si and glass platforms. This is a well-established technology for low-cost and high-volume production of portable wireless electronic applications, but yet with limited involvement into the microfluidic landscape. However, its multilayer fabrication approach allows monolithic integration of complex structures. As a result, three-dimensional microfluidic channels and cavities have been already reported [14]. The compatibility with ink printing techniques has enabled the development of highly integrated devices that incorporate electrochemical detection as well as all the electronic

components for signal and data processing [15]. However, the integration of optical detection means in such platforms still constitutes a serious issue due to the overall opacity of ceramic material. To mitigate this issue, two strategies have been implemented: (1) the integration of optical fiber and (2) the implementation of transparent windows for localized optical analysis [16]. Recently, Microresist Technology Inc. (Berlin, Germany) developed a promising and organically modified ceramic Ormocer as an optically transparent and UV curable ceramic material to fix this issue.

2.2. Polymers: thermoset and thermoplastic materials

In the last 15 years, polymers have played the leading role from prototyping and manufacturing perspectives. This is because they offer a broad range of physical and surface chemical properties through adaptable formulation and enriched chemical modification. Based on their adhesiveness intrinsic properties, or activated bonding strategy, complex and 3D multilayered systems have been implemented. Considering both the cost of raw materials and manufacturing perspectives, several polymer interfaces, but none of all, are amendable to mass production processes (e.g., hot-embossing, injection-molding, and roll-to-roll). For the aforementioned thermoset subcategory, the most common materials are thermal and UV curable materials, Respectively, polydimethylsiloxane (PDMS) and the so-called SU-8 photoresist [17, 18]. On the thermoplastic side, the most popular materials are polycyclo-olefin (PCO), polycarbonate (PC), polytetrafluoroethylene (PTFE) and polystyrene (PS). Polycyclo-olefin offers high moldability and low water uptake [19, 20]. Polycarbonate [21] has excellent material toughness properties while polytetrafluoroethylene [22] and polyimide [23, 24] feature excellent chemical resistance, electrical, and thermal properties, respectively. Polystyrene has become more and more involved for cellular-based microfluidic systems, due to its wide applications in cell biology. Indeed, PS has provided decades of conclusions and sensitive protocol establishment, and numerous cell behaviors and functions have been determined [25]. Currently, PDMS and a dozen thermoplastic materials cover the vast majority of microfluidic research activities. The intensive use of PDMS is devoted to rapid prototyping and proof-of-concept demonstrations. However, despite its intrinsic drawbacks, mainly related to its cellular toxicity, molecular adsorption and absorption, gas permeability, and bonding issues, PDMS still appears as the preferred material in laboratories. Its industrial transfer is generally not overly recommended due to the cost concern. On the other hand, although thermoplastic interfaces offer the solution for product development, the entry cost in terms of required mold making as well as equipment setting limits its development. In addition, only well-established groups have significant infrastructures and facilities to afford it. To end this section, recently introduced soft thermoplastic elastomers (sTPE) are discussed. These materials bridge the gap between PDMS and classical rigid thermoplastic materials such as PC, PCO, and PS [48–52]. The sTPE materials are low-cost and commercially available polymers, offering transparency, biocompatibility, and flexibility compared to PDMS, and they can be rapidly thermoformed like currently used thermoplastics. Additionally, they display enhanced molding properties; indeed, low-pressure molding can be performed (e.g., <1–2 bar), lowering down the thermomechanical properties of the required molds. Due to their elastomeric properties, they can easily be peeled off from the mold. Assembling and bonding procedures are more convenient, and cohesive sealing can be done without thermal and/or plasma treatment. This material promotes a seamless integration, promoting therefore a realistic transfer path from prototyping to feasible and realistic industrialization. The sTPE material could become a mainstream platform for microfluidic technology and POC applications.

2.2.1. Thermoset

Thermosets are liquid or solid materials at room temperature. Their molecular polymer chains cross-link through a process called curing, when the polymer is heated or exposed to light or to others radiations. The curing step involves an irreversible chemical reaction. Therefore once formed, the thermoset parts cannot be reshaped anymore. Typical examples of thermoset polymers used in microfabrication are PDMS and SU-8 materials.

2.2.2. PDMS

First introduced in the late 90s, PDMS [26] is the most common microfluidic substrate, a large portion (40–45%) of research papers published on microfluidic devices utilizes this material [27]. Devices' molds are formed via traditional micromachining and photolithography means and a mixture of two liquid polymer components are mixed together and then casted, cured, and finally peeled off from the master. Due to its elasticity, integration of microvalves or micropumps is possible, and complex 3D system and numerous point-of-care devices have been developed for research applications. PDMS displays excellent optical properties for convenient fluorescence detection and fluid imaging. Due to its gas permeability, PDMS appears as a material of choice for many but not all cellular studies.

On the other hand, four important properties of PDMS have negative impacts: (1) channel deformation due its high mechanical compliance, (2) evaporation and absorption, (3) leaching-out of uncrosslinked oligomers, and (4) hydrophobic recovery [25]. The compliance issue is particularly true considering cell culture experiments that require accurate control over shear force on the endothelial monolayer, the inability to account for mechanical deformation bias in data analysis and subsequent data interpretation [28]. Oxygen permeability in PDMS is three orders of magnitude higher as compared to PS, and may in fact produce a hyperoxic microenvironment leading to cellular stress [29, 30]. Water vapor resulting from the permeable PDMS also leads to problematic shifts on volumes, concentrations, and chemical balances [31]. PDMS is also largely prone to bulk absorption of hydrophobic compounds. Regehr et al. [32] have shown significant depleted estrogen levels in culture media, leading to inhibition of protein-1 activator. Similarly, important shifts were identified in response to fluoxetine over transfected human embryonic kidney (HEK) in between PDMS and polystyrene interfaces [33]. Finally, absorption not only affects fundamental cellular biology on chip, but also drug discovery and high-throughput screening applications. However, the PDMS abilities to provide rapid (1–4 h), easy, low-cost (50–200 $/kg) and straightforward accessibility make it a major leading player for prototyping uses at academic level. However, its commercial applications are generally avoided. A noticeable exception emerges from this mainstream Eration, indeed around the multilayered PDMS pressure-actuated crossed-channel valve architecture initially developed by Quake et al. [34]; the Fluidigm company (San Francisco, CA, USA) has developed several impressive applications ranging from qPCR platforms to mass cytometer as depicted in Figure 3(a) and (b). Such system can run 48 samples in 770 reaction chambers. Each chip shown in Figure 3(c) ranges at a cost from $ 400 to 800. This price highlights the intrinsic and overall PDMS difficulties and its lengthy processing steps to tackle for low-costs microfluidic system for single use perspective. We refer the readers to the following reviews for extended discussion related to microfluidic products development and the ambiguous positioning of PDMS material in the community for applications development [25, 35, 36].

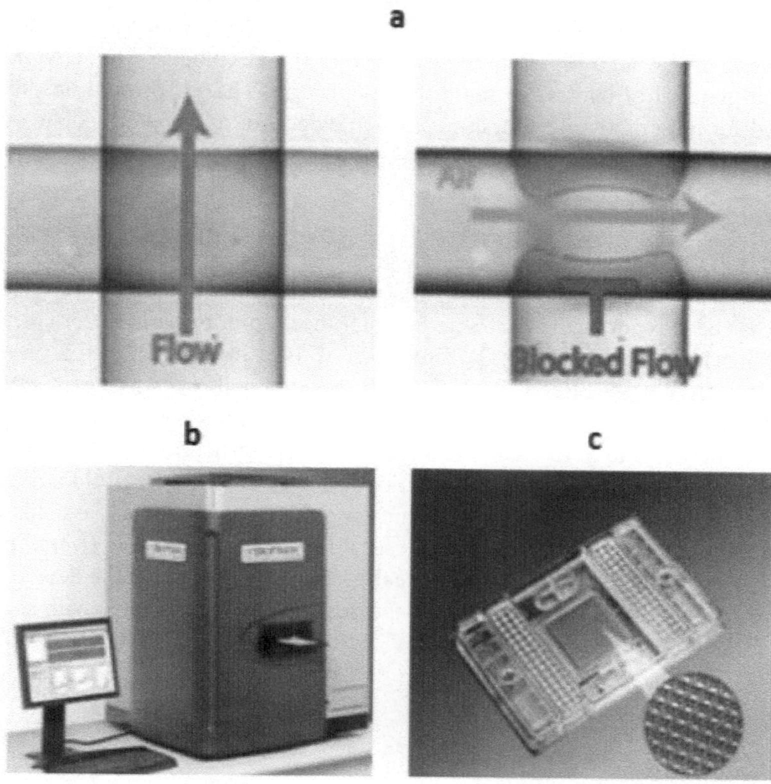

Figure 3. *(a) Bilayer PDMS pneumatic valving microfluidic system, liquid flow inside the vertically oriented top channel and the bottom channel (air flow) is pressure-actuated for clogging/or liquid motion of the fluidic layer. (b) and (c) Biomark HD system for digital qPCR from Fluidigm, which run 48 samples in 770 reactions/sample chip.*

2.2.3. SU-8: an epoxy-based material

Using classical means of photolithography or stereolithography, SU-8 microfluidic devices can be built in a complex 3D structure out of an initial liquid resist. SU-8 contains eight epoxy groups which undergo a very strong crosslinking upon exposure to UV light. Such composition and process lead to highly mechanically, thermally, and chemically stable materials. However, large internal stress exists due to the process, providing thus an overall brittle characteristic, and therefore making it difficult in handling and transferring part. SU-8 can be deposited and patterned in a range of thicknesses from nanometer to millimeter films, using lithography (either UV and e-beam), the lateral feature resolution extends from the macroscopic surface down to submicrometer. SU-8 can be structured also from the laser ablation approach. However, as for photolithography means, it appears that both methods have limited throughput [37]. Indeed, the cycle for resin preparation and others processes are lengthy. It seems that SU-8 is limited mainly to academic uses. For in-depth descriptions of the SU-8 interface for microfluidic applications, we refer the reader to a couple of comprehensive and exhaustive reviews [18, 38]. Beyond its use for direct microfluidic device fabrication, SU-8, due to its high mechanical strength and its capabilities to create high aspect ratio structures [39] and complex 3D networks [40], is definitely a material of choice for mold master making. The SU-8 mold has been intensively used for hot-embossing of thermoplastic devices, or as an intermediate system for injection-molding [41].

2.2.4. Alternative thermoset material

Besides PDMS soft thermoset material, several attempts have been introduced in order to provide alternatives for soft thermoset materials and mainly the use of fluorinated-based polymers has been reported. The attractiveness of those solutions arises from the inertness of the perfluorinated compounds. De Simone et al. [42] have developed a photocurable soft perfluoropolyethers (PFPE). It exhibits low toxicity and low surface energy and displays enlarged chemical resistance. For multiphasic microfluidic environment, it might be a material of choice due to the fact that such Teflon-like structure is both oleophobic and hydrophobic. More recently, high aspect ratio (up to 6.5) PFPE microfluidic devices have been fabricated by a direct photolithographic process. Through a mask-assisted photopolymerization approach, the authors have successfully developed a rapid overall process of around 5 min and demonstrated important sealing capabilities, indeed the device can withstand a pressure up to 3.8 bar [43]. Finally, the devices have been tested with some model reactions employing organic solvents.

2.2.5. Thermoplastic polymers

2.2.5.1. Rigid thermoplastics

According to the aforementioned drawbacks of PDMS, intensive use of thermoplastic materials, such as polycarbonate and polystyrene, is increasing. These materials are amendable for rapid thermoforming manufacturing technology in the CD and biology industries. Those platforms have been clearly identified as materials of choice for microchip research and subsequent technology transfer. Complementary polymethylmethacrylate (PMMA) and polyimide, due their favorable position in the semiconductor industry, complete the set of foreseen thermoplastic candidate. More recently, polycyclo-olefin polymer has become another popular substrate for microfluidics. It displays high chemical resistance, low water absorption, and excellent optical transparency in the near UV range. The materials are moldable polymers when heated above their glass temperature, offering thus recyclable and reshaping perspectives. They also provide a subsequent bonding pathway. They are optically clear and commercially available, and they display slightly better solvent compatibility than PDMS. However, they are incompatible with most organic solvents, such as ketones and hydrocarbons. Currently used thermoplastic interfaces are rigid

and stiff materials with Young modulus in the range of Giga Pascal. Consequently, they are not as convenient as PDMS to achieve conformal contact. Therefore, sealing strategies involve pressurized solvent-assisted and thermal approaches to melt surfaces. In addition, the following thermoplastics: perfluoroalkoxy (Teflon PFA) and fluorinated ethylenepropylene (Teflon FEP) can be used for extremely inert microfluidic devices. Ultimately, they feature nonstick and antifouling properties. PFA has been used for high-quality immobilization of *Escherichia coli*, *Pseudomonas putida,* and *Bacillus subtilis* cells in highly dense microarray patterns [44, 45]. For more classically used interfaces, Zhang et al. [46] reported sealing and chemical surface modification of an integrated monolithic PMMA microdevice for DNA purification and amplification of *E. coli*. Also, a highly integrated polystyrene microfluidic chip coupled to electrospray ionization mass spectrometry for on-chip protein digestion and online analysis was developed [47]. One of the most important challenges faced when targeting for molding of thermoplastic parts is the realization of the master cavity featuring submicron resolution, high aspect ratio, and densely packed structure areas. Most of the materials are crystalline and/or semi-crystalline, and they display high coefficient of thermal expansion (CTE) and high shrinkage parameters compared to amorphous ones. Additionally, due to the fact that they are rigid and often brittle, they represent sensitive challenges from the manufacturing perspectives. The high shear force resulting from the pressurized environment generates asymmetric and/or random pull-off of plastic edges due to friction in demolding, which is downstream detrimental for sealing and bonding. The characteristics of the thermoplastic materials induce high specifications on the thermo-mechanical properties of the master molds. Currently, the impacts on the master mold making related to the realization of such thermoplastic parts are only metallic (stainless, nickel-cobalt alloy, and aluminum), and from prototyping perspectives, only few epoxy molds can be employed. Even if elegant CNC laser machining and electroplating processes are available for mold making, the implementation of such master are expensive and are limited in terms of resolution, pattern density, and aspect ratio. The work suggest an overall in-depth consideration when starting to envisage a microfluidic system development, undoubtedly a holistic approach should be taken. The overall chip constraints in terms of fabrication, sealing, packaging, thermomechanical loads, biological compatibility (both in terms of physicochemical surface properties and reagent integration), biomicrofluidic functions, microfluidic statistical, and robust properties have to be considered.

2.2.5.2. Soft thermoplastic elastomers: sTPE

The sTPE is a class of material in which elastomeric properties of elastomer rubber (e.g., PDMS) are embodied with the ease of processing of thermoplastic materials such as PMMA, PCO, and PC. The sTPE thus bridges the gap in between thermoplastics and elastomers, enhancing the advantages of each material. Moreover, the range of sTPE Young modulus extends continuously from 0.1 MPa to 1–5 GPa. The combination of elastomeric and thermoplastic properties makes these materials potential substitutes to PDMS and/or hard TP polymers that are commonly employed in microfluidics [48]. Unlike PDMS, sTPE can be used in the form of extruded sheets that provide off-the-shelf availability without the need of performing any precompounding step. Extruded films can be stored over a long period of time (e.g., several years) without any notable degradation, making it possible to use the material on demand at any time (Figure 4(a)) [49, 50]. The sTPE materials are available at low cost, and they display optical transparency and biocompatibility for proteins, nucleic acids, and cell and tissue engineering and diagnostics [49–53]. sTPEs are block copolymers comprising different monomer sequences that are distributed randomly or statistically in domains through diblock or triblock architectures [51]. For styrenics-based sTPE materials with low PS content (10–12%), thermodynamic incompatibility between blocks induces nanophase separation and self-assembly of PS domains into nanometric clusters (typically 10–30 nm in diameter) that are distributed in a three-dimensional fashion within a hexagonal symmetry in the rubber matrix of ethylene-butylene (EB). This morphology provides the basis of the material performance: rigid PS domains act as junction points that stabilize the polymer matrix while the EB-dominant phase offers elastomeric properties. Moreover, size and cluster distributions promote the sTPE surface to be uniform and homogenous at the microfluidic device level [49–53]. As block copolymer materials, sTPE exhibits two glass transition temperatures corresponding to the EB soft block ($T_{g,EB} \sim -60$ to -75°C) and to the styrene rigid block ($T_{g,EB} \sim -90$ to 105°C), respectively. The negative value of $T_{g,EB}$ predicts liquid-like behavior of the materials.

Indeed, like PDMS, which similarly displays a negative glass transition temperature, the selected sTPE forms a spontaneous and conformable close contact with flat substrates, generating tight air and water sealing [49–53]. Additionally, the soft blocks provide bonding capabilities above their glass transition, which implies that even at room temperature, the polymer chains can be reorganized according to the contact surface. The bonding strength is variable and dependent on time and temperature; and irreversible bonding was obtained at room temperature [50, 52]. Depicting the enhanced molding capabilities of sTPE interface, Brassard et al. [51] have demonstrated the rapid and reliable patterning of open through-hole microstructures in sTPE material using a method based on hot-embossing (Figure 4(b)). The sTPE-based 3D microfluidic patterning device was then used for the immobilization of up to 96 different biological probes in a 10×10 array format of 50 μm×50 μm spots (Figure 4(c)).

Additionally, for novel tissue engineering biomaterial platform, high molding performances have been confirmed in a newly reported process. We reported a rapid microfabrication of a biocompatible sTPE sheet in an overall 3 min process operating within an ultralow applied pressure (1.6 bar) [52]. Smooth muscle cells contact guidance studies have been conducted over an array of 4-μm-patterned grooves [52]. For reader's information, contributing to the establishment of the biocompatibility, the bonding and the microfabrication performance of sTPE, which are highly dependent on block copolymers formulation (molecular weight of each block of each diblock (DB) and triblock (TB) and the DB/TB ratio) and also to the additives composition (tackifiers and processing agents); we underline the Flexstene sTPE materials performance (InfineFlex Inc., San Diego, US and Blackholelab Inc., Paris) for the fabrication of adjacent micropillar arrays with different heights for cellular studies [53]. We also demonstrated that sTPE can be used as a rapid technique for the fabrication and assembly of pneumatically driven valves in a multilayer microfluidic device using a simple SU-8 mold material for embossing purposes (Figure 4(d)) [54]. The quality of the obtained soft thermoplastic valve shows a robust behavior with an opening–closing frequency of 5 Hz. Finally, more recently [49], we demonstrated the implementation of a sTPE CD-like microfluidic system for genomic assay. This device integrates all required molecular assay steps, from cellular lysis to gDNA polymerase chain reaction amplification, amplicons digestion, and microarray hybridization on a plastic support. The

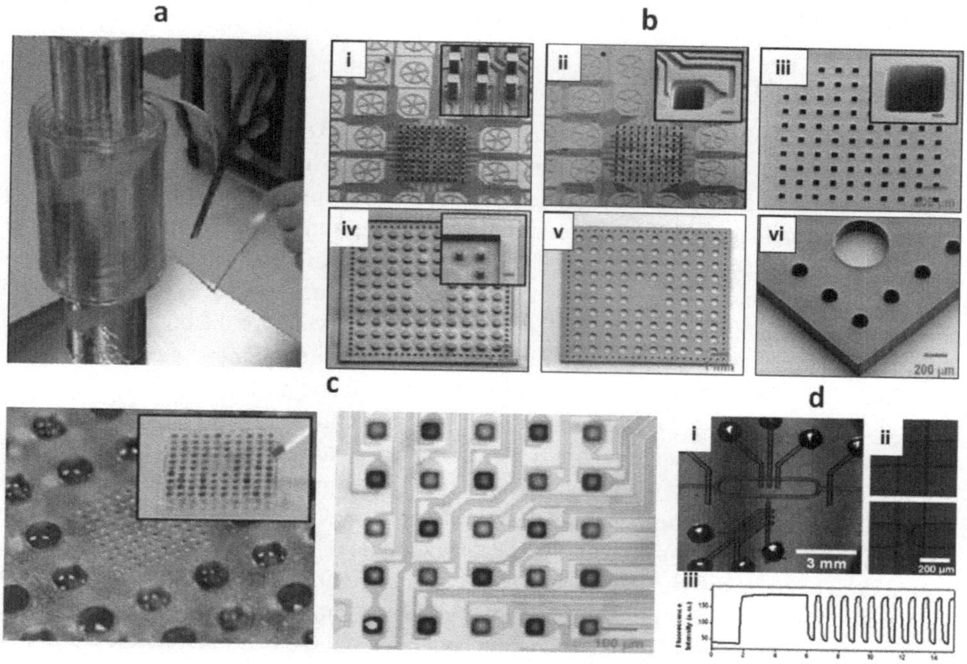

Figure 4. *(a) Photograph of an extruded flexestene foil on a roll from which pieces can be cut conveniently before use. (b) Series of SEM images illustrating the fabrication of the microfluidic flexestene device. (i) SU-8 embossing mold used for the fabrication of the bottom flexestene membrane, (ii) Upper and (iii) lower side of the bottom sTPE membrane, (iv) SU-8 mold used for the fabrication of the top sTPE membrane, (v) overview, and (vi) close-up view of the top flexestene membrane. Scale bars in the insets of (i), (ii), (iii), and (iv) correspond to 50, 20, 10, and 200mm, respectively. The images shown in (iv) and (v) were assembled from several SEM micrographs to achieve the desired field of view. (c) Photographs of an assembled 3D microfluidic immobilization after filling with solutions of a red and green dye (left), optical microscope image of the resulting red-green pattern obtained on the central region of the microfluidic device (right). (d) high-throughput fabrication method of sTPE multilayered microfluidic devices, (i) Optical micrograph of the whole micromixer made of two layers of sTPE material bonded on a poly(cyclo)olefin polymer substrate. The device size is smaller than a centimeter square, and each valve measures 200 μm×200 μm, (ii) detailed view of a valve at both open and closed positions; and (iii) curve representing the valving cycle at 1.2 Hz, fluorescence intensity under the deflected membrane is registered, it is maximal when the valve is open and minimal when closed.*

low-temperature, pressure-free assembly and bonding of sTPE material on the flat polyclo-olefin thin substrate offer a pertinent solution for simple and efficient loading and storage of the required on-CD board elements. This was demonstrated through the integration and the conditioning of microbeads, magnetic discs, and dried enzyme species. This work highlights a seamless strategy that promotes a feasible path to transfer from prototyping toward realistic industrialization. This work aims to establish the full and pertinent potential for sTPE centrifugal system as a mainstream microfluidic diagnostic platform for clinical molecular diagnostics, water and food safety, besides other applications.

2.3. Paper, biodegradable and hydrogel materials

2.3.1. Hydrogels

Hydrogels display a molecular architecture analog to extracellular matrix, with water uptake properties up to 80% of its total mass. Hydrogels are highly porous and thus an excellent matrix for cellular biology studies. However, direct tissue engineering from bulky hydrogels are challenging due to the restricted depth for nutriments diffusion of around hundreds micrometers [55]. Microfluidic technology, through both top-down and bottom-up approaches, has demonstrated its abilities to tackle this fundamental issue. From a top-down approach, microchannels are fabricated inside the hydrogel while for the bottom-up, hydrogels filled microchannels cavities. Matrigel™ and collagen are the mostly used animal-derived hydrogels. They promote excellent cellular adhesion and proliferation [55]. Recently, Bang et al. [57] engineered a 3D neural circuit in a microfluidic Matrigel hydrogel system. They had grown, aligned, and organized 3D networks of axon bundles at an average speed of 250 $\mu m.d^{-1}$ for a period of 6 *in vitro* days. Alternatively, alginate and agarose plant-derived hydrogels and synthetic ones, such as polyacrylamide or polyethylene glycol (PEG), can be used [58]. Even though synthetic hydrogels slightly lack cellular adhesion compared to animal-derived ones, they nevertheless promote higher flexibility and enriched formulation adaptability for fine tuning objectives. Hydrogel composition, structure, morphology, and rigidity have been used in a high-throughput manner in droplet-based microfluidics [59]. Recently, agarose hydrogels have been integrated in a microfluidic system for *E. coli* purification and concentration, and finally for fluorescence immune detection. Authors reported that 90% recovery efficiency could be achieved with a million-fold volume reduction from 400 μL to 400 pL. For concentration of 1×10^3 cells mL^{-1} bacteria, approximately ten million-fold enrichment in cell density was realized. Urine and blood clinical isolates were further tested and validated [60]. We refer the readers to follow the review for further reading [61].

2.3.2. Biodegradable materials

Biodegradable polymers for tissue engineering and drug delivery purposes display degradation time ranging from 24 h to several months. They offer, in a microfluidic format, a promising opportunity for microstructured tissue scaffolds. Commonly used biodegradable matrices are polycaprolactone, poly(lactide-co-glycolide), and polyglycolic acid (PGA). Their degradation and mechanical properties are tunable, and they display minimal changes in the systemic immune responses. Their degradation products such as glycolic acid for PGA are through metabolite response absorbed by the hosted living body. Curing of these materials takes place as their dimer version polymerizes via a ring-opening reaction under appropriated heating and catalytic steps. An excellent review related to biodegradable material properties and their microenvironment integration has been recently published [62]. Various technologies, such as printing, soft-lithography, stereolithography, hot-embossing, and injection-molding methods, have been used toward integration [63, 64].

2.3.3. Paper

Paper is the most recently introduced microfluidic material. Its cellulose matrix acts to wick liquids while specifically, hydrophobically, patterning areas to avoid liquid motion.

The patterned barriers define the shape (i.e., width and length) while the thickness of the paper accounts for the height. Hydrophilic wicking regions thus serve as channel networks opened to air. Paper as chip material is one of the cheapest materials, and it can be easily stacked in 3D devices [65]. The fabrication approaches can be divided into two groups. Lithographic-based methods, where particular coated polymer areas are removed, thus are forming the channels. Second, the printing and the cutting approaches allow direct hydrophobic barrier definition without exposure of the effective channels to any reagents. Fundamentally, paper-based microfluidic systems are not suitable for large-volume samples and their applicable detection perspectives are limited owing to the intrinsic cellulose matrix properties. Recently published by Mace et al. [66], paper-based diagnostic devices and their manufacturing perspectives are commented in detail. Wax printing approaches and assembly of 3D vertical flow assays are further discussed therein. Concerning printing and cutting approaches, the reported channel resolution is quite low and limited to 200 μm. Paper-based technology, displays undoubtedly tremendous limitations considering heterogonous component integration such as valves and other reagent storage issues. Recently, Thom et al. [67] have performed the integration of light-emitting diodes (LEDs) onto paper microfluidic device for the fluorescence detection of β -d-galactosidase. From medium to low complexity bioassay integrations, paper-based technology appears to be a promising pathway for portable and low-cost platform in the future, and thus a material of choice for optimized and multiplexed lateral flow assays in the health care segment. We refer the readers to follow recently published review for further reading on paper-based microfluidic system for bioanalytical applications [68, 69].

3. CONCLUSION

This chapter presents an overview of materials for microfluidic applications and their applications in recent research. The large range of materials dedicated to microfluidics is a key component for successful microfluidic applications. The optimal selection of an adequate material platform for a targeted application is of tremendous importance and represents significant technical challenges. In a concomitant manner, this decision has to be taken accordingly an exhaustive list of requirements essentially related to the biocompatibility, the overall thermomechanical properties, the latter inherent to bonding and reagent integration. Beyond the traditional proof-of-concept works developed at the academic research level, another higher level of concern exists when real applications and medical research are envisioned. Two important issues therefore need to be addressed. First, a reevaluation of the biocompatibility and the overall stability of the intrinsic microfluidic performance when handling real samples. The second aspect involves the scaling-up of each microfabrication, bonding, conditioning, and other packaging needs, and their interdependences and costs. The gap currently is wide in these aspects, and it is one of the most severe limitations for microfluidic applications. Therefore dedicated efforts are needed to tackle this issue. The introduction of sTPE highlights a seamless strategy that promotes a feasible path transfer from prototyping toward realistic industrialization, working from the earliest research steps to the end with a unique polymer interface. Beyond, the polymer materials presented in this chapter, there is tremendous space considering the introduction of other functional polymers in the microfluidic applications. We envision that new research activities focused on conductive, piezo, and magnetically doped polymers among other polymers not only provide a fantastic opportunity for further progress and advancement but also for a new field of research and IP development. The considerations extend surely over new blends of material development for specific goals and needs at large. The high level of multidisciplinary skills required in the field is challenging for the academic community; however, such multidisciplinary nature that extends from biology/medicine, microfabrication, microfluidic materials, and electronic and optical through sensors also provide unique solutions. Definitively, the development of innovative materials will bring innovations, and our community has to act proactively in this direction for the success of microfluidic research and the real benefit for progress in health-related biomicrofluidic applications.

REFERENCES

1. S. C Terry, J. H. Jerman and J. B. Angell, A gas chromatographic air analyzer fabricated on a silicon wafer, IEEE Trans. Electron. Dev., 26(12), 1880–1886, 1979.
2. R. Tiwari and S. Chandra, Low-temperature silicon-to-silicon anodic bonding using sodium-rich glass for MEMS applications, J. Electr. Mater., 43(2), 555–566, 2014.
3. K. Y. Fong, M. Poot and H. X. Tang, Nano-optomechanical resonators in microfluidics, Nano. Lett., 15(9), 6116–6120, 2015.
4. A. F. Coskun, A. E. Cetin, B. C. Galaterra, D. A. Alvarez, H. Altung and A. Ozcan, Lens free optofluidic plasmonic sensor for real-time and label-free monitoring of molecular binding events over a wide field-of-view, Sci. Rep., 4(6789), 2014.
5. A. Rival, D. Jary, C. Delattre, Y. Fouillet, G. Castellan, A. Bellemin-Comte and X. Gidrol, An EWOD-based microfluidic chip for single-cell isolation, mRNA purification and subsequent multiplex qPCR, Lab Chip, 14, 3739–3749, 2014.
6. V. Marx, PCR: paths to sensitivity, Nat. Methods, 11, 241–245, 2014.
7. J. Tamayo, P. M. Kosaka, J. J. Ruz, Á. San Paulo and M. Calleja, Biosensors based on nanomechanical systems, Chem. Soc. Rev., 42, 1287–1311, 2013.
8. E. Biavardi, S. Federici, C. Tudisco, D. Menozzi, C. Massera, A. Sottini, G. G. Condorelli, P. Bergese and E. Dalcanale, Cavitand-grafted silicon microcantilevers as a universal probe for illicit and designer drugs in water, Angew. Chem. Int. Ed. Engl., 53(35), 9183–9188, 2014.
9. M. C. Breadmore, R. M. Tubaon, A. I. Shallan, S. C. Phung, A. S. Abdul Keyon, D. Gstoettenmayr, P. Prapatpong, A. Alhusban, L. Ranjbar, H. H. See, M. Dawod and J. P. Quirino, Recent advances in enhancing the sensitivity of electrophoresis and electrochromatography in capillaries and microchips (2012–2014), Electrophoresis, 34(1), 29–54, 2013.
10. R. A. Saylor and S. M. Lunte, A review of microdialysis coupled to microchip electrophoresis for monitoring biological events, J. Chromatogr. A., 1382, 48–64, 2015.
11. D. J. Guckenberger, T. E. de Groot, A. M. D. Wan, D. J. Beebe and E. W. K. Young, Tutorial review micromilling: a method for ultra-rapid prototyping of plastic microfluidic devices, Lab Chip, 15, 2364–2378, 2015.
12. F. Z. Fang, X. D. Liu and L. C. Lee, Micro-machining of optical glasses—a review of diamond-cutting glasses, Sadhana Acad. Proc. Eng. Sci., 28(5), 945–955, 2003.
13. E. M. Chan, A. P. Alivisatos and R. A. Mathies, High-temperature microfluidic synthesis of CdSe nanocrystals in nanoliter droplets, J. Am. Chem. Soc., 127 (40), 13854–13861, 2005.

14. G. A. Grob, T. Thelemann, S. Schneider, D. Boskovic and J. M. Kohler, Fabrication and fluidic characterization of static micromixers made of low temperature cofired ceramic (LTCC), Chem. Eng. Sci., 63, 2773–2784, 2008.

15. S. Gomez-de Pedro, M. Puyol, D. Izquierdo, I. Salinas, J. M. de la Fuente and J. Alonso-Chamarro, A ceramic microreactor for the synthesis of water soluble CdS and CdS/ZnS nanocrystals with on-line optical characterization, Nanoscale, 4(4), 1328–1313, 2012.

16. P. Couceiro, S. Gómez-de Pedro and J. Alonso-Chamarro, All-ceramic analytical microsystems with monolithically integrated optical detection cells, Microfluid. Nanofluid., 18(4), 649–656, 2015.

17. S. K. Sia and G. M. Whitesides, Microfluidic devices fabricated in poly(dimethylsiloxane) for biological studies, Electrophoresis, 24(21), 3563–3576, 2003.

18. S. Arscott, SU-8 as a material for lab-on-a-chip-based mass spectrometry, Lab Chip, 14, 3668–3689, 2014.

19. A. B. Azouz, S. Murphy, S. Karazi, M. Vázquez and D. Brabazon, Fast fabrication process of microfluidic devices based on cyclic olefin copolymer, Mater. Manuf. Process., 29, 93–99, 2014.

20. K. Tsougeni, K Ellinas, H Archontaki and E Gogolides, A microfabricated cyclo-olefin polymer microcolumn used for reversed-phase chromatography, J. Micromech. Microeng., 25, 015005, 2015.

21. D. Ogonczyk, J. Wegrzyn, P. Jankowski, B. Dabrowski and P. Garstecki, Bonding of microfluidic devices fabricated in polycarbonate, Lab Chip, 10, 1324–1327, 2010.

22. M. Salim, G. Mishr, G. J. S. Fowler, B. O'Sullivan, P. C. Wright and S. L. McArthur, Non-fouling microfluidic chip produced by radio frequency tetraglyme plasma deposition, Lab Chip, 7, 523–525, 2007.

23. H. Mogi, Y. Fukushi, S. Koide, R. Sano, T. Sasaki and Y. Nishioka, Ascorbic acid fuel cell with a microchannel fabricated on flexible polyimide substrate, IEEJ Trans. Sens. Micromach., 134, 366–371, 2014.

24. A. Zulfiqar, A. Pfreundt, W. E. Svendsen and M. Dimak, Fabrication of polyimide based microfluidic channels for biosensor devices, J. Micromech. Microeng., 25, 035022, 2015.

25. E. Berthier, E. W. K. Young and D. Beebe, Engineers are from PDMS-lad, biologists are from polystyrenia, Lab Chip, 12, 1224–1237, 2012.

26. E. Kim, Y. Xia and G. M. Whitesides, Polymer microstructures formed by moulding in capillaries, Nature, 376, 581, 1995.

27. H. Becker, Famous last words, Lab Chip, 11(13), 2133–2134, 2011.

28. J. W. Song and L. L. Munn, Fluid forces control endothelial sprouting, Proc. Natl. Acad. Sci., 108, 15342–15347, 2011.

29. J. S. Gewandter, R. J. Staversky and M. A. O'Reilly, Hyperoxia augments ER-stress-induced cell death independent of BiP loss, Free Radic.. Biol. Med., 47, 1742–1752, 2009.

30. Y. Tang, E. A. Scheef, Z. Gurel, C. M. Sorensen, C. R. Jefcoate and N. Sheibani, CYP1B1 and endothelial nitric oxide synthase combine to sustain proangiogenic functions of endothelial cells under hyperoxic stress, Am. J. Physiol.: Cell Physiol., 298, C665–C678, 2010.

31. E. Berthier, J. Warrick, H. Yu and D. Beebe, Managing evaporation for more robust microscale assays, Part 1, Volume loss in high throughtput assays, Lab Chip, 8, 852–859, 2008.

32. K. J. Regehr, M. Domenech, J. T. Koepsel, K. C. Carver, S. J. Ellison-Zelski, W. L. Murphy, L. A. Schuler, E. T. Alarid and D. Beebe, Biological implications of polydimethylsiloxane-based microfluidic cell culture, Lab Chip, 9, 2131–2139, 2009.

33. X. Su, E. W. K. Young, H. A. S. UnderKofler, T. J. Kamp, C. T. January and D. J. Beebe, Microfluidic cell culture and its application in high-throughput drug screening: cardiotoxicity assay for hERG channels, J. Biomed. Screen., 16, 101–111, 2011.

34. M. A. Unger, H. P. Chou, T. Thorsen, A. Sherer and S. R. Quake, Monolithic microfabricated valves and pumps by multilayer soft lithography, Science, 288, 113–116, 2000.

35. M. I. Mohammed, S. Haswell and I. Gibson, Lab-on-a-chip or chip-in-a-lab: challenges of commercialization lost in translation, Procedia Technol., 20, 54, 2015.

36. C. D. Chin, V. Linder and K. Sia, Commercialization of microfluidic point-of-care diagnostic devices, Lab Chip, 12, 2118–2134, 012.

37. G. Chantal and M. Khan, Laser processing for bio-microfluidics applications (Part II), Analyt. Bioanalyt. Chem., 385(8), 1362–1369, 2006.

38. A. Bertsch and P. Renaud, Special issue: 15 years of SU-8 as MEMS material, Micromachines, 6(6), 790–792; 2015.

39. R. Daunton, A. J. Gallant and D. Wood, Manipulation of exposure dose parameters to improve production of high aspect ratio structures using SU-8, J. Micromech. Microeng., 22(7), 075016, 2012.

40. H. O. Moser and C. Rockstuhl, 3D THz metamaterials from micro/nanomanufacturing, Laser Photon. Rev., 6, 219–244, 2012.

41. A. Shamsi, A. Amiri, P. Heydari, H. Hajghasem, M. Mohtashamifar and M. Esfandiari, Low cost method for hot embossing of microstructures on PMMA by SU-8 masters, Microsys. Technol., 20(10), 1925–1931, 2014.

42. J. P. Rolland, R. M. Van Dam, D. A. Schorzman, S. R. Quake and J. M. DeSimone, Solvent-resistant photocurable liquid fluoropolymers for microfluidic device fabrication, J. Am. Chem. Soc., 126, 2322–2323, 2004.

43. A. Vitale, M. Quaglio, S. L. Marasso, A. Chiodoni, M. Cocuzza and R. Bongiovanni, Direct photolithography of perfluoropolyethers for solvent-resistant microfluidics, Langmuir, 29(50), 15711–15718, 2013.

44. G. Stojkoviča, M. Kriveca, A. Veselb, M. Marinšeka and P. Žnidaršič-Plazla, Surface cell immobilization within perfluoroalkoxy microchannels, Appl. Surf. Sci., 320, 810–817, 2014.

45. W. H. Grover, M. G. von Muhlen and S. R. Manalis, Teflon films for chemically-inert microfluidic valves and pumps, Lab Chip, 8, 913–918, 2008.

46. Y. Zhang, K. T. L. Trinh, I. S. Yoo and N. Y. Lee, Integrated monolithic PMMA microdevice, for DNA purification and amplification of Escherichia coli pathogen, Sensor. Actuat. B. Chem., 202, 1281–1289, 2014.

47. X. Hu, Y. Dong, Q. He, H. Chen and Z. Zhu, Fabrication of a polystyrene microfluidic chip coupled to electrospray ionization mass spectrometry for protein analysis, J. Chromatogr. B., 990, 96–103, 2015.

48. G. Holden, N. R. Legge, R. Quirk and H. E. Schroeder, Thermoplastics elastomers, 2nd ed. Hanser/Gardner, Cincinnati, 1996.

49. E. Roy, M. Mounier, G. Steward, L. Malic, C. Liviu, M. Madou, M. Bergeron and T. Veres, From cellular lysis to gDNA micro-array detection, an integrated point of care thermoplastic CDlab, Lab Chip, 15, 406–416, 2015.

50. E. Roy, M. Geissler, J. C. Galas and T. Veres, Prototyping of microfluidic systems using a commercial thermoplastic elastomer, Microfluid. Nanofluid., 11, 235–244, 2011.

51. D. Brassard, K. Li, M. Geissler, C. Miville-Godin, E. Roy and T. Veres, 3D thermoplastic elastomer microfluidic devices for biological probe immobilization, Lab Chip, 11, 4099, 2011.
52. M. Guillemette, E. Roy, F. Auger and T. Veres, Rapid isothermal substrate microfabrication of a biocompatible thermoplastic elastomer for cellular contact guidance, Acta. Biomater., 7, 2492–2498, 2011.
53. J. Wein, S. Jian, B. Wang, Y. Tan, X. Tu, E. Roy, B. Ladoux and Y. Chen, Fabrication of adjacent micropillars arrays with different heights for cell studies, Microelectr. Eng., 158, 22–25, 2016
54. E. Roy, J. C. Galas and T. Veres, Thermoplastic elastomers for microfluidics: towards a high-throughput fabrication method of multilayered microfluidic devices, Lab Chip, 11, 3193–3196, 2011.
55. N. W. Choi, M. Cabodi, B. Held, J. P. Gelghorn, L. J. Bonassar and A. D. Stroock, Microfluidic scaffolds for tissue engineering, Nat. Mater., 6, 908–915, 2007.
56. M. W. Tibbitt and K. S. Anseth, Hydrogels as extracellular matrix mimics for 3D cell culture, Biotechnol. Bioeng., 103, 655–663, 2009.
57. S. Bang, S. Na, J. Kim and N. L. Jeon, Engineering-aligned 3D neural circuit in microfluidic device, Adv. Healthc. Mater., 5, 159–166, 2016.
58. G. Y. Huang, L. H. Zhou, Q. C. Zhang, Y. M. Chen, W. Sun, F. Xu and T. J. Lu, Microfluidic hydrogels for tissue engineering, Biofabrication, 3(1), 012001, 2011.
59. M. Chau, H. Thérien-Aubin, Y. Li, Y. Wang, D. Velasco, E. Tumarkin, A. Ramachandran and E. Kumacheva, Microfluidic generation of composite biopolymer microgels with tunable compositions and mechanical properties, Biomacromolecules, 15, 2419–2425, 2014.
60. Y. Li, X. Yan, X. Feng, J. Wang, W. Du, Y. Wang, P. Chen, L. Xiong and B. F. Liu, Agarose-based microfluidic device for point-of-care concentration and detection of pathogen, Anal. Chem., 86(21), 10653–10659, 2014.
61. V. van Duinen, S. J. Trietsch, J. Joore, P. Vulto and T. Hankemeier, Microfluidic 3D cell culture: from tools to tissue models, Curr. Opin. Biotech., 35, 118–126, 2015.
62. D. I. Dan Cho and H. Jung Yoo, Microfabrication methods for biodegradable polymeric carriers for drug delivery system applications: a review microelectromechanical systems, J. Microelectromech. Syst., 24(1), 10–18, 2015.
63. P. A. Gunatillake and R. Adhikari, Biodegradable synthetic polymers for tissue engineering, Eur. Cell. Mater., 5, 1–16, 2003.
64. Y. Lu and S. C. Chen, Micro and nano-fabrication of biodegradable polymers for drug delivery, Adv. Drug Deliv. Rev. 56, 1621–1633, 2004.
65. H. Liu and R. M. Crooks, Three-dimensional paper microfluidic devices assembled using the principles of origami, J. Am. Chem. Soc., 133(44), 17564–17566, 2011.
66. C. R. Mace and R. N. Deraney, Manufacturing prototypes for paper-based diagnostics devices, Microfluid. Nanofluid., 16, 801–809, 2014.
67. N. K. Thom, K. Yeung, M. B. Pillion and S. T. Phillips, "Fluidic batteries" as low-cost sources of power in paper-based microfluidic devices, Lab Chip, 12, 1768–1770, 2012.
68. D. M. Cate, J. A. Adkins, J. Mettakoonpitak and C. S. Henry, Review, recent developments in paper-based microfluidic devices, Anal. Chem., 87, 19–41, 2015
69. DY. Xia, J. Si and Z. Li, Fabrication techniques for microfluidic paper-based analytical devices and their applications for biological testing: a review, Biosens. Bioelectron., 77, 774–89, 2016.

2 CHAPTER

Microfluidics for Ultrafast Spectroscopy

Adrien A. P. Chauvet

GAP-Biophotonics, Geneva University, Geneva, Switzerland

ABSTRACT

Ultrafast laser technologies became one of the essential tool in the characterization of molecular compounds. Being comprised of spectroscopists, laser scientists, chemists and biologists, the "ultrafast community" is often disconnected and consequently unaware of the developments in microfluidic systems. The challenges of studying limited amount of precious liquid sample by means of ultrafast spectroscopy remains silent and, while no commercial systems are available, each research group is developing its own "home-made" options. This chapter will therefore contribute in filling up the gap that exist between the two communities, that of the ultrafast spectroscopy and that of microfluidics by revealing the importance of this analytical tool as well as the advantages of applying microfluidic technics to it. In this goal, the chapter will focus of the recently developed microfluidic flow-cell. With a minimal volume of about 250 μL, the flow-cell enables the study of precious protein complexes that are simply not available in larger quantities. The multiple advantages of the microfluidic flow-cell will be illustrated by the analysis of the cytochrome bc1. In particular, the study will describe how the capabilities of the microfluidic flow-cell enabled the resolution of the ultrafast electronic and nuclear dynamics of specific embedded chromophores.

Keywords: *Microfluidics; Ultrafast Spectroscopy; Liquid Sample*

1. INTRODUCTION

The aim of this chapter is to address the gap that exists between two research communities: ultrafast spectroscopy and microfluidics. Indeed, the development of pulsed laser systems in the last few decades has ushered in new techniques in ultrafast spectroscopy. These techniques have opened new doors for the study of fundamental photo-chemical and photo-physical behavior of a variety of photosynthetic protein complexes.[1] For methodological reasons, i.e. samples being rare and the use of highly specialized equipment, there existed a pressing need to apply microfluidics systems in ultrafast spectroscopy, two fields that unfortunately developed separately and whose researchers are rarely knowledgeable in both areas. For the microfluidic readers who may be unfamiliar with this literature, I introduce, in the first part of this chapter, the basic concepts of ultrafast pump-probe spectroscopy. In so doing, I highlight the relevance of this technique in gaining understanding protein dynamics and therefore biological properties and functions. It is in fact due to the laborious procedures that are involved in the purification processes of these proteins complexes that most biological samples are only available in sub-milliliters quantities. In the second part of this chapter, I will therefore expose some of the most common solutions that are in use in order to manipulate liquid samples in ultrafast spectroscopy. Finally, in the third part of this chapter, I describe one of the latest applications to the field of ultrafast spectroscopy—what is now called microfluidics—as it handles micro-liters volumes of a given sample. I illustrate the benefits of the application of such a microfluidic system through an analysis the of the cytochrome bc_1. I conclude with a discussion on the areas of the present microfluidic flow-cell in need of further research and investigation.

2. THE NEED FOR MICROFLUIDICS IN ULTRAFAST SPECTROSCOPY

In this section I will introduce the field of ultrafast spectroscopy to which the microfluidics systems will be applied. I describe the basics of the technique and its relevance to the current fundamental research efforts in biochemistry and biophysics. In the last section of this chapter, I cover the practical challenges that emerge from such studies, which justify the need for microfluidics.

2.1. Ultrafast transient absorption spectroscopy

Transient absorption is a spectroscopic technique whose aim is to resolve the relaxation dynamics of an excited molecule "simply" by looking at its spectral modifications. The technique involves two principal light beam: a pump and a probe. While the role of the pump is to promote the molecule to a particular excited state, the probe is used to "look" at the state in which this molecule is in. The fact that these light beams are not continuous but are strains of pulses allows one to excite the molecule for a brief instant (the duration of the pulse) and to probe its state at a later time. The light pulses are first produced in a cavity (oscillator) that is built around a doped crystal such as titanium doped sapphire crystals (Ti:S) as shown in Figure 1.[2]

The crystal, when excited, serves as a photon tank that will amplify any pulses passing through, via stimulated emission. The cavity then enters into a "mode locked" state as soon as the enclosed pulse strains are in resonance with the dimension of the cavity itself. The energy trapped within the cavity builds up until the pulses are intense enough to leak through one of the cavity end-mirror. Typical Ti:S cavity produces strains of ~30 fs pulses centered around ~800 nm at 80 MHz repetition rate; each pulse being about tenth of nano-Joules. The extracted pulses are then amplified in a similar fashion by passing in a second cavity comprised of a second Ti:S crystal.[2] The frequency is however decreased

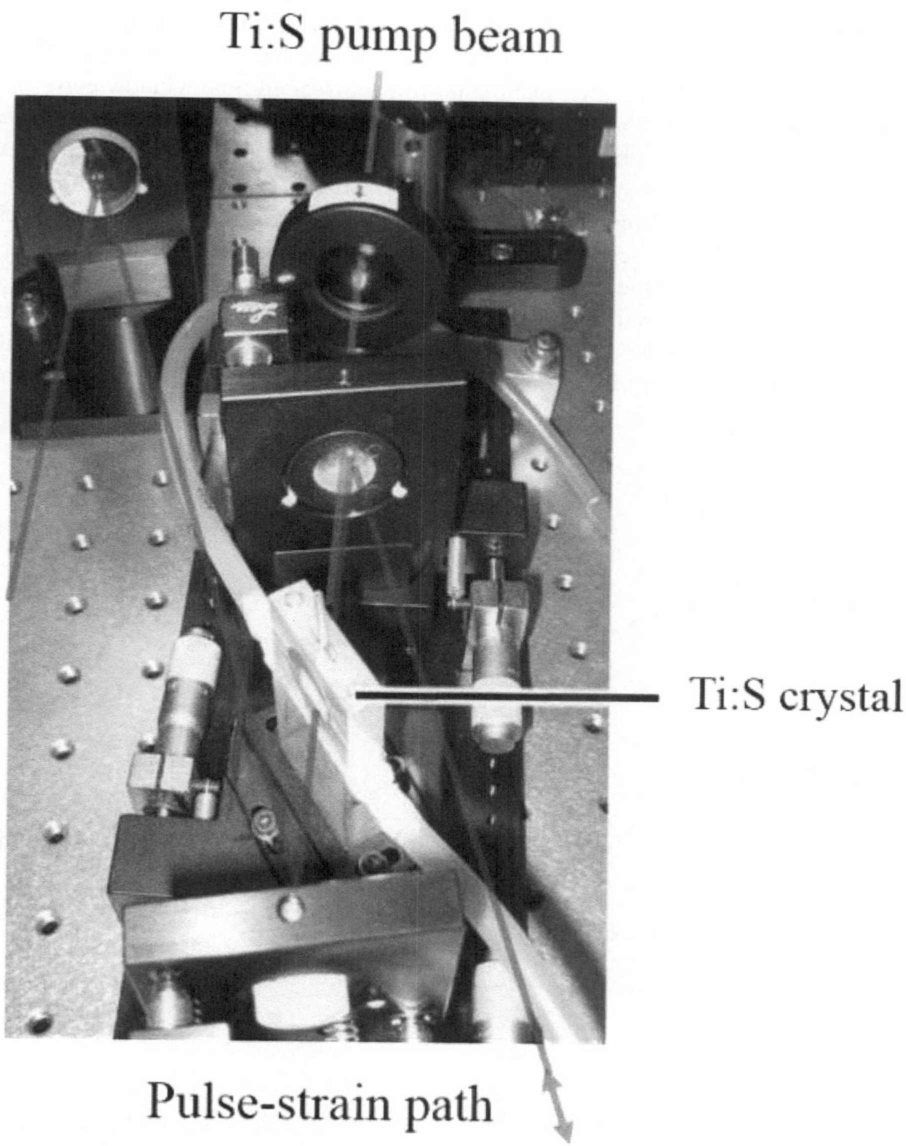

Figure 1. *Ti:S crystal cavity.*

to few kHz in order to reach few milli-Joules per pulses. It is these amplified pulses that are then split into pump and probe pulses.

As illustrated in Figure 2, varying the time delay that separate both pump and probe enables to follow the evolution of the transient excited states over time. This delay is introduced by physically increasing the path of one of the pump or probe arm of the setup via a precision stage. In a typical configuration, a 30 cm stage provides with a temporal window of about 4 ns (round-trip). The time resolution with which we can follow the molecular dynamics is then given by the duration of the light pulses themselves. A regular transient absorption setup is today capable of producing amplified strains of ~40 fs pulse with a kHz repetition rate. Such femto-to-nano-second time window corresponds to the dynamics of energy and electron transfer within and between molecules as well as local structural modifications.[3] Furthermore, the high repetition rate and the development in matters of laser stability and detection system gives the ability to monitor the absorption changes of a single molecule out of a thousand (corresponding to changes of ~10^{-4} OD) in less than a second of accumulation time.[4] With such setups, it is for example possible to trigger the charge separation in the Photosystem I protein complex and to follow the liberated electron as it progresses from one side of the protein to the other.[4] Another example is the monitoring of the heme-ligand dissociation and rebinding dynamics that results from the absorption of a photon such as it is the case of various types of cytochromes.[5, 6]

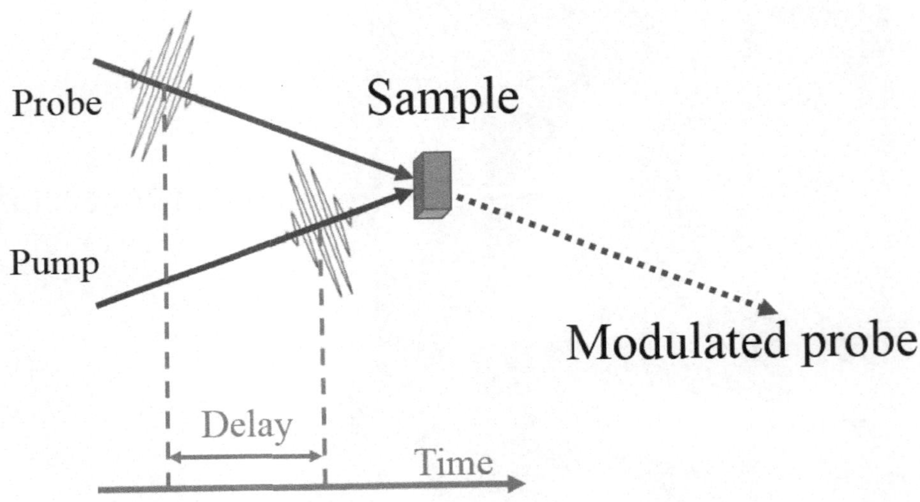

Figure 2. *Pump-probe experimental scheme.*

The use of non-linear optics, such as in the famous and various kind of optical parametric amplifiers (OPA, Non-collinear OPA and multi-pass OPA)[7] as well as the different pulse shaping devices (in either transmission or reflection),[8] gives the ability to tune both the pump and the probe to the desire wavelength, therefore enabling to excite and to follow a particular molecular transition. For example, in the study of the bacterial reaction centers complexes, being comprise of multiple types of pigments that are spectrally distinct, careful tuning of the pump allows to preferably excite one pigment while living the others in their fundamental state.[9]

It is therefore out of the development in laser technology and specifically in tunable table-top pulsed lasers that the field of ultrafast spectroscopy came to know the success it knows today.

2.2. Studying biological samples

As implied in the previous section, transient spectroscopy is best suited for the study of compounds that have distinct spectral feature. Fortunately, most organic compounds are made of either aromatic amino acids or incorporates chromophores within their protein structure, each having distinct spectral features. The technics consequently became in the past decade a common analytical tool for biologists and chemists. Ultrafast transient spectroscopy is indeed used for a broad range of investigations: being sensitive to changes in absorption spectrum of the proteins, it is possible to collect data on local conformational deformations, electronic transitions and (low) vibrational modes of oscillation within molecules, intra and inter molecular energy and electron transfers, etc. [10] In the field of solar energy conversion for example, which is one of today's essential topic in our energy savvy societies, this technique allowed to better understand the conversion processes from light to consumable energy. In particular, the study of photosynthesis showed how the specific arrangement of pigment within larger protein structure either favor the absorption and passing of the photon energy, as it takes place in antennae systems, or favors the generation of a charge separated state, as it happens in photosynthetic reaction centers, [4, 9] which results in the liberation of a high energy, and therefore usable, electron. Other chromophores can also serve as electron docking sites and electron carriers.[6] The knowledge gained from such studies is then applied for medical and industrial purposes and used to optimize specific molecular reactions. From these examples, it is possible to understand why the study of biological samples by means of ultrafast transient spectroscopy, among other spectroscopic techniques, became and remains one of the standard analytical tool for the fundamental understanding of a broad range of molecular dynamics.

2.3. The challenges of the application

As discussed above, ultrafast transient spectroscopy is suitable for the study of biological samples. These samples are usually made of purified and solvated proteins. Once a particular molecule is excited, the deposited energy ultimately dissipates into the solvent (so long as the changes are not permanent). It is to remember that the typical repetition rate of the pulses are in the order of the kHz and at this rate the sample is excited about every milliseconds. The risk is that the photo-induced molecular modifications live for a time that is comparable, which will then results in a rapid saturation of the sample. In other words, the excited molecule might not have sufficient time to relax to its fundamental state before the coming of the next light pulse. Saturation thus takes place and as the molecules gets overexcited: they are unable to release the deposited energy quickly enough and end up by "burning". In order to avoid such consequences, the sample is usually flown in front of the laser beams. The condition is that the flux is high enough so that the sample is refreshed for each laser pulse.

The second major constraint is related to the quantity of the sample available. The sample being made of purified proteins, it is then concentrated in order to reach an optical density that is suitable for spectroscopic analysis. Ideally, transient absorption spectroscopy requires an optical density of about 0.6 for the transition of interest, which in the case of heme protein corresponds to a concentration of hundreds of micro molars. The samples are consequently limited in terms of volumes and sub-milliliters quantities already requires months-long of successive growth of the organisms and protein purification cycles.[11, 12]

The third constraint concerns the susceptibility of the sample in respect to its solvent and atmosphere in which it is enclosed. Indeed some biological samples are hydrophobic and require to be dissolved in various chemicals in order to avoid aggregation and the subsequent scattering of the light pulses, such as it is the case for solvated porphyrins. Many samples are also sensitive to oxygen and therefore require the atmosphere to be controlled. For example, myoglobin is able to effectively bind a variety of diatomic molecule. It however has a high affinity for oxygen, so much that it is not possible to study its deoxygenated state unless in anaerobic conditions.[13]

In conclusion, ultrafast transient spectroscopy is today one of the standard analytical tool for whoever desire to study the local structural, electronic and vibrational photo-induced dynamics. In particular, solvated chromophores and chromophore-containing proteins are well suited for the techniques as they can be specifically targeted via their absorption band. However, these liquid samples are often limited in quantities and have to be flown in order to avoid any photo-damages. From these challenges arise the need for microfluidics so as to flow the limited sample volumes. Furthermore, in order to perform ultrafast spectroscopic measurement, the probe pulses must pass through the sample, therefore through the flow-cell in which it is enclosed. The cell consequently requires adequate windows that do not alter the signal-to-noise ratio nor the temporal and spectral resolution of the apparatus. Additionally, due to the properties and high sensitivity of certain samples, the cell must be resistant to the solvent while providing control of the atmosphere. In such cases, the sample must be hermetically confined within the microfluidic system which therefore has to also play the role of an anaerobic chamber.

3. MOST COMMON SOLUTIONS AVAILABLE

In this section I will review different techniques that are commonly employed to flow the sample in front of the laser beam. I will discuss the advantages and inconveniences of each in terms of their compatibilities with the requirements of ultrafast spectroscopic laser systems.

3.1. Flow cell

The main idea is to flow the sample in between two transparent plates. These windows are usually made of quartz so as to permit the broadband (near UV-Visible-near IR) beams to pass through. This type of cell allows for small path lengths and thin windows, down to 0.02 mm such as the one shown in Figure 3, therefore reducing scattering of the excitation beam through the quartz. The thin windows also have the advantage to minimally alter the pulse duration (limited group velocity dispersion), therefore allowing for an optimal time resolution. Furthermore, the cell is steady and its stability allows for optimal signal to noise levels.

The quartz cells, by themselves, are commercially available.[14] However, due to their size, these cells already enclose about one milliliter of sample. Furthermore they must be connected to a pump, typically a peristaltic or flow-through pump, in order to generate the flow. Altogether, the flow cell system requires few tenth of milliliters and is consequently not adequate for precious samples that are simply not available in such quantities.

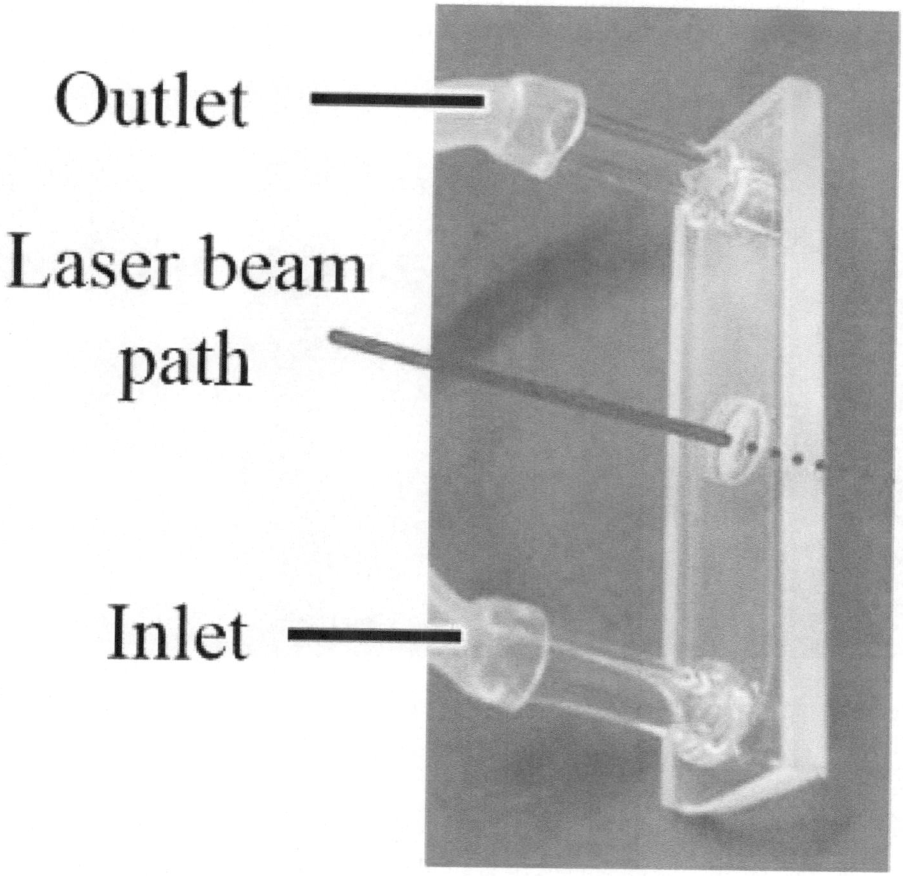

Figure 3. *Flow cell from Starna Cell [14]*

3.2. Liquid micro-jet

The term micro here comes from the diameter of the jet itself, which produces a couple of centimeters-long of regular flow. The advantage of expelling the sample as a jet is to remove the constraints of having it to pass through windows, i.e. avoiding any additional absorption, scattering and group velocity dispersion. It therefore allows pump probe experiment in all spectral region including UV and X-ray region. Associated with an adequate sample collector, as shown in Figure 4, it is also applicable in vacuum chambers and becomes suitable for photo-electron spectroscopy.[15]

The inconvenience of having to produce a constant flow rate is that the sample must pass through a sophisticated HPLC pump,[15-17] which consequently requires sample volumes that are larger than our targeted sub-milliliter. Note also that while passing in either the air or in vacuum, the sample's solvent is subjected to evaporation. In such conditions, recycling of the sample results in a change of the sample's concentration and temperature over the course of the experiment.[16] Furthermore, the high speed at which the sample go through the nozzle induces charging of the liquid and or of the nozzle, which might alter the measurement.[17]

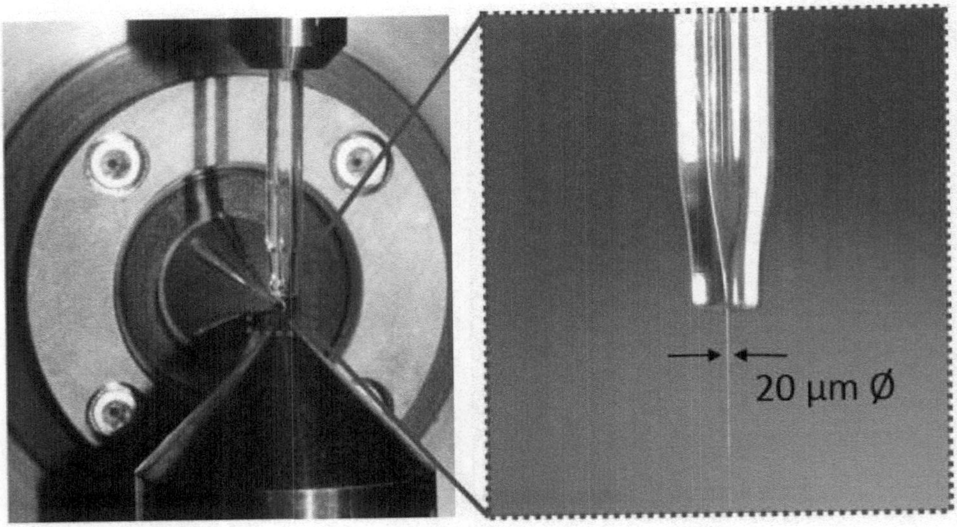

Figure 4. *Micro-jet implemented for photo-electron spectroscopy. Courtesy of José Ojeda, EPFL.*

3.3. Spinning /moving cell

The idea behind the spinning-cell is similar to that of the flow cell as the sample is housed between two circular glass plates that are spaced by the desirable optical path-length.[18] While rotating, the sample either creates a rim at the edge of the cell, or at lower speed, the solution remains at the bottom of the cell and is constantly mixed due to friction with the glass as shown in Figure 5. The advantage is that it typically requires minimal amounts of sample (~0.3 mL) as well as to provide control over the initial atmosphere in which the sample in enclosed since the sample is hermetically confined.

However, because the rotation of the glass plate causes the excitation beam to sweep a large surface, the cleanness of the plates is directly related to the noise. It is consequently burdensome to clean. Furthermore, the fact of having a moving piece of glass in which the beam is focused renders the alignment of the cell crucial, and any slight asymmetry of the glass plates has consequences on the transmitted probe beam. Also, since the glass plates are typically few cm in diameter, the required minimal thickness of the glass lowers the time resolution. Another inconvenience is that once the cell is set it does not allow access to the enclosed sample and each experiments therefore requires its own sample preparation.

In conclusion, the proposed solutions used to flow the sample in front of the laser beam have each distinct advantages and inconveniences, and none ideally respond to all the requirement, i.e. is suitable for sub-milliliter sample volumes; provides a close atmosphere;

grant access to the sample once closed; affecting minimally the signal-to-noise ratio and resolution of the apparatus. Through these few examples I hope to have convinced the reader of the need for a development in the application of microfluidic system to the field of ultrafast spectroscopy.

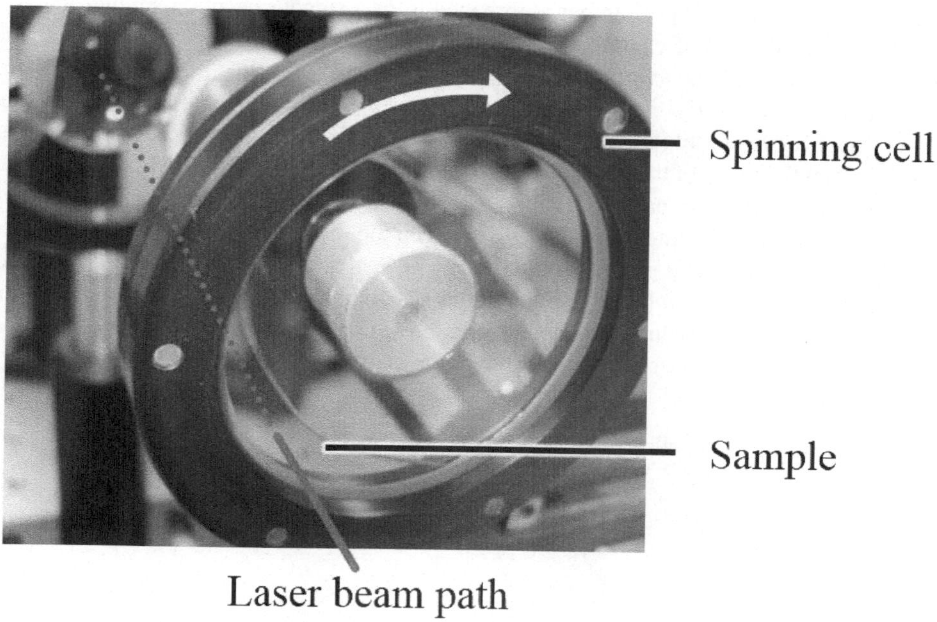

Figure 5. *Spinning cell used if few-kHz spectroscopy setups.*

4. RECENT IMPROVEMENT AND APPLICATION: THE MICROFLUIDIC FLOW CELL

In this section I will describe the recently developed microfluidic flow-cell in order to illustrate the importance and advantages of applying microfluidic systems to the field of ultrafast spectroscopy. I will show that recent improvements in the field of microfluidics have the capabilities to solve the previously mentioned constraint all at once. I will thus discuss the advantages of the system in light of the other technics. Following the technical properties I will illustrate the flow-cell's effectiveness through a study done on rare *bc*-cytochrome and conclude with an appeal for further development.

4.1. The microfluidic flow-cell

The microfluidic flow-cell, [19] as illustrated in Figure 6, is composed of three main elements that are connected via flexible tubing of 1-mm diameter:

The decantation chamber is a home-made polymer cylindrical chamber as shown in Figure 6. The 0.5-mm diameter inlet and outlet are at the bottom of the chamber in order to minimize turbulences that are created at high flow rates. The chamber requires a minimal amount of ~50 μL of liquid sample in order to have a continuous flow between the inlet and outlet. Any excess of sample fills up the chamber and allows the bubbles that might be enclosed in the closed circuit to rise to the surface. The bubbles are naturally trapped by the chamber while passing through at low flow-speed. At high flow-speed however, larger sample volume are required in order to avoid the suction of air due to the liquid's turbulences. The top of the chamber is threaded to fit a standard septum screw cap. This allows for the addition of chemicals to the enclosed solution while keeping the confined atmosphere protected and avoiding evaporation of the solvents.

The capillary window is made of a square quartz silica capillary bought from Composite Metal Services Ltd (CMS). It has a path-length of 0.5 mm with 0.1-mm thin walls. Knowing that the focus of the laser beam is about 100 um in diameter, the window can easily be set within the beam path. The capillary is fixed at the center of a xyz-rotation mount that allows for fine adjustment of the angle between the incident beam and the window.

The turbisc pump is a design from CSEM.[20] In short, the flow is created by direct friction between a grooved barrel and the liquid. The inner volume that the pump contains is about 100 μL only. Because the housing and the seal are respectively made out of Polyetherimide and of Polyetheretherketone, the pump is relatively resistant to chemicals.

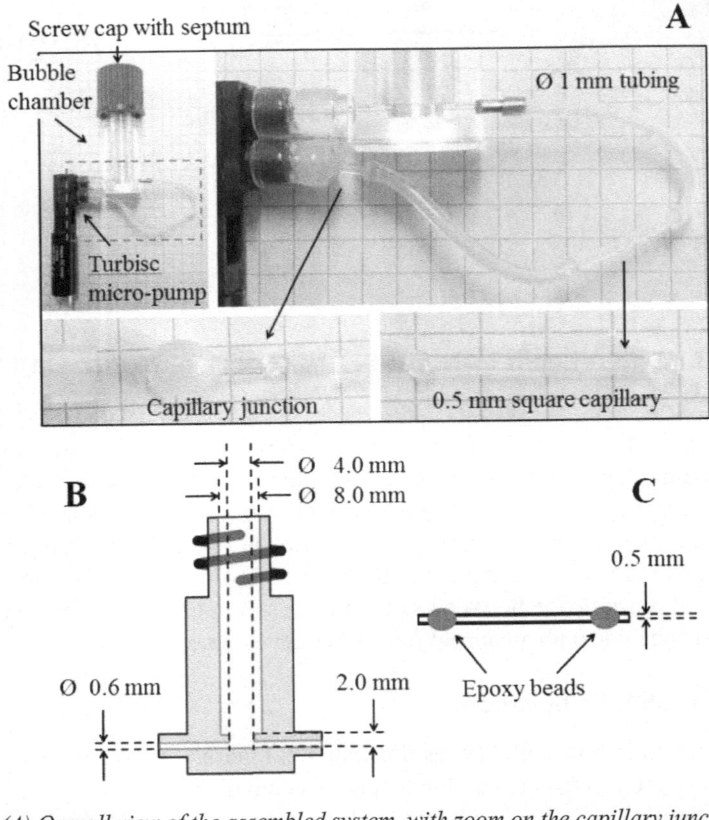

Figure 6. *(A) Overall view of the assembled system, with zoom on the capillary junctions. (B) Side view of the bubble chamber and (C) of the capillary window. Reproduced from Ref. [19].*

When the pump, the chamber and the capillary are connected, the minimal volume of sample needed for good working conditions is about 250 μL only. This configuration includes a total tubing (1-mm inner diameter) length of ~6 cm and a sufficient amount of sample in the cuvette to avoid the suction of bubbles due to turbulences with a minimal flow of ~0.1 ml/sec. The flow is proportional to the voltage applied to the pump and under the same configuration the maximum flow rate was measured to be ~0.36 mL/sec, as shown in Figure 7.

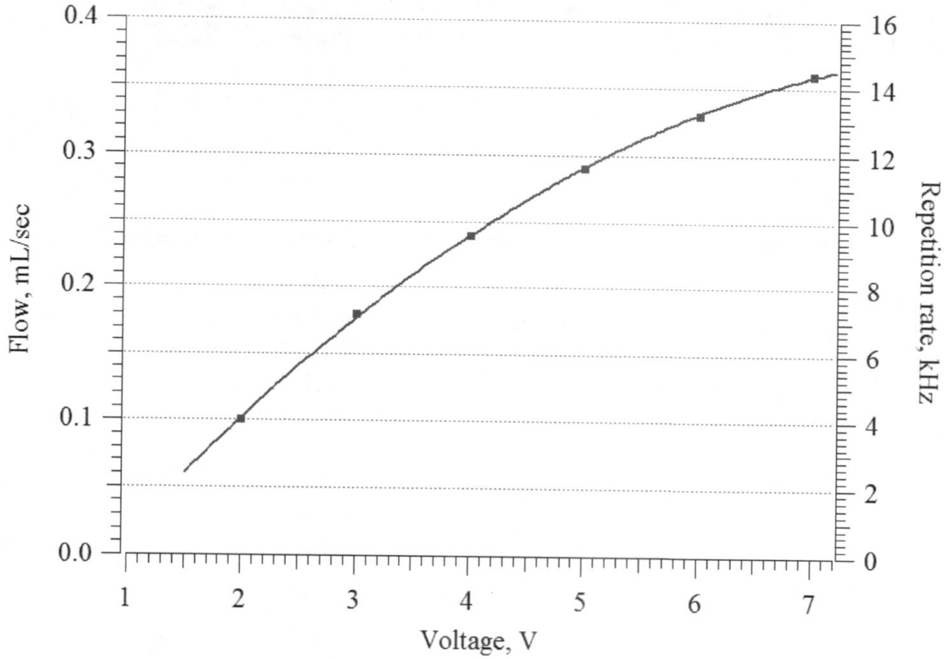

Figure 7. *Measured flow-rate (dots) and its best fit (line, 2nd order polynomial) while the pump is connected to ~6 cm of 1-mm diameter tubing and ~1-cm long 0.5x0.5 mm² square capillary. In order to ensure fresh sample at each laser shot, while assuming a typical laser focus size of 100 μm, the expected maximum repetition rate of excitation is indicated as a reference only (right axis).*

While assuming a typical laser focus diameter of 100 μm within the 0.5x0.5 mm² square capillary, the flow, when assumed to be uniform, is expected to effectively refresh the sample within the laser focus for each laser shot at an excitation rate of up to ~14 kHz. In practice, it is to remember that the flow is impeded on the edges of the capillary and consequently the value of 14 kHz has to be taken as an upper limit only. Taking into account that the inner volume of the pump is only about 100 μL, it represents one of the best (if not the actual best) compromise between flow-rate and required volume. The assembled microfluidic flow cell, in working conditions is shown in Figure 8.

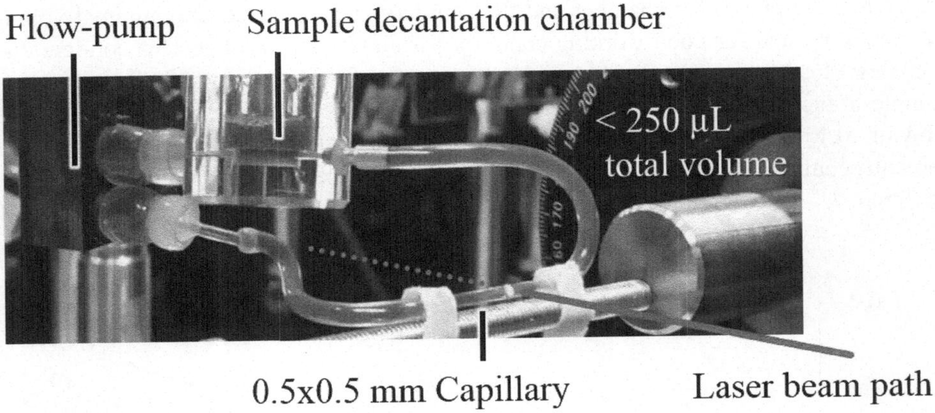

Figure 8. *Microfluidic flow-cell in action*

4.2. Example of application: the study of cythochrome bc1 complexes

In order to illustrate the applicability of the microfluidic flow-cell as well as some of its advantages I present here a unique analysis, that of the Cytochrome (cyt) bc_1 complex.[6] The cyt bc_1 complex is a key player in mitochondrial and bacterial respiratory chains.[21] It is the main actor in the protonmotive Q cycle and results in the formation of a proton gradient across the membrane via a series of embedded hemes as shown in Figure 9.[22-

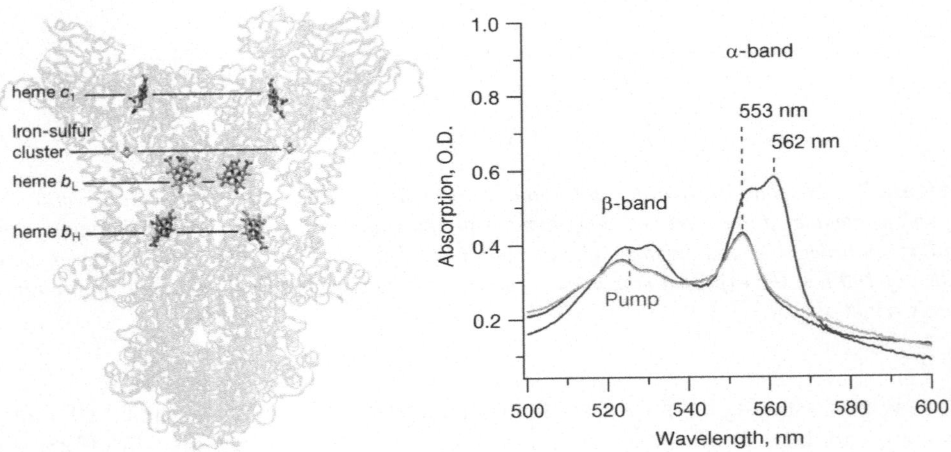

Figure 9. *(Left) Structure of the bc$_1$ complex[25] with the protein backbone being shaded for clear visualization of the key actors in the proton-coupled-electron mechanisms. (Right) Static absorption spectrum of the sequential reduction and oxidation of the cyt bc$_1$ dimer complex: after pre-reduction by ubiquinol and before (red, c$_1$-hemes reduced) and right after (blue, b- and c$_1$-hemes reduced) the addition of dithionite, and at the end, after oxidation of the complexes by oxygen (green, c$_1$-hemes reduced). The 523-nm excitation pump is indicated as reference. Reproduced from Ref. [6] with permission from the Royal Society of Chemistry.*

24] The generated potential gradient serves as the driving force for the synthesis of ATP, the universal energy transporter in living organisms. The understanding of the protein complex is therefore of primary importance. However the sample is rare and mainly because of the limited quantity available after each purification process, the heme dynamics were until then never studied by means of ultrafast spectroscopy.

Thanks to the microfluidic flow-cell, and more particularly to the access it provides to the sample via the septum, the reduction and oxidation of the hemes of interests could be controlled chemically. All data were therefore taken from the same sample preparation within the same experiment, in the same experimental conditions. The different signals that emerge from the reduced states of the hemes could then be directly compared: either the c_1-hemes can be exclusively reduced, or both c_1- and *b*-hemes can be reduced together. Direct subtraction of the two data set enabled the extraction of the sole signal form the *b*-hemes, as shown in Figure 10.

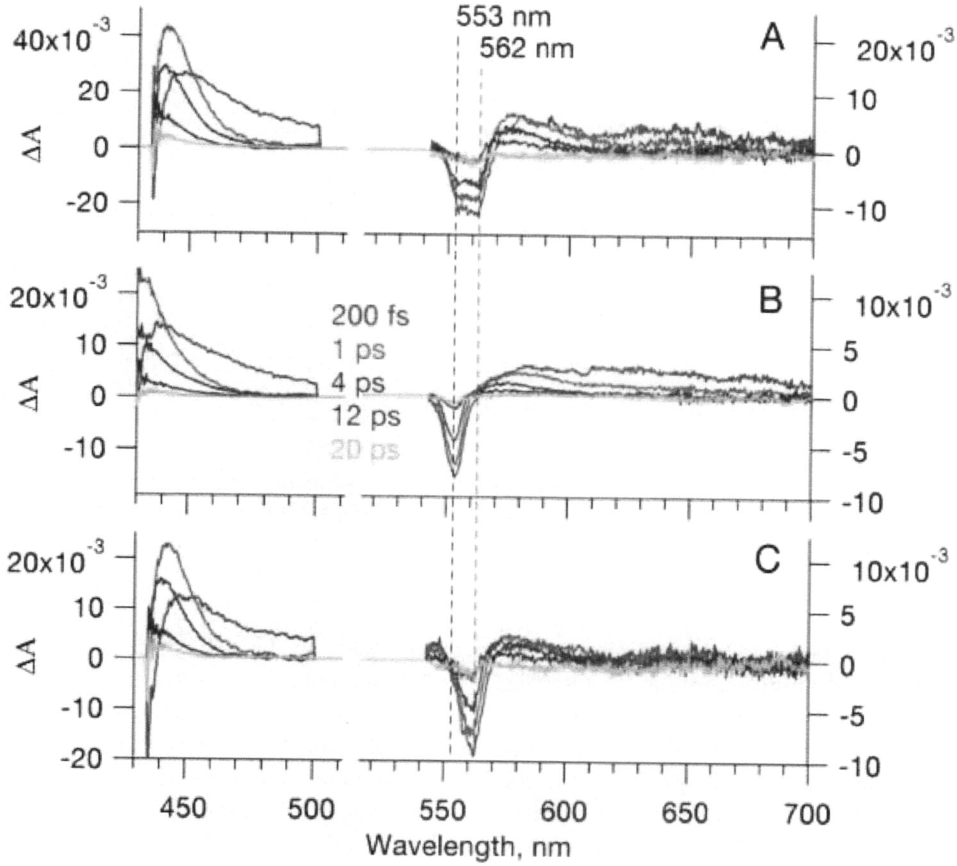

Figure 10. *(A) Spectra at selected time delays while both b- and c_1-hemes are reduced. (B) Spectra at selected time delays while only the c_1-hemes are left reduced. (C) Difference (A-B) corresponding solely to the signal of the ferrous b-heme, as discussed in the text. Note that the vertical scales below and above 515 nm differs by a factor of two. Reproduced from Ref. [6] with permission from the Royal Society of Chemistry.*

The dynamics of each heme type could be analysed separately by means of singular value decomposition and global fitting as described in detail elsewhere.[6] The resulting decay associated spectra, shown in Figure 11, revealed the clear differences between the *b*- and c_1-hemes behaviours within the bc_1 protein complex: while the c_1-hemes undergo photo-dissociation of their axial ligand as a result of ultrafast laser excitation, the *b*-hemes were shown to undergo photo-oxidation with a high (> 0.4) quantum yield that is beyond all expectations.

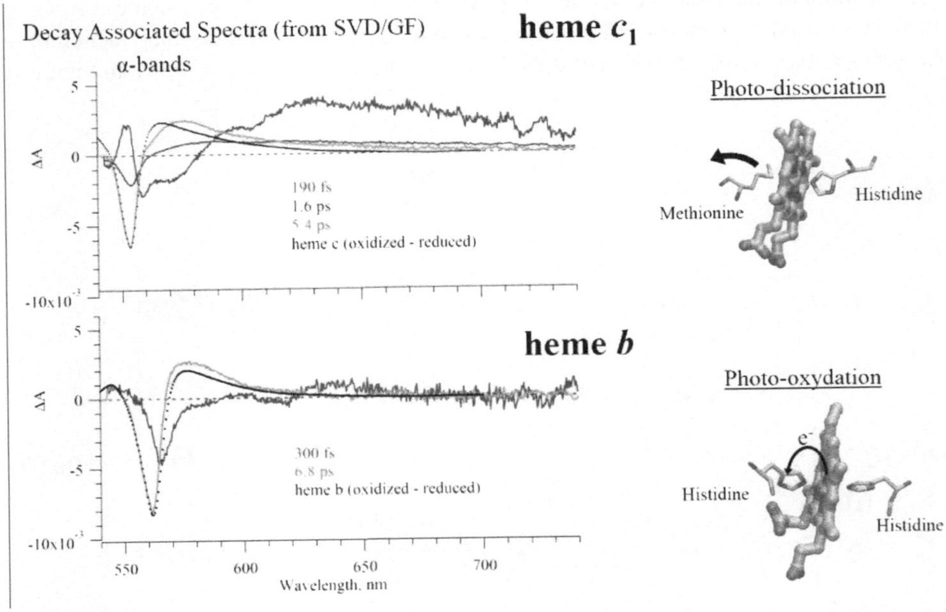

Figure 11. *Photo-dissociation of the c_1-heme (left) and photo-oxidation of the b-heme (right) upon light excitation of the cyt bc_1 complex. Reproduced from Ref. [6] with permission from the Royal Society of Chemistry.*

The *b*-heme's high electronic reactivity makes sense in the light of cyt bc_1 having to efficiently fulfil its role in the Q-cycle: it favours the reduction and oxidization of the ubiquinone and ubiquinol, respectively.[22] The *b*-hemes have to efficiently "process" the electrons, which demand them to easily loose or gain electron. Similar electronic-reactivity would in fact be counterproductive in soluble cytochromes as they would less efficiently keep their electrons from being scavenged by other solutes. The hydrophilic environment of the bc_1 core on the other hand preserves the *b*-hemes from unwanted solvated electron carrier and their high electronic reactivity is then an advantage. In contrast to the high electronic reactivity of the *b*-hemes, the high photo-dissociation quantum yield of the c_1-hemes can be understood as being an efficient "heat sink" that protects the reduced state of the heme against light excitations.

Overall, this study illustrate that, even though the *b*-hemes in cyt bc_1 and in other cyt *b* have similar ligation to their protein backbone; specific structural constraints and amino-acid arrangements result in clearly different responses, and therefore functions. While cytochromes were known to serve only as electron carriers, this study demonstrates that with the appropriate environment, light-induced charge separation can readily be initiated within single heme structures. The use of the microfluidic flow-cell therefore not only enabled the study of this rare protein complex, but allowed to shine light on the relevance of local heme-bonding and structural environment in initiating larger chemical reactions. This particular case study is only one example of how the field of ultrafast spectroscopy can benefit from the application of microfluidics technologies.

4.3. Call for development

As I mentioned, the described microfluidic flow-cell, beside the numerous advantages it provides in respect to the other systems commonly employed, is certainly not perfect, which leaves room for improvement.

For example, in terms of sample volumes, most of the liquid is contained in the pump that is used to generate the flow and in the tubing. Miniaturization of both would allow to use even smaller sample volumes. Already the small turbisc pump that is employed uses a newly developed friction based technologies[20] that is able to flow even viscous samples. The samples studied in ultrafast spectroscopy are however water-like and, not being in need of the actual viscous sample capabilities of the pump. I can therefore imagine that a simplified version of the pump would be sufficient and require even smaller volumes. Concerning the actual 1 mm diameter tubing, that links the pump with the capillary, it could be replaced by other microfluidics technologies that uses micro-channels that are specifically designed for spectroscopy.[26] The goal through these proposed improvements is to reduce the amount of liquid used, keeping in mind that reducing further the diameter of the channels might impede the overall flow rate.

Another field of improvement that I see is that of the control over the inner atmosphere of the cell in which the sample is enclosed. As I mentioned, the actual pumping is done by friction between a grooved barrel and the sample. The spinning motion is however done via a shaft and under high pressure difference between the inner and outer part of the cell, either the sample can leak through or air can be sucked in via the shaft's joints. In order to palliate to this issue, I can imagine that the shaft would be replaced, not by another mechanical interaction, but remote electromagnetical interaction so as to preserve the inner atmosphere of the sample. Such systems are already implemented for applications in biomedical for example.[27] Furthermore, the microfluidic flow-cell uses epoxy beads in order to hermetically fit the square capillary in the cylindrical tubing. Better seal could be achieved if the capillary themselves were to be made with initial beads on each sides such as it is the case for the larger commercially available flow-cells.[14]

At last, I would like to precise that, due to friction between the sample and the capillary, the capillary does not provide with a flow that is homogeneous. Sample that is the closest to the window has consequently lower flow rate and might not be refreshed for each laser pulse, being therefore subjected to photo-damage. One solution would be to employ the newly developed Electro-Osmotic flow systems that are used to generate homogeneous flows.[28]

In conclusion, the microfluidic flow-cell that I propose is specifically designed to fits the requirement imposed by the ultrafast spectroscopy of quantity-limited and sensitive sample, while it remains cost effective and easy to use. As an illustration of the flow-cell's effectiveness, I showed how its implementation enabled the study of the cytochrome bc_1. More precisely, I was able to resolve the ultrafast electronic and nuclear motions that precedes some of the larger physiological function of the protein. The microfluidic flow-cell not only facilitates but also opens the door to the study of a whole range of samples that cannot be purified in large quantities such as it is the case for most organic compound that are extracted from living organisms.

The advantages of this microfluidic flow-cell over other pre-existing solutions are clear; as clear as there is room for development.

5. GENERAL CONCLUSION

The implementation of the microfluidic flow-cell to the field of ultrafast spectroscopy can be considered as one of the first attempts to bridge two communities, i.e. that of microfluidics and that of ultrafast spectroscopy. In this chapter I have first described the technique of ultrafast spectroscopy in order to show its requirements in terms of sample as well as the importance of such analytical tool when applied to the study of biological protein complexes. In particular, ultrafast transient spectroscopy became one of the essential approach for whoever desires to understand the local electronic and nuclear modifications that are at the origin of the larger physiological functions of proteins. I then exposed the advantages and inconveniences of the different techniques that are commonly used in order to flow liquid samples in front of the laser beams. The aim of this discussion is to better appreciate how the application of microfluidics technologies is able to responds to the challenges raised by the technique. In this aim, the recently developed microfluidic flow-cell is adequate as it requires only about 250 μL while generating flow rate that are suitable for high repetition rate laser systems. Its steady window and decantation chamber allow for an optimal time and spectral resolution. By providing direct access to the sample while running a single experiment and monitoring chemical changes in "real time", the microfluidic system enables studies that were otherwise not possible. The advantages of microfluidics over other usual systems are numerous but as it represents only one of the first attempt, there is ample room for improvement. In this regard, one of the goal of this chapter is to serve as an initial step in an effort to bridge the microfluidics community with that of ultrafast spectroscopy in order to foster new ideas, new applications and new perspectives.

Acknowledgments

Thanks to Professor S. Savikhin (Purdue University, USA) as well as to Professor M. Chergui (EPFL, Switzerland) from whose laboratory of Ultrafast Spectroscopy the different pictures are taken. The described microfluidic flow cell as well as the cyt bc_1's study have been funded by the Swiss NSF via the NCCR:MUST, by the FP7 Marie Curie COFUND, by the Excellence Initiative of the German Federal and State Governments (EXC 294, BIOSS) and by the Deutsche Forschungsgemeinschaft (RTG 1976).

REFERENCES

1. Berera R, van Grondelle R, Kennis JTM. Ultrafast transient absorption spectroscopy: principles and application to photosynthetic systems. Photosynth Res. 2009;101:105-18.
2. Koechner W. Solid-State Laser Engineering. 6th ed: Springer; 2006.
3. Rosspeintner A, Lang B, Vauthey E. Ultrafast Photochemistry in Liquids. Annu Rev Phys Chem. 2013;64:247-71.
4. Chauvet A, Dashdorj N, Golbeck JH, Johnson TW, Savikhin S. Spectral Resolution of the Primary Electron Acceptor A0 in Photosystem I. J Phys Chem B. 2012;116(10):3380-6.
5. Vos MH, Battistoni A, Lechauve C, Marden MC, Kiger L, Desbois A, et al. Ultrafast Heme-Residue Bond Formation in Six-Coordinate Heme Proteins: Implications for Functional Ligand Exchange. Biochemistry. 2008;47:5718-23.
6. Chauvet AAP, Al Haddad A, Kao W-C, van Mourik F, Hunte C, Chergui M. Photo-induced dynamics of the heme centers in cytochrome bc_1. Phys Chem Chem Phys. 2014.
7. Cerullo G, De Silvestri S. Ultrafast Optical Parametric Amplifiers. Rev Sci Instrum. 2003;74(1).
8. Weiner AM. Femtosecond Pulse Shaping Using Spatial Light Modulators. Rev Sci Instrum. 2000;71(1929).
9. Chauvet A, Sarrou J, Lin S, Romberger SP, Golbeck JH, Savikhin S, et al. Temporal and Spectral Characterization of the Photosynthetic Reaction Center from Heliobacterium modesticaldum. Photosynth Res. 2013;116:1-9.
10. Rosspeintner A, Bernhard Lang B, Vauthey E. Ultrafast Photochemistry in Liquids. Annu Rev Phys Chem. 2013;64:247-71.
11. Mitra S. Sample Preparation Techniques in Analytical Chemistry. Winefordner JD, editor: John Wiley & sons, INC.; 2003.
12. Berg JM, Tymoczko JL, Stryer L. The Purification of Proteins is an Essential First Step in Understanding their Function. In: Moran S, Hadler GL, Zimmerman P, editors. Biochemistry. 5th edition ed: W H Freeman; 2002.
13. Monni R, Al Haddad A, van Mourik F, Auböck G, Chergui M. Tryptophan-to-Heme Electron Transfer in Ferrous Myoglobins. PNAS. 2015;112(18):5602-06.
14. Starna Cells. Linear Flow Cells, Type 48 series http://www.starnacells.com/d_cells_s/flow/linear/T048.html2015.
15. Arrell CA, Ojeda J, Sabbar M, Okell WA, Witting T, Siegel T, et al. A Simple Electron Time-of-Flight Spectrometer for Ultrafast Vacuum Ultraviolet Photoelectron Spectroscopy of Liquid Solutions. Rev Sci Instrum. 2014;85(103117).
16. A. K. Charge Transfer to Solvent Dynamics in Iodide Aqueous Solution Studied at Ionization Threshold: Freie Universität Berlin; 2015.
17. 1Duffin AM, Saykally RJ. Electrokinetic Power Generation from Liquid Water Microjets. J Phys Chem. 2008;112:17018-22.
18. Savikhin S, Wells T, Song P-S, Struve WS. Ultrafast Pump-Probe Spectroscopy of Native Etiolated Oat Phytochrome. Biochemistry. 1993;32:7512-8.
19. Chauvet A, Tibiletti T, Caffarri S, Chergui M. A microfluidic flow-cell for the study of the ultrafast dynamics of biological systems. Rev Sci Instrum. 2014;85: 103118.
20. Lisibach A, Casartelli E, Schmid N, editors. Flow Investigation in a Disk Micropump. ASME 2010 3rd Joint US-European Fluids Engineering Summer Meeting and 8th Interna-

tional Conference on Nanochannels, Microchannels, and Minichannels; 2010; Montreal, Quebec, Canada.

21. Berry EA, Guergova-Kuras M, Huang L, Crofts AR. Sructure and Function of Cytochrome bc Complexes. Annu Rev Biochem. 2000;69:1005-75.
22. Crofts AR. Proton-coupled electron transfer at the Qo-site of the bc1 complex controls the rate of ubihydroquinone oxidation. Biochem Biophys Acta. 2004;1655:77-92.
23. Hunte C, Solmaz SRN, Palsdóttir H, Wenz T. A Structural Perspective onMechanism and Function of the Cytochrome bc1 Complex. Results Probl Cell Differ. 2007;45:253-78.
24. Trumpower BL. Cytochrome bc1 Complexes of Microorganisms. Microbiol Rev. 1990;54(2):101-29.
25. Solmaz SRN, Hunte C. Structure of Complex III with Bound Cytochrome c in Reduced State and Definition of a Minimal Core Interface for Electron Transfer. J Biol Chem. 2008;283(25):17542-49.
26. Horrocks HM, Tosatto L, Dear AJ, Garcia GA, Iljina M, Cremades N, et al. Fast Flow Microfluidics and Single-Molecule Fluorescence for the Rapid Characterization of α-Synuclein Oligomers. Anal Chem. 2015;87(17):8818-26.
27. Al-Halhouli AT, Kilani MI, Büttgenbach S. Development of a novel electromagnetic pump for biomedical applications. Sensor Actuat A-Phys. 2010;162:172-6.
28. Stone HA, Stroock AD, Ajdari A. Engineering Flows in Small Devices: Microfluidics Toward a Lab-on-a-Chip. Annu Rev Fluid Mech. 2004;36:381-411.

3 CHAPTER | Microfluidics for Soft Electronics

Babak Taghavi, Jiantong Li, Mikael Östling and Shi Cheng
KTH, Royal Institute of Technology, Sweden
Ericsson AB, Sweden

ABSTRACT

Microfluidics- based soft electronic systems have the potential to assist conventional rigid devices and circuits to achieve extreme levels of elasticity in wearable electronics and other applications. The goal of employing microfluidics-based approach among other existing methods is to enhance users comfort through fulfillment of wearable's mechanical performance requirements such as flexibility, twistability, and stretchability. This chapter presents a brief survey of different solutions for developing elastic electronics and a thorough review of the progress in microfluidics-based approaches. This chapter mainly focuses on the description of the fabrication process, design, and measurement steps of different antennas and complex systems realized using microfluidic interconnects.

Keywords: *Elastic antennas; liquid metal; microfluidics; stretchable electronics; wearable technology*

1. INTRODUCTION

In recent years, wearable electronics have received increasing attention in both academic research and industrial development. Wearable electronics are being utilized in various applications, including healthcare and medical electronic devices [1] as well as consumer electronics and robotics [2]. Human body surfaces are huge compared to the sizes of electronic components. Large-area coverage is sometimes a must-have feature, where tiny wearable gadgets are certainly not a good fit. Consider an electronic system equipped with multiple sensors, each placed at a fingertip for pressure sensing, and a radio frequency (RF) transmitter associated with an antenna to wirelessly send out the sensing data. Each sensor is connected to the core transmitter circuit by long, in the order of 10 cm, interconnects. Additionally, an antenna with a size of few centimeters is needed for RF communication. Therefore, the overall size of such an integrated electronic device can be as large as 50 –100 cm^2. Using conventional rigid printed circuit board (PCB) techniques to develop this kind of systems would inevitably lead to a clumsy gadget that will most likely give an uncomfortable experience to the wearer. As wearable devices intimately interact with users' body and skin, their physical shapes and impact on haptic feelings of wearers are very important design considerations. Human body is soft and our skin is stretchable, especially on joints, thus any external gadget that comes into seamless contact with our body should feature certain degree of flexibility and stretchability in order to facilitate natural motions [3]. Researchers have been experimenting different materials and assembly methods to create more human-friendly wearable systems. Flexibility can be achieved by implementing electronics on thin flexible substrates, ranging from polyimide- based Kapton films such as circuit boards to thin film transistors (TFTs) on device level. Such flexible electronic devices and systems can withstand moderate folding and twisting, but not stressing. Excessive straining often results in non-reversible mechanical deformation and permanent electrical failure. Achieving stretchability with maintained electronic functions is significantly more challenging. Radically new fabrication techniques and design strategies are required to realize soft electronics, capable of reaching extreme levels of reversible bendability, twistability, and stretchability.

Significant advances have been made in the emerging field of soft electronics in the past two decades. Great potential has been shown in numerous new appealing applications, e.g., body-worn healthcare/medical sensing systems, ergonomic units, and tissue engineering constructs, where conventional rigid, bulky electronics are facing insurmountable obstacles to fulfill the demands. One of the most promising concepts is soft, rubbery electronic devices, implemented in the form of elastic structures, materials, or a combination of both [4]. Wavy silicon ribbons to realize stretchable silicon integrated circuits (ICs) have been reported [5]. Also, integration of TFTs on prestressed substrates has been proposed [6].

Unlike complex, miniaturized semiconductor- based active circuits, many passive components such as interconnects, sensors, and antennas, which occupy the major areas in large-area electronic devices and systems, can only be shrunk at the expense of efficiency, sensitivity, and functionalities. Elasticity of such large passive devices, thereby, dominates the overall flexibility and stretchability of integrated electronics. Various approaches to implement stretchable interconnects have been presented [7,8]. One typical example is meandered metallic interconnects built on flex foils, operating at different frequency ranges [9]. Flexible and stretchable interconnects, realized by incorporating low melting temperature solder alloy into microstructured channels in polydimethylsiloxane (PDMS), were first proposed by Siegel, et al. [10]. Various properties of different solder metals and alloys, including melting temperature, electrical conductivity, Brinell hardness, and material costs, were also investigated. Later, room temperature liquid metal alloy-, eutectic gallium and indium (EGaIn), filled microfluidic channels acting as elastic direct current (DC) interconnects have been demonstrated [11]. Additional appealing characteristics of EGaIn, such as good rheological properties, excellent wetting on almost any surfaces, and low toxicity, make it well suited for microfluidics- based soft electronics that might be in proximity contact with human body in practical use [12]. The first stretchable electronic devices, operating at radio frequencies, are a liquid metal unbalanced loop antenna and a fluidic ultra-wideband planar inverted cone antenna (PICA). Both antennas were implemented by utilizing liquid metal alloy, galinstan, incorporated into microstructured, elastomeric channel matrices [13,14]. Similar concept was also proposed to demonstrate a stretchable dipole antenna. Modified inhomogeneous elastomer substrate was subsequently used to significantly improve elasticity of stretchable antennas [15]. A hybrid integration approach for combining active circuits on flexible PCBs and large-area stretchable passive electronics for RF radiation sensing was later reported by Cheng and Wu [16]. This integration strategy was afterwards expanded to create multi-layer, self-contained, large-area, elastic wireless strain sensor [17]. Standalone patch antenna using serpentine channels filled with EGaIN was also published [18]. More recently, a reconfigurable fluidic serpentine antenna with varying resonant frequency in response to strain was presented [19]. Moreover, successful development of new fabrication techniques for microfluidics- based soft electronics has been demonstrated [20].

The following sections present and discuss recent advances in the emerging field of microfluidics -based soft electronics. This chapter focuses on ultracompliant, elastic electronic devices and systems, with the capability of wireless communications and remote sensing. Following sections start with an elaborate description of the fabrication process, followed by presentation of several application examples of microfluidics- based RF elastic electronics. Measurement results on foldable, twistable, stretchable antennas and integrated devices under different mechanical deformations are also shown. This chapter ends with a glimpse of some ongoing research activities at the Department of Integrated Devices and Circuits, KTH Royal Institute of Technology, Sweden, among which new advanced processing techniques and nanomaterials, e.g., inkjet-printed, high precision, graphene resistors and conductors, are being investigated to enable novel sensing functions. Furthermore, feasibilities of utilizing elastic, microfluidic RF energy harvesters for wireless charging of soft energy storage units or remote powering active stretchable electronics are discussed.

2. FABRICATION PROCESS

Microfluidic channel design procedure can be concisely described in three major steps, which are soft lithography, reverse molding, and liquid metal injection. Similar to PCB routings, depending on the complexity and architecture of circuits and antennas, a single or a multi-layer configuration can be selected to develop an electronic system. First, fabrication processes of single-layer liquid metal- filled channel matrices embedded in PDMS elastomers are presented in this section. Subsequently, processing steps for implementing multi-layer microfluidics-based soft electronic devices are described in detail. Finally, heterogeneous integration method for connecting active circuits assembled on flexible PCBs to microstructured fluidic channels encased in elastomeric substrates is discussed.

2.1. Single-layer microfluidic, elastic channels

Single- layer microfluidic channels are developed on bare silicon wafers, which are used as the substrate for molding and formation of elastic layers. SU8 photoresist is employed as a reverse mold for fluidic channels, and its thickness determines the height of microfluidic channels. SU8 photoresist is spin coated on a silicon wafer as shown in Figure 1(a).

As discussed in the Introduction, microfluidic channels act the same as conductor traces on conventional PCBs; therefore, their patterns are designed in the similar way as routing the traces on PCBs. The designed patterns are then used to build the corresponding lithography mask. After transferring the pattern to the photoresist and removing unexposed parts, the mold is ready for casting the channels. Cross- sectional view of the developed mold on silicon is displayed in Figure 1(b). The prepared mold is then thermally stabilized at 150 °C for 30 minutes to enhance the adhesion between silicon and the remaining SU8 layer. PDMS is used as the soft substrate because of its stretchability, transparency, and relatively easy preparation process. PDMS is widely used in bioelectronics and microfluidic applications such as lab-on-a-chip. It is prepared by mixing PDMS prepolymer and cross linker (Elastosil RT601A and B, Wacker Chemie, Munich, Germany) with 9:1 mass ratio. As shown in Figure 1(c), PDMS should be poured on top of the silicon to form the substrate. In addition to the patterned structures, a thin layer of PDMS is prepared on the blank silicon wafer to form a cover layer for the reversely molded PDMS. After degassing, both patterned and blank PDMS layers are cured at 70 °C for 30 minutes to solidify the PDMS layers. Once the PDMS layers are ready, it is time to peel them off from the silicon carriers as illustrated in Figure 1(e). Before proceeding to the next step, numerous openings should be punched on the patterned structure so that the channels can be accessed for liquid metal injection through those openings. The PDMS layers are bonded together using corona discharging (ETP, Chicago, IL, USA) activation afterwards. At this stage, fabrication of the microfluidic channels is completed and the subsequent step is to inject galinstan into the channels as seen in Figure 1(g). Galinstan is a liquid metal alloy, consisting of 68.5%, Ga; 21.5%, In; and 10%, Sn. It is in liquid state from –19 to 1300 °C, and it also features a good electrical conductivity of 3.46×10^6 S/m and non-toxicity in comparison with mercury. Its excellent wettability on almost any surface simplifies the channel filling process and ensures good

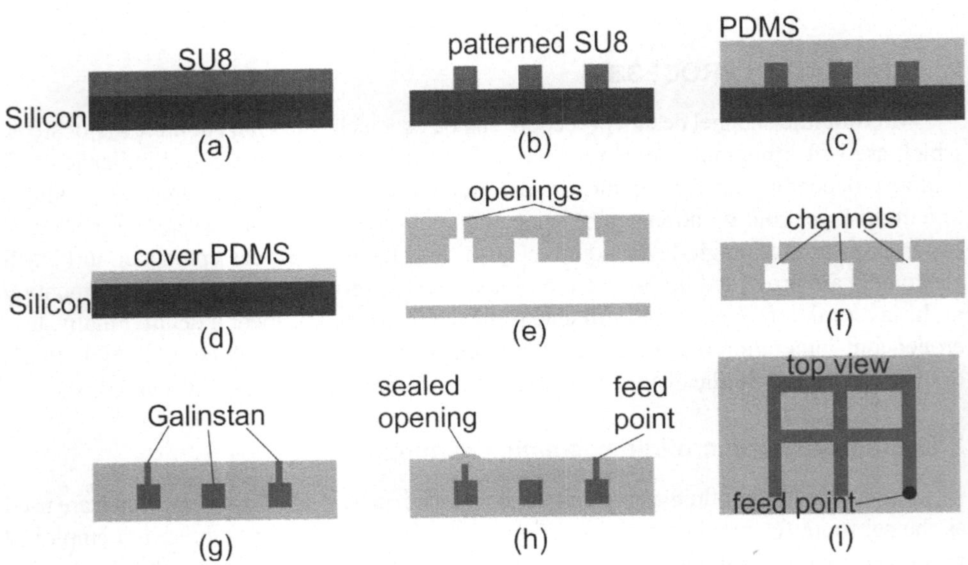

Figure 1. *Single-layer microfluidic soft electronics fabrication process.*

electrical continuity despite of mechanical deformation. After filling the channels, a visual inspection or a simple electrical conductivity check should be performed to ensure that all electrical connections are established as expected. All openings except feeding points for electrical measurement and interconnections with active circuits are sealed with uncured PDMS droplets to prevent any liquid metal leakage. Cross- sectional and top view of the resulting single-layer microfluidic channels filled with liquid metal in a PDMS substrate are shown in Figure 1(h) and (i).

2.2. Multi-layer microfluidic, elastic channels

Some applications necessitate multi-layer structures. Additional channel layers make it possible to implement more sophisticated structures that are not feasible to build in a single-layer configuration. As depicted in Figure 2, fabrication steps resemble the previously presented processes for constructing a single-layer structure. Extra channel layers, however, require more lithography and bonding steps. In most applications, as shown in Section 2.3, feeding points on the same side of an elastomer substrate are often needed to facilitate circuit assembly or characterizations. For assessing the bottom liquid metal filled channels from the top surface of the PDMS substrate, an additional inlet is punched through the top and middle layers after bonding. In the liquid metal injection phase, top layer channels are filled and then openings 2 and 5 are temporarily sealed by small pieces of Scotch tapes and opening 1 is encapsulated with uncured PDMS. Afterwards, the structure is flipped to the other side and galinstan is injected into the channel and both openings 3 and 4 are permanently sealed with uncured PDMS droplets. Later, the bonded substrate is flipped over and the temporary sealants are removed. After verifying all connections, the sample is cured at 85 °C for about 30 minutes.

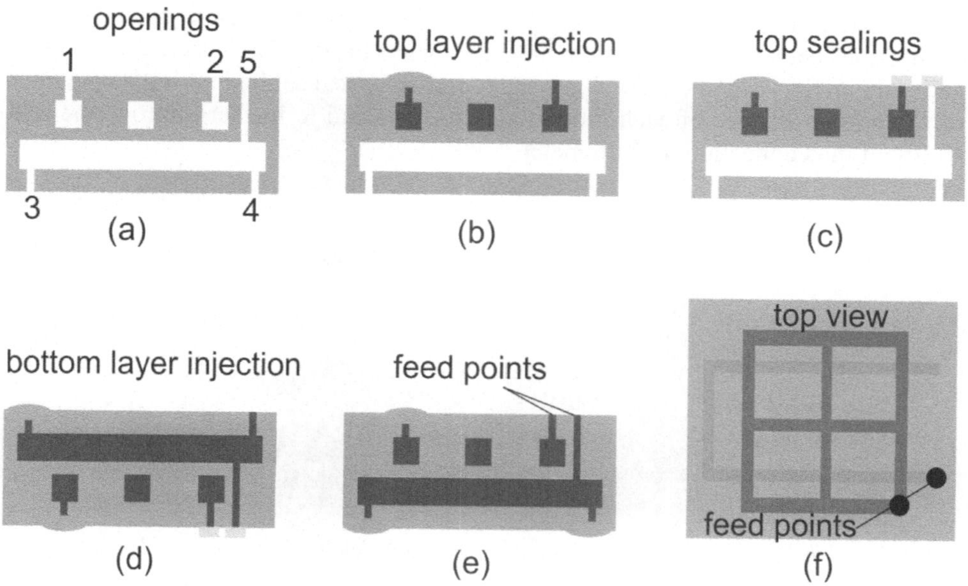

Figure 2. *Processing steps of multi-layer microfluidic soft electronics.*

2.3. Strain isolation cells

Active circuits play an essential role in modern electronic systems. Purely passive microfluidic soft electronics without any active components can only be used in very limited application scenarios in practice. A pressing issue is how to realize active electronics in bendable, flexible, and stretchable manners. Implementing active circuits using microfluidic technologies would be ideal, but to our best knowledge, none of the existing microfluidic techniques support fabrication of high performance, basic active components, such as diodes and transistors, operating at gigahertz frequency range. A compromising solution is to heterogeneously integrate off-the-shelf packaged IC chipsets with passive microfluidic electronics. Here the biggest challenge lies in how to create reliable interconnections between rigid IC chips associated with discrete components and liquid metal filled elastic channels. Figure 3 illustrates a hybrid integration strategy, where IC chipsets as well as external lumped components are first assembled on a small piece of flexible PCB. Such a quasi-flexible circuit contains complex routing and most connections with high reliability. The subsequent step is to implement interconnects between flexible circuits and microfluidic elastic channels via solid metallic pins. Since much fewer connections are needed at this step, the overall reliability of such interconnects are dramatically improved. In order to address the issue of mechanical mismatch between elastomers and rigid components, strain isolation cells, the so-called local stiffening cells (LSCs), are introduced to encapsulate PCBs and metal pins with locally thickened PDMS. Although such cells are a bit more rigid than any other areas on the PDMS elastomer and inevitably degrade the overall elasticity of integrated devices to some extent, they can effectively minimize mechanical deformation around active circuits and interconnect pins when the entire devices are being folded, twisted, or stressed. Reliable galvanic contacts between metal pins and liquid metal alloy are crucial. Soldering tin plating on metallic pins is employed to ensure good electrical connections between solid and liquid metals, thanks to superior wetting properties of galinstan on tin. As depicted in Figure 3, a semi-spherical solder ball is also mounted on the bottom surface of each tin plated pin at one end to further improve the reliability of the connection to fluid metal.

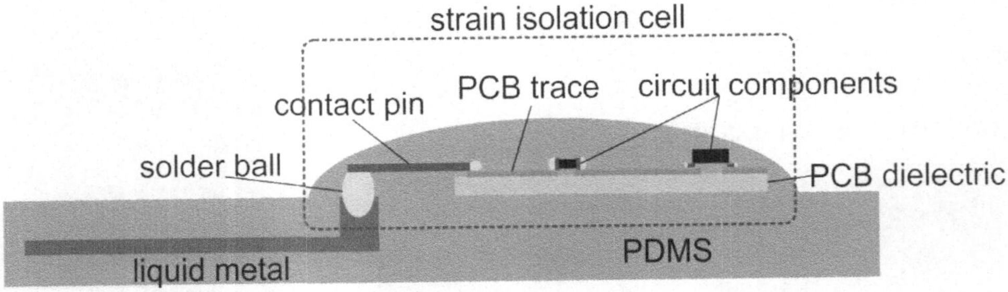

Figure 3. *Hybrid integration of flexible active circuits and fluidic, elastic passive electronics.*

3. MICROFLUIDICS- BASED ELASTIC ANTENNAS AND HETEROGENEOUSLY INTEGRATED RF ELECTRONICS

Fabrication methods introduced in Section 2 are applied to realize different elastic RF-integrated electronic devices and systems. This section presents the design, fabrication, and measured results on standalone single-layer foldable, twistable, stretchable antennas; a fluidic soft RF radiation sensor; and a multi-layer, microfluidic, reversibly stretchable, large-area wireless strain sensor.

3.1. Liquid metal stretchable unbalanced loop antenna

An unbalanced loop antenna, operating at 2.4 GHz Industrial, Scientific, and Medical (ISM) band, is designed, fabricated, and characterized, using single-layer microfluidic channels filled with liquid metal. Schematic of the resulting loop antenna prototype as well as a cross- sectional view of microfluidic channels are shown in Figure 4. The antenna consists of a loop radiating element and a semi-circular, meshed ground plane. The loop is grounded at one end and isolated from the ground plane at the other end for coaxial cable feeding.

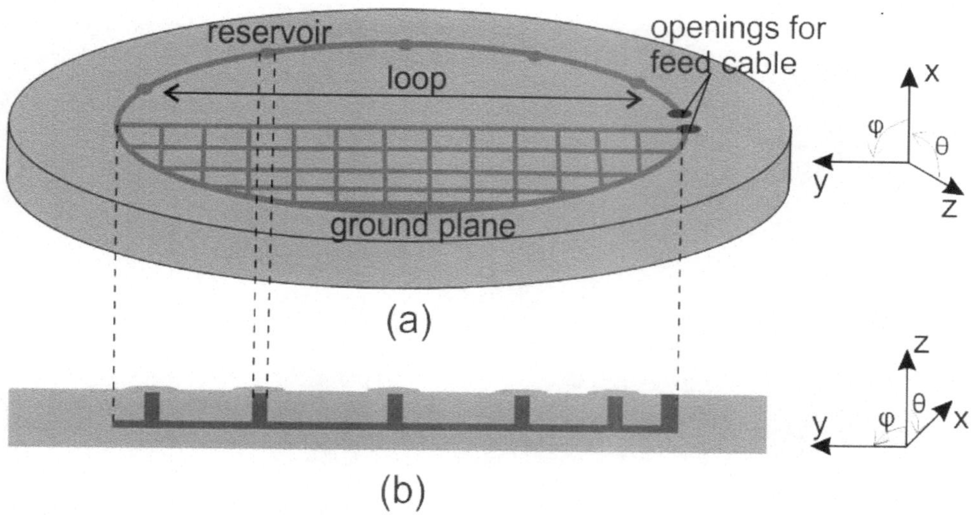

Figure 4. *Schematic of liquid metal stretchable unbalanced loop antenna: (a) overlook view, (b) cross- sectional view.*

The resonant frequency (f_{res}) of a loop antenna is a function of the loop length and can be approximated by the following Equation 1:

$$f_{res} \approx \frac{c}{2L_{loop}\sqrt{\varepsilon_{eff}}}, \tag{1}$$

where c is the velocity of light in free space, L_{loop} is the loop length, and ε_{eff} is the effective permittivity, which is about 1 for very thin PDMS layers. According to the equation, to obtain f_{res} = 2.4 GHz, the loop should be 5.6 cm in length. Numerical simulations are performed in Ansoft HFSS to verify antenna port impedance and radiation characteristics. As seen in Figure 4, numerous small cylinders are aligned along the loop-shaped channel, in which extra liquid metal alloy can be stored and released to guarantee reliable galvanic connections while the antenna is mechanically deformed. It should be noted that the presence of such liquid metal reservoirs results in slightly increased electrical length of the loop antenna so that a minor decrease in resonant frequency compared with the value calculated from the equation can be observed both in simulations and measurements. For example, if the loop length, L_{loop}, equals 56.4 mm, the resonant frequency, f_{res}, derived according to the equation should be 2.7 GHz. However, considering the effect of reservoirs, f_{res} = 2.4 GHz is obtained in numerical simulations. Once the simulation results meet the requirements, the antenna layout is exported to print a mask on transparency for subsequent soft lithography process. The printed plastic mask is then used to transfer the antenna patterns to the SU8 photoresist and to fabricate the antenna following the fabrication steps described in Section 2.1. The resulting unbalanced loop antenna prototype is illustrated in Figure 5.

Figure 5. *Photograph of the fabricated fluidic stretchable loop antenna.*

In order to verify simulated results, the port impedance and radiation characteristics are measured using a vector network analyzer in an anechoic chamber. All experiments are first carried out with the antenna in its relaxed state. Measurement results along with the corresponding simulations are shown in Figure 6.

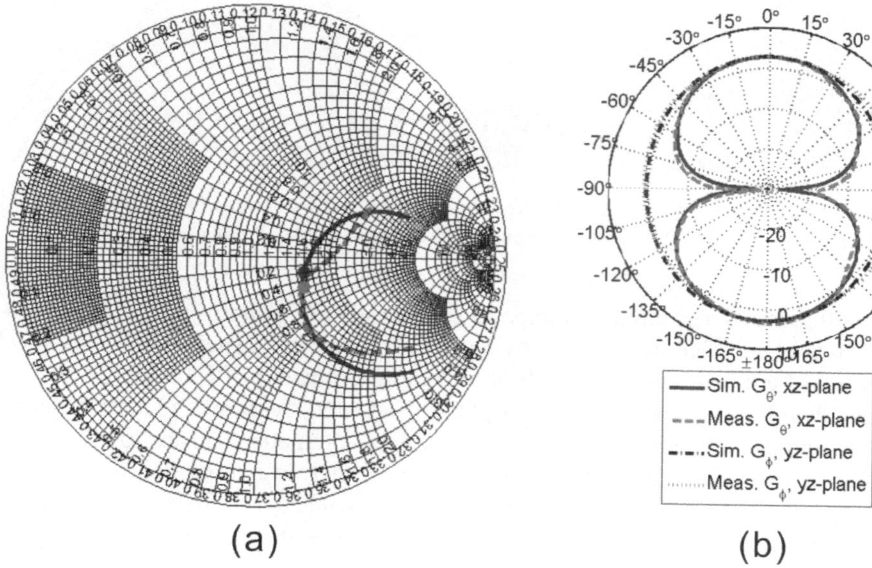

Figure 6. *(a) Simulated (solid line) and measured (dotted line) reflection coefficient (S11) and (b) radiation patterns of the liquid metal- based stretchable unbalanced loop antenna. The coordinate system is depicted in Figure 4.*

Mechanical performance of the microfluidic elastic antenna is tested under various deformations as shown in Figure 7. It is observed that the unbalanced loop antenna prototype can withstand omnidirectional straining, as well as severe folding and twisting, and return to its original shape after the removal of external forces.

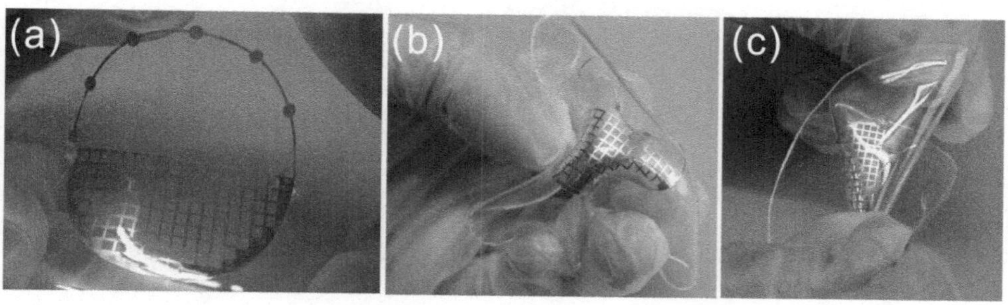

Figure 7. *Photographs of the unbalanced loop antenna in (a) omnidirectionally strained, (b) twisted, and (c) folded states.*

Electrical properties of the fluidic soft antenna in response to varying strains applied along its *x*- and *y*-axis are also characterized. Photographs of the elastic antenna in its relaxed and stressed states are displayed in Figure 8.

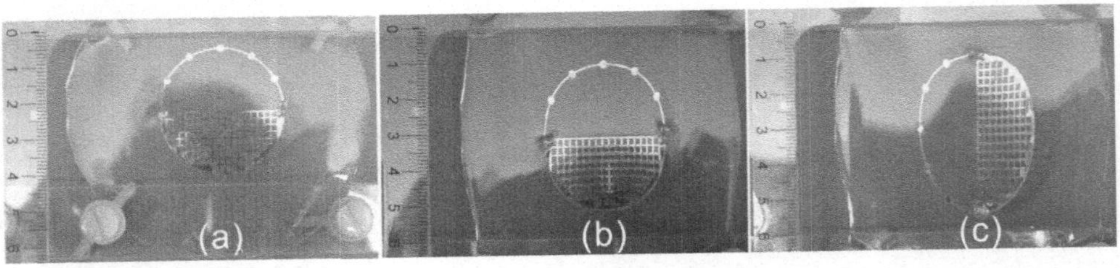

Figure 8. *Photograph of the resulting elastic loop antenna in its (a) relaxed state, and with 40% stain along (b) x- and (c) y-axes. The corresponding coordinate system is shown in Figure 4.*

Figure 9 presents measured port impedance and radiation patterns of the loop antenna in different states. The decrease in the antenna resonant frequency can be explained by an increased antenna size as a result of stretching. Antennas often exhibit wider radiation bandwidth than impedance bandwidth. This is why, measured radiation efficiency remains relatively stable despite being strained. No dramatic electrical performance degradation is detected in the experiments. It implies that no interruption in electrical connections occurs during antenna deformation. This is mainly because of the superior wettability of galinstan on the inner surfaces of elastomeric channels as well as its good electrical conductivity.

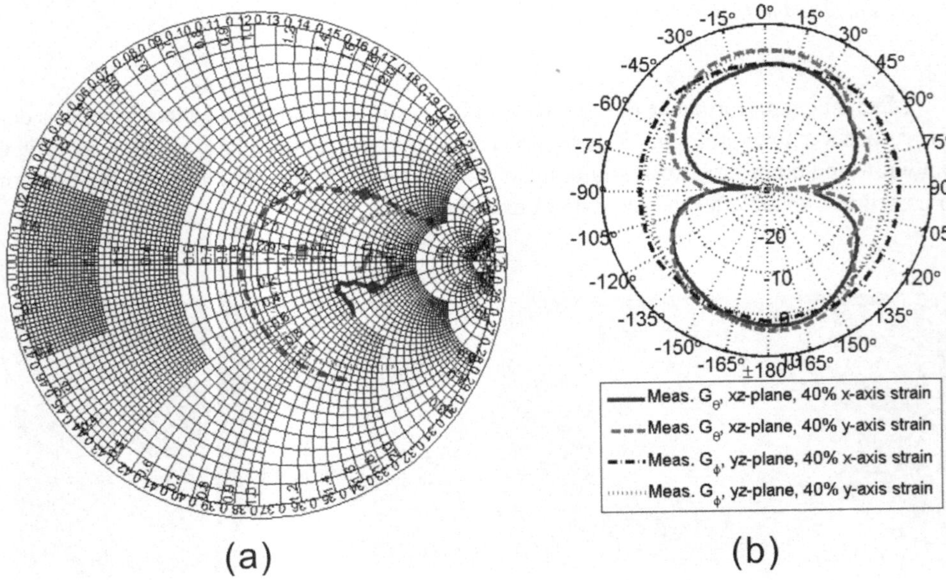

Figure 9. *Measured (a) port impedance and (b) radiation patterns of the stretchable unbalanced loop antenna. The coordinate system is seen in Figure 4.*

3.2. Elastic electromagnetic radiation sensor

Our daily life has been greatly benefiting from revolutionary mobile communication technologies rapidly developed in the past decades. Billions of portable devices, equipped with wireless links, are new parts of our everyday life. People and electronic equipment and systems are connected anywhere at any time. Rising concerns on potential health issues caused by the exposure to RF radiation has been expressed by both scientists and the public. It is, thereby, worth monitoring the level of ambient electromagnetic (EM) wave radiation to assess possible hazards and improve signal coverage quality in real life. In order to make it more portable, an EM radiation sensor can be developed as a wearable device. As discussed in the Introduction Section 1, better user experience can be achieved by constructing such a radiation sensor in ultracompliant, soft, and elastic manners. An RF power detector circuit assembled on a flex foil and encapsulated in an LSC is designed and interconnected to a microfluidics- based stretchable antenna to form a hybrid device, *cf.* Figure 10. Target operating frequency of the proposed radiation sensor is 900 MHz. A uniplanar rectangular unbalanced loop antenna is chosen for its ease of fabrication. The overall dimensions of the loop obtained from full wave simulations conducted in Ansoft HFSS are 86.9 mm × 88.0 mm × 1.0 mm. Ambient RF radiation is picked up from free space and directed to the input of the RF power sensing module via an elastic, fluidic antenna. The injected RF energy is then proportionally converted to DC voltages at the RF power detector IC output. The upper and lower sensing range of RF radiation energy is determined by the dynamic range of the power detection IC chipset. The chosen IC chip features a dynamic range from approximately –60 dBm to 0 dBm that is sufficient in most application scenarios. Apart from the core IC, three coupling and decoupling surface mounted device (SMD) capacitors along with an input matching inductor are needed. The entire power sensing active circuit mounted on a flexible PCB is powered externally, using a pack of four AA batteries connected serially. A light emitting diode (LED) is soldered at the voltage output of the power detector for visualization, which will be switched on when the received RF energy exceeds the pre-defined threshold. More details about the integrated RF radiation sensor can be found in Figure 10.

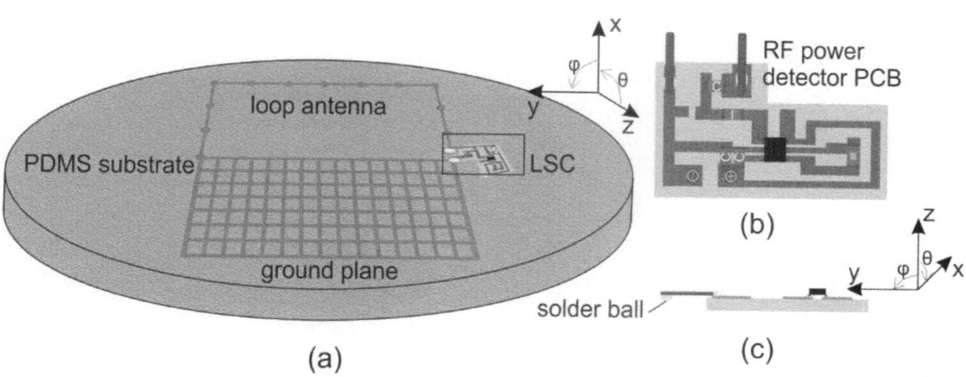

Figure 10. *Schematic view of the elastic EM sensor: (a) overlook view of the integrated sensor, (b) top view, and (c) cross view of the flexible active circuit.*

Prior to connecting the microfluidic antenna to the active circuit, electrical performance of both the antenna and the detector circuit is evaluated separately. Figure 11 shows measured reflection coefficient of the loop antenna in different stretched modes.

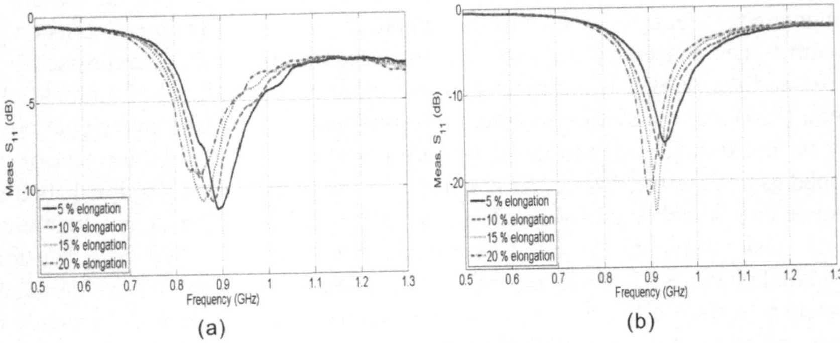

Figure 11. *Measured reflection coefficient of the microfluidic elastic rectangular loop antenna in response to varying strain applied along its (a) x- and (b) y-axis. The coordinate system is presented in Figure 10.*

The output DC voltage of the standalone active circuit in response to varying RF input power is characterized. In this experiment, continuous wave (CW) signal with incremental power level generated from a signal synthesizer is fed into the RF power detector input through an RF coaxial cable, and the LED indicator at the DC voltage output is temporally disconnected. The output voltage of the RF power detector as a function of input power is shown in Figure 12, in which fairly linear behavior is achieved over the entire dynamic range.

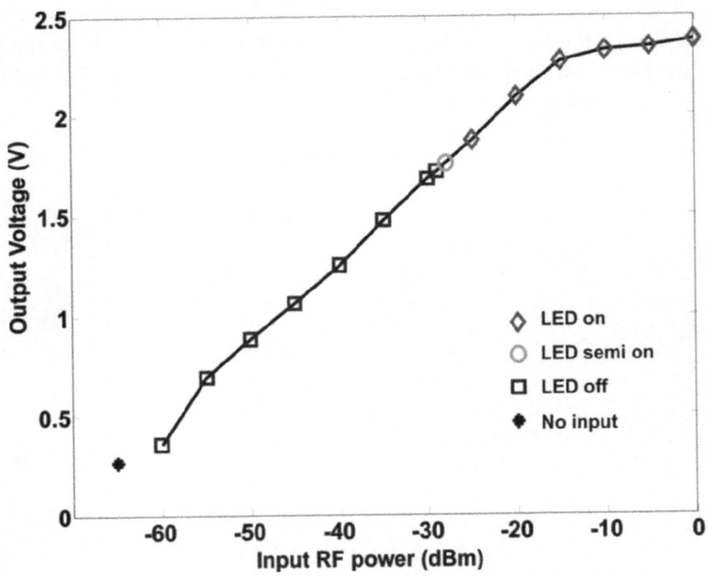

Figure 12. *RF power module output voltage and LED status versus input RF power.*

Subsequently, the PCB is assembled on the PDMS substrate using the proposed hybrid integration method. The fabricated elastic EM radiation sensor is shown in Figure 13.

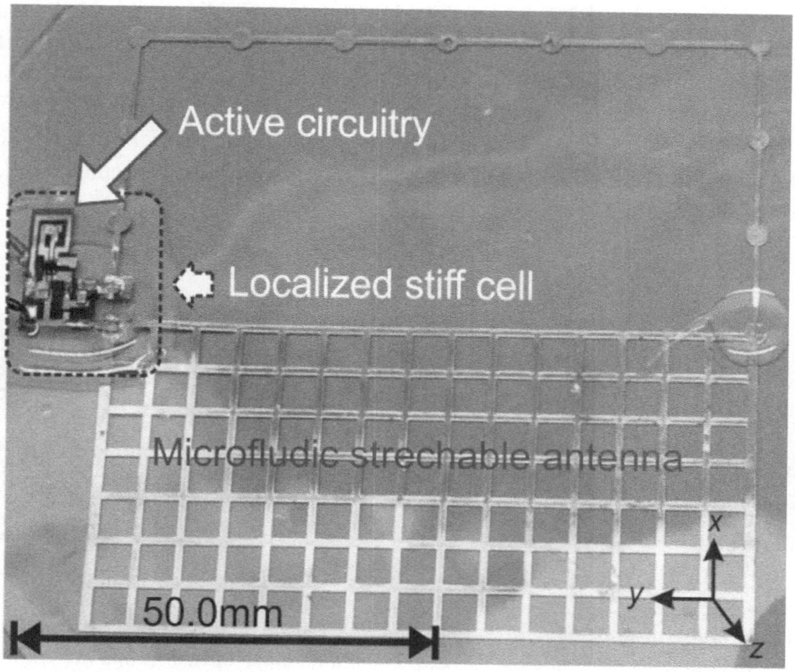

Figure 13. *Photograph of the resulting elastic EM radiation sensor prototype.*

The measurement setup illustrated in Figure 14 is utilized to demonstrate RF radiation sensing capability of the integrated sensor device in ordinary office environment. The distance from the dedicated radiation source comprising a standard horn antenna and a signal generator to the elastic RF radiation sensor is approximately 5 m.

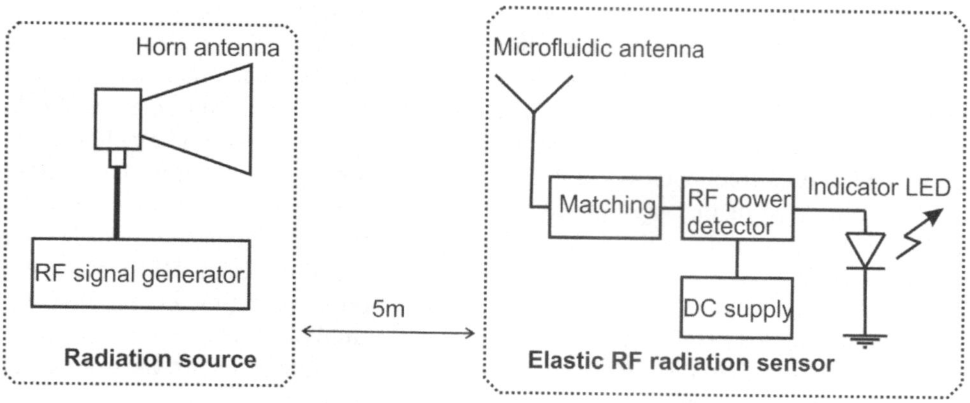

Figure 14. *Schematic of the demonstration setup for the implemented elastic RF radiation sensor.*

The transmitter sends out a CW signal at 900 MHz with variable power levels into the free space. The LED indicator on the integrated stretchable sensor device is on when being exposed to high levels of radiation. The ON/OFF state of the LED indicator only depends on the level of the RF power received by the microfluidic stretchable loop antenna. Increased distance between the radiation source and the sensor would result in greater free space loss. Higher power is, therefore, required from the RF signal generator to turn on the LED in this case. In addition to measurements in its relaxed state, the stretchable EM radiation detector is also tested in its folded, twisted, and strained modes. Figure 15 shows that the implemented microfluidics- based soft radiation sensor prototype remains functioning regardless of deformation.

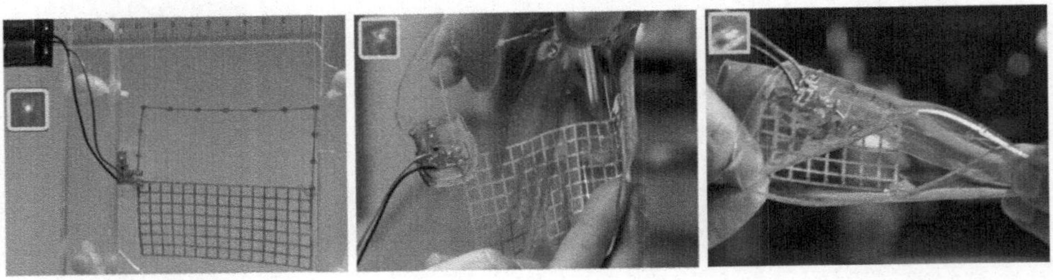

(a) (b) (c)

Figure 15. *Photographs of the elastic EM radiation sensor operating under deformation: with 15% elongation along its (a) x- and (b) y-axis, (c) omnidirectional stressing, and (d) severe twisting. The corresponding coordinate system can be found in Figure 10.*

3.3. Fluidic, elastic ultra-wideband antenna

Antennas with broad operational frequency bands are sometimes required. In ultra-wideband (UWB) systems, an extremely broad frequency range from 3.1 GHz to 10.6 GHz needs to be covered using a single antenna. Implementing a UWB antenna in highly flexible, bendable, and stretchable formats would be of great interest. Such an antenna can also be deployed in other applications that only require a fractional bandwidth of UWB, as it is expected to feature robust electrical characteristics in some narrower frequency ranges under deformation. A non-resonant antenna, PICA, is chosen because of its uniplanar configuration and ease of fabrication. The lower and upper operational frequencies of a PICA are determined by the height of the inverted cone and the size of the gap between the radiator and the ground, respectively. Its tapered shape gradually transforms the free space impedance of 377 Ω to 50 Ω over a wide frequency range. Full wave simulations performed in Ansoft HFSS suggest a radius of 10 mm (*R*) and a gap size of 300 μm (G) for covering the operational frequency range of 3–11 GHz. Schematic views of the liquid metal elastic PICA are seen in Figure 16. Similar to the previously presented microfluidic antennas, its ground and radiating element are realized in meshed configurations to separate the top and bottom PDMS membranes.

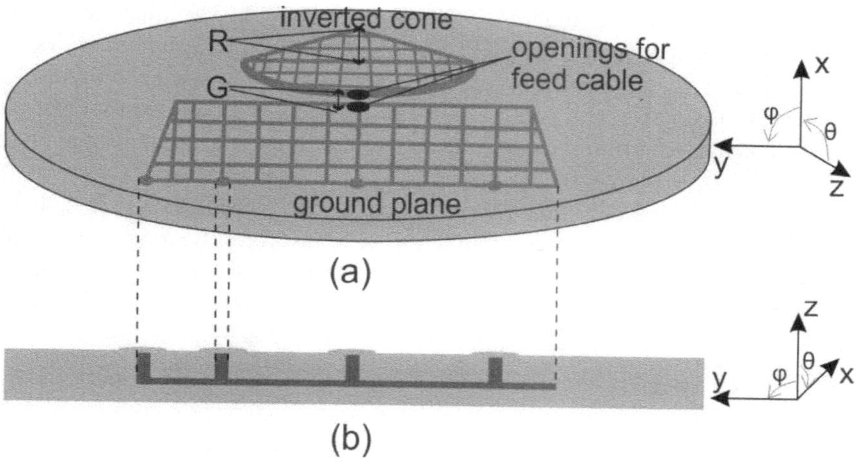

Figure 16. *Schematic of the microfluidics- based stretchable ultra-wide band antenna: (a) overlook view and (b) cross view of microfluidic channels in a PDMS elastomer.*

Mechanical properties of the fabricated antenna are first evaluated. Figure 17 shows that the resulting PICA prototype is reversibly stretchable, foldable, and twistable. After removing external forces, the antenna always returns to its original shape without any visible mechanical damages.

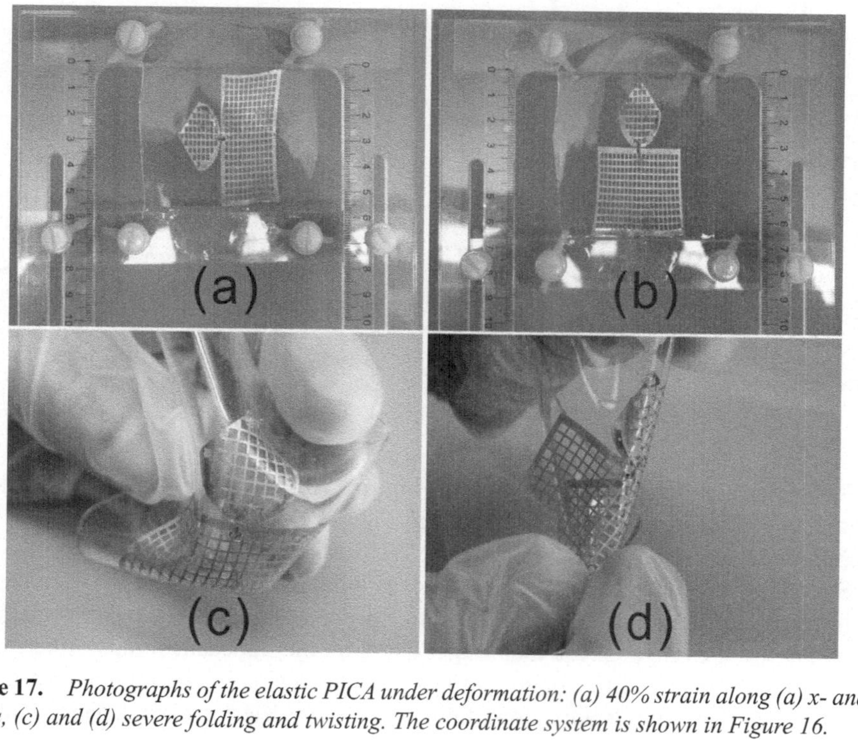

Figure 17. *Photographs of the elastic PICA under deformation: (a) 40% strain along (a) x- and (b) y-axes, (c) and (d) severe folding and twisting. The coordinate system is shown in Figure 16.*

Relative stretchability as high as 100% should be within reach according the data sheet of the PDMS elastomer in use. Openings in the top PDMS membrane as well as the rigid feed cable set an upper limit of approximately 50% on stretchability. Low temperature test is also performed, using a household freezer with a lowest temperature of –24 °C. No mechanical failure can be found after cooling the antenna prototype at –24 °C for 2 hours.

Port impedance and radiation characteristics of the microfluidics- based stretchable PICA are all measured at room temperature. Measured reflection coefficient, S_{11}, of the relaxed antenna is in line with the corresponding simulated result and good impedance matching (S_{11}<–10 dB) is achieved between 3 GHz and 11 GHz, as seen in Figure 18.

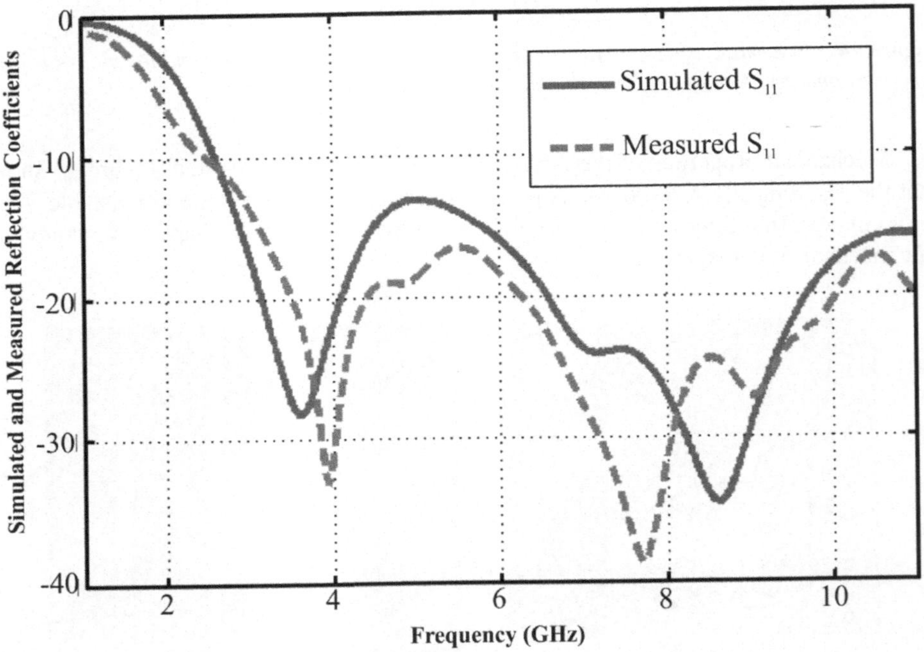

Figure 18. *Simulated and measured S11 of the stretchable ultra-wide band PICA in its original state.*

Reflection coefficients of the stretchable fluidic PICA are also characterized in various flexed states. Experimental data presented in Figure 19 indicate that robust impedance matching is achieved even if the microfluidic PICA is strained.

Stretching the antenna along its *x*-axis leads to increased height of the inverted cone and lowered first resonant frequency. Port impedance at higher frequencies remains stable in spite of stressing, as it is mainly governed by the impedance transformer formed by the tapered shape and the small gap between the antenna radiating element and the ground plane. Excessive strains along the *x*-axis might deteriorate impedance matching performance to some extent because of significantly increased gap size and severe deformation of the impedance transformer. Elongation along the *y*-axis only slightly impacts measured reflection coefficients because of relatively small changes in the tapered shape and the gap size. Radiation patterns of the relaxed and flexed elastic PICA are measured at 2.5 GHz and 5 GHz. Figure 20 illustrates the measured data at 2.5 GHz.

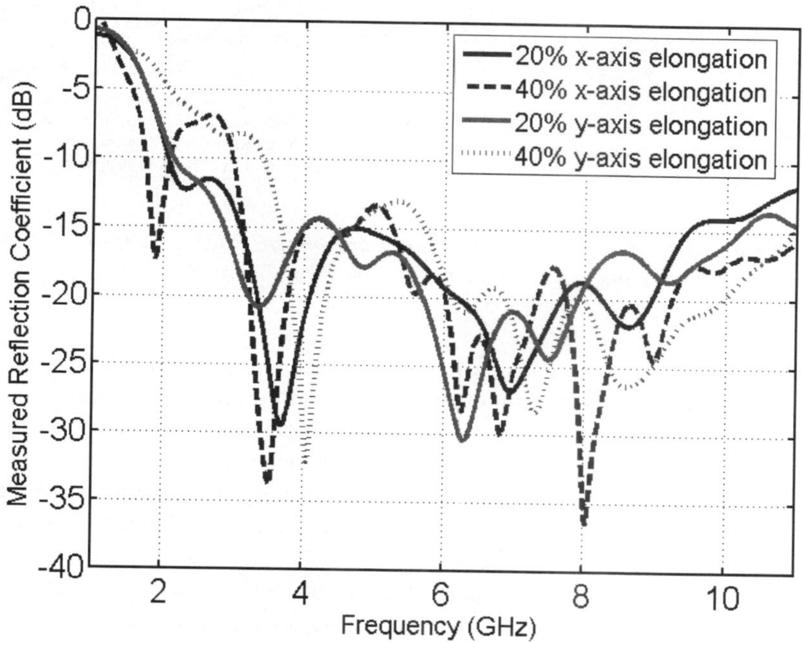

Figure 19. *Measured S11 of the flexed PICA. The coordinate system can be found in Figure 16.*

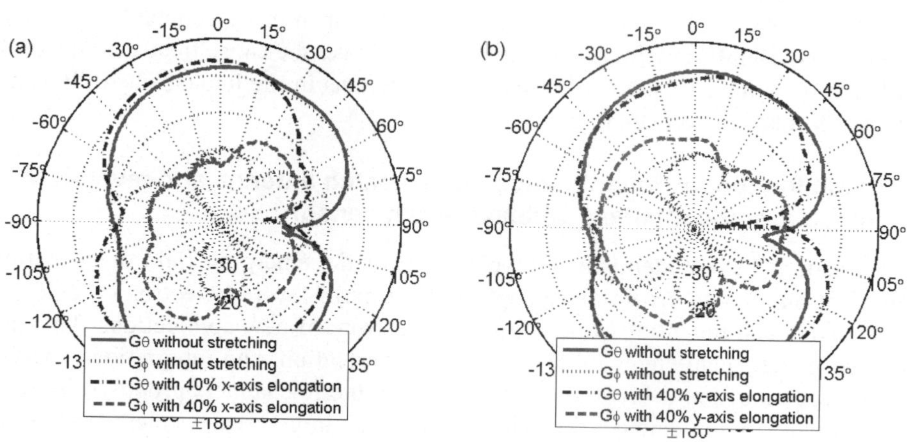

Figure 20. *(Continued)*

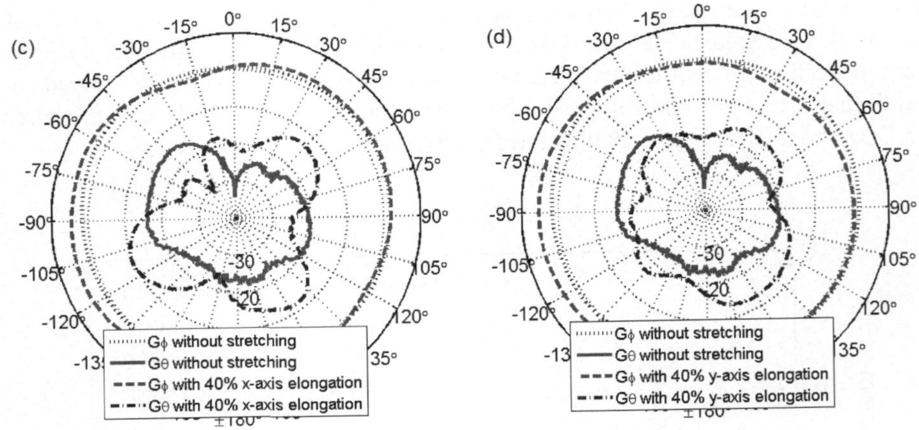

Figure 20. *Measured radiation patterns of the relaxed and stressed PICA in the (a), (b) xz-, and (c), (d) yz-plane. The corresponding coordinate system is depicted in Figure 16.*

In comparison with antenna gain, variations on measured radiation patterns as a result of stressing are greater. Broad beam coverage similar to an ordinary monopole antenna is achieved in both relaxed and flexed cases. Maximum gain of approximately 2.2 dBi is measured at 2.5 GHz. Detailed results on radiation characteristics of the resulting elastic PICA can be found in Cheng et al. [14]. Strong ripples are seen in measured radiation patterns at 5 GHz. The major cause for these fast ripples is frequency- dependent disturbance introduced by the feed cable in the experiments.

Radiation efficiency is another important figure of merit for terminal antennas featuring omnidirectional radiation patterns. A custom-designed reverberation chamber is employed for rapid radiation efficiency measurements. Experimental results at different frequencies are plotted in Figure 21.

Although measured radiation efficiency decreases a bit at 3 GHz during stretching, it still remains above 70%, an acceptable figure, better than many commercially available antennas. No significant efficiency drop is detected over the entire frequency range. This further verifies the robustness of electrical connections in the implemented PICA under different forms of deformation.

3.4. Multifunctional, microfluidics- based patch antenna and reversibly stretchable, self-contained, wireless strain sensor

A natural further step in the development of the emerging field of microfluidic soft electronics is to create more sophisticated multi-layered electronic devices and systems that can potentially open up a full spectrum of new applications. This, however, poses new challenges in device design, fabrication, and integration. The fabrication process of multi-layer microfluidic, elastomeric channels proposed in Section 2.2 is utilized to implement a reversibly stretchable, large-area, wireless strain sensor. This section details the design, fabrication, assembly, and characterizations of the sensor device.

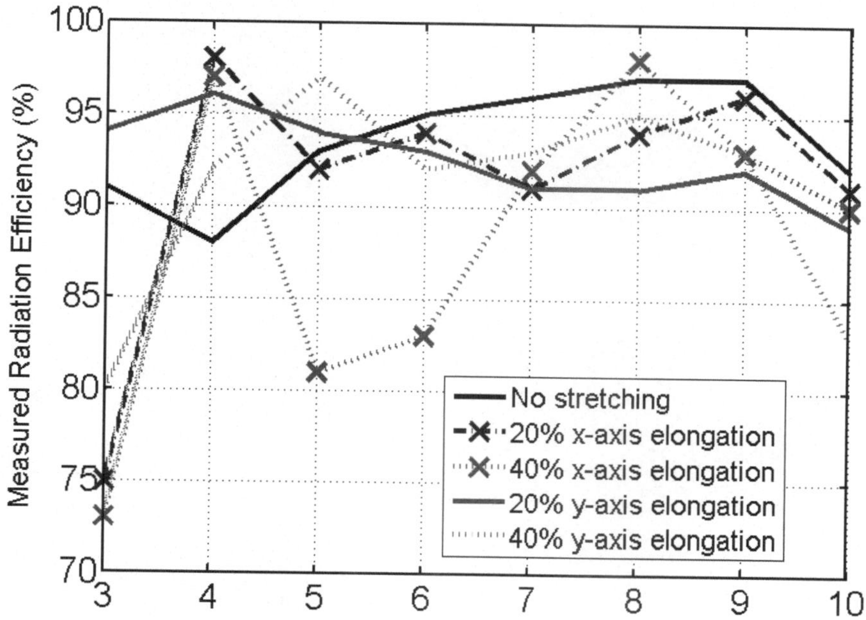

Figure 21. *Measured radiation efficiency of the PICA in its relaxed and stretched states. The coordinate system is presented in Figure 16.*

The integrated wireless strain sensor contains an RF transmitter on a flexible PCB and a multifunctional, microfluidic, stretchable patch antenna for both wireless communication and strain sensing. The fluidic patch antenna is constructed in a double-layer configuration, *cf.* Figure 22. The principle for using an elastic antenna for strain sensing is simple. Because the resonant frequency of the patch antenna in use is largely determined by its electrical length along the x-axis, elongation along this direction would result in greater electrical length so that it decreases its resonant frequency. In the case of a patch antenna with narrow bandwidth (or high quality factor, Q), considerable impedance mismatch will occur at its original resonant frequency as a result of strain along its x-axis. When feeding such an antenna with a 50 Ω source at a constant power level, significantly decreased radiation will be measured in the far-field. Higher strain sensing sensitivity can be attained by reducing antenna bandwidth. But decreasing bandwidth would put high demands on fabrication precision. A tradeoff between sensitivity and fabrication yield has to be made. In addition to the multifunctional antenna, the integrated sensor also contains a simplified RF transmitter, which can be externally powered by a stack of two AA batteries connected serially. Details of this integrated electronic sensor device can be found in the schematic views presented in Figure 22.

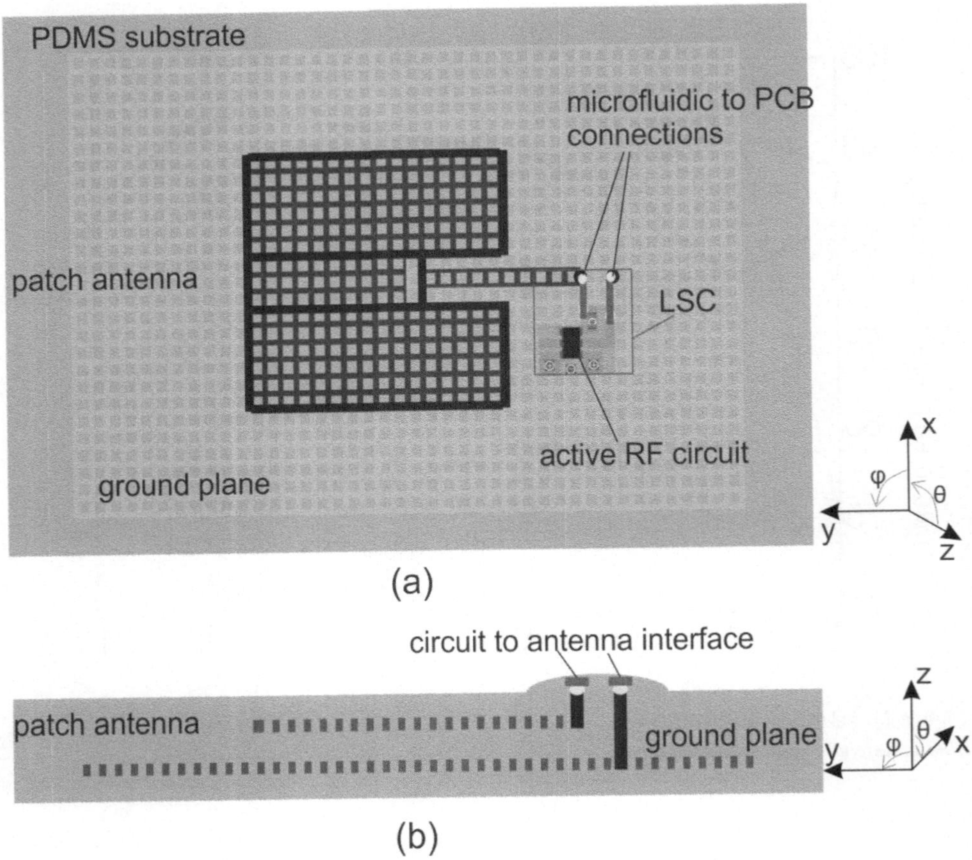

Figure 22. *Schematic view of the proposed elastic wireless strain sensor: (a) top view and (b) cross view of microfluidic channels.*

The designed patch antenna is fed by a short piece of microstrip. The spacing between the patch and the ground plane determines the bandwidth of the antenna. The greater the spacing is, the wider bandwidth the antenna features. Impedance bandwidth is also affected by dielectric losses of the substrate. Lossy dielectric materials would increase antenna bandwidth at the expense of radiation efficiency. Poor radiation efficiency should be definitely avoided, as it significantly shortens remote sensing range. Resembling microfluidic antennas reported in the previous sections, both the antenna patch and ground plane are realized in the meshed configuration. The active circuit assembled on a flex foil comprises a 1.5 GHz voltage controlled oscillator (VCO) associated with a few discrete components. The effective strain sensing area is as large as the overall size of the integrated sensor, approximately 110.0 mm × 80.0 mm. It makes the resulting wireless sensor well suited for remote, large-area strain sensing.

Mechanical properties should be taken into account in the design as well. Thinner PDMS substrates are preferred because of higher elasticity. A very thin substrate would, however, constrain the spacing between the patch and the ground plane, and lead to too narrow antenna bandwidth. Parametric analysis in numerical simulations shows that good compromise values for the overall thickness, and the spacing are 2.5 mm and 1.5 mm. A microfluidic, flexible, stretchable wireless strain sensor is implemented, following the fabrication and integration process described previously. Two openings on the top surface of the PDMS elastomer are utilized for hybrid integration with RF transmitter. A photograph of the fabricated sensor device is presented in Figure 23.

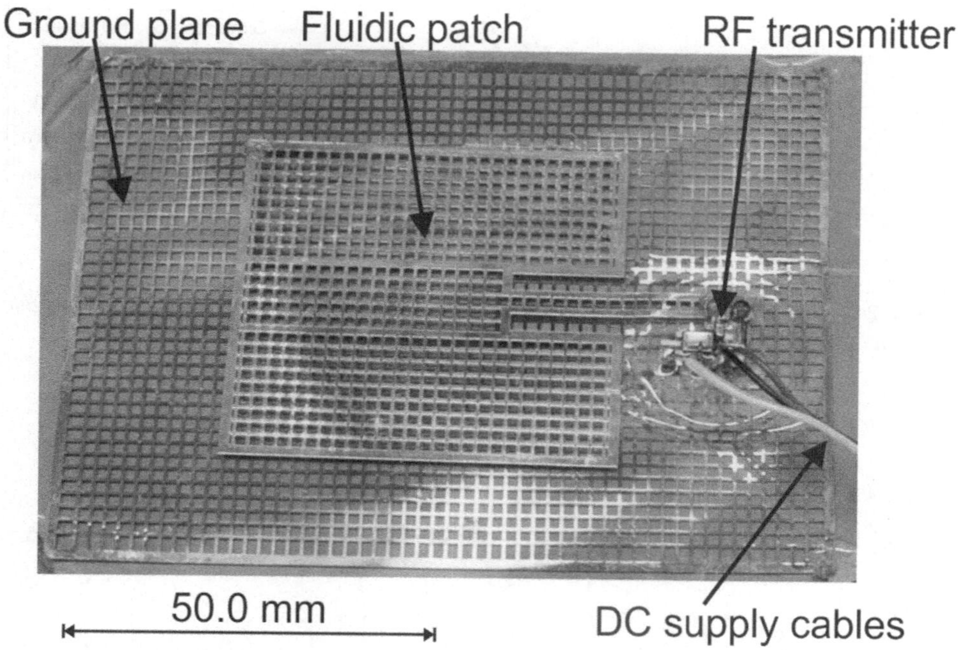

Figure 23. *Photograph of the resulting microfluidics- based stretchable wireless strain sensor.*

Prior to the final integration, the standalone patch antenna and RF transmitter circuit are tested individually. Reflection coefficients of the fluidic soft patch in response to varying strains are seen in Figure 24.

Experimental results indicate a dramatic resonant frequency decrease in response to incremental strain along the x-axis. Significant mismatch losses are observed at the oscillation frequency, 1.46 GHz, of the VCO during stretching. Measured radiation patterns (including mismatch losses) depicted in Figure 25 show greater than 10 dB realized antenna gain drop when the patch antenna is stressed from its original state to 15% elongation along the x-axis. Front-to-back ratio of radiation patterns is better than 10 dB. It should be noted that the meshed ground plane acts as an effective radiation shield and excellent front-to-back ratio of higher than 10 dB implies good forward radiation.

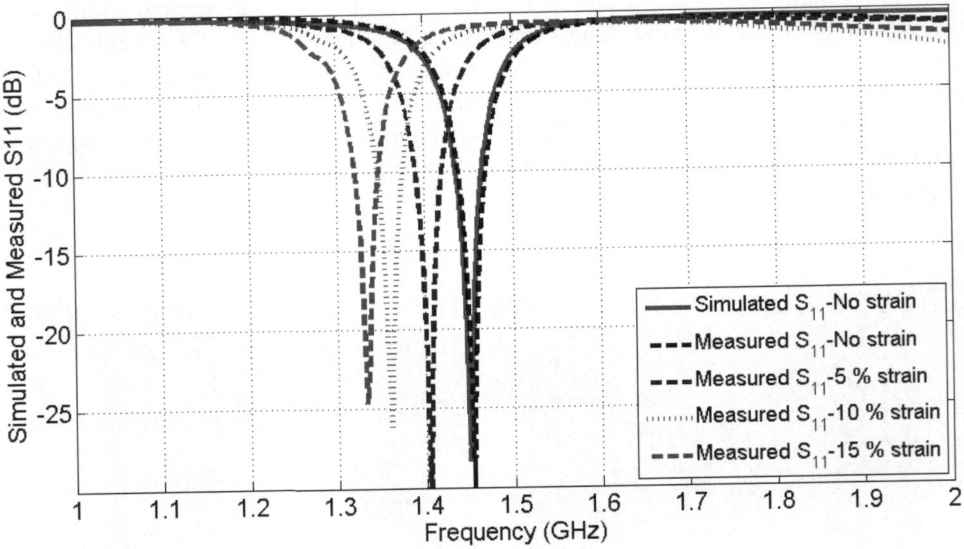

Figure 24. *Simulated and measured S11 of the standalone fluidic patch antenna in its original state, and with varying strain along x-axis. The coordinate system is seen in Figure 22.*

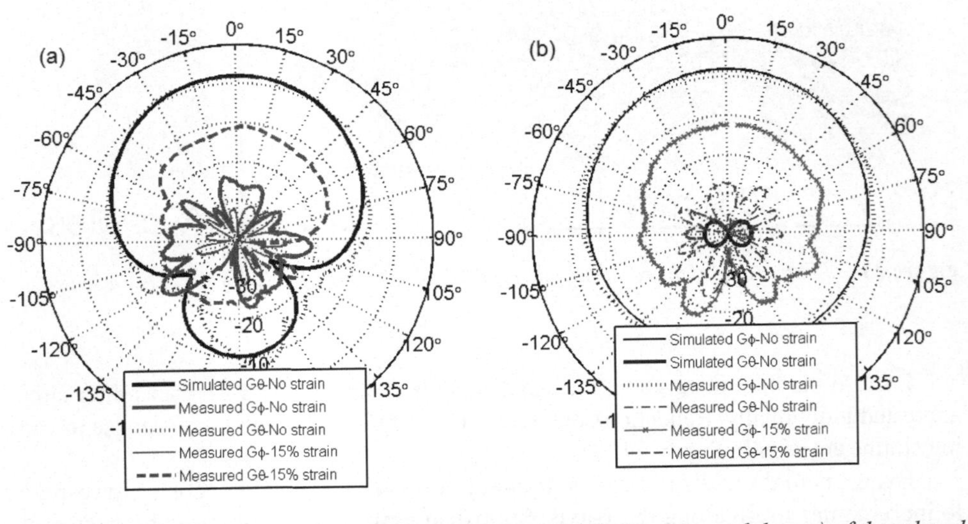

Figure 25. *Simulated and measured radiation patterns (including mismatch losses) of the relaxed and flexed microfluidic patch antenna in the (a) xz- and (b) yz-plane. The coordinate system is shown in Figure 22.*

After antenna measurements, the active circuit and the elastic antenna are heterogeneously integrated. On the other end of the wireless strain sensing system, a custom-designed receiver is developed to receive and interpret sensing data. The system diagram of the entire test setup for the elastic wireless strain sensor is illustrated in Figure 26.

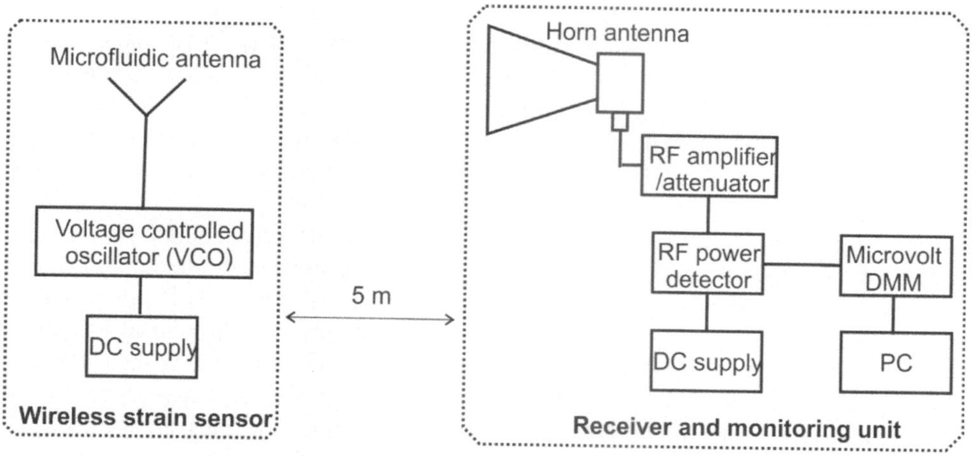

Figure 26. *Illustration of the setup for demonstrating the microfluidic, reversibly stretchable, large-area wireless strain sensor.*

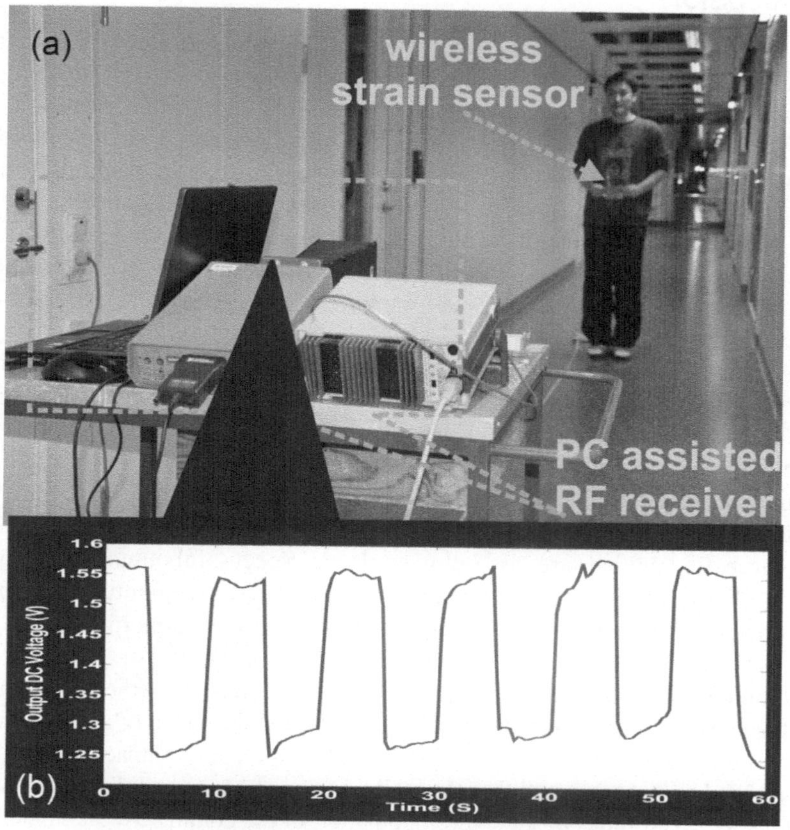

Figure 27. *(a) Photograph of the demonstration of the elastic, wireless strain sensor. (b) Voltage readouts at a custom-designed PC-aided receiver.*

As shown in Figure 26, the receiver consists of numerous sub-modules for RF power detection and is assisted by a personal computer (PC) for data readout and plotting. The output of the RF power detector is a DC voltage, proportional to the input RF power level. This voltage is measured by a high resolution digital multi-meter (DMM) controlled by the PC. The receiver unit is calibrated with varying RF power around 1.5 GHz, using a high precision synthesizer prior to the actual measurements.

The microfluidic, reversibly stretchable wireless strain sensor is tested in an ordinary office environment, *cf.* Figure 27. The distance between the integrated sensor and the receiver station is roughly 5 m. Repeated strain with a period of approximately 10 seconds is manually applied along the x-axis of the sensor. The plot in Figure 27 displays the data readout at the PC, which is in line with the manual stretch. Reversibility can also be clearly seen in the plot, as the measured voltage always returns to the original level after the removal of external strain. Such an integrated sensor can be deployed as a comfortable wearable electronic device for real-time monitoring of body motion.

4. CONCLUSION

An overview in the emerging field of ultracompliant, soft electronics that can be twisted, compressed, and stressed into almost any arbitrary shapes has been presented. Different strategies for realizing bendable, flexible, stretchable electronic devices and systems have been briefly discussed. Microfluidics- based soft electronics, one of the most promising approaches for implemented elastomeric electronics, has been addressed in detail. Fabrication techniques, hybrid integration concepts, as well as various electronic devices built in microfluidic, elastic formats are the focus of this chapter. Appealing application examples, ranging from single-layer standalone fluidic stretchable antennas to multi-layer, self-contained, large-area, wireless sensors, have been demonstrated.

Research efforts in incorporating new materials and novel technologies into soft electronics are ongoing. Figure 28 shows an example of a hand-shaped, soft, rubbery sensor for multi-point, remote pressure sensing. Emerging nanomaterials and fabrication processes, e.g., pressure-activated conductive rubber and inkjet-printed graphene -based flexible circuits, are introduced to create novel pressure sensors and active RF electronics. Feasibilities of implementing microfluidic, elastic EM energy harvesting devices are also being explored. Successful outcomes will pave the way for realizing batteryless, ultracompliant, wearable and implanted electronics that can be powered by ambient EM energy generated from billions of wireless devices around us.

Microfluidic flexible, compressible, stretchable electronics together with other members in the new category of soft electronics are reshaping the world of electronics so as to the future of our daily life. The number of cell phones already outpaced the world population in 2014. It is anticipated that the total number of connected devices will grow to approximately seven times of the entire population in the coming 5 years, among which billions of soft electronic devices will be found in many new applications, including healthcare/medical monitoring, tissue engineering, bio-inspired robotics, and so on.

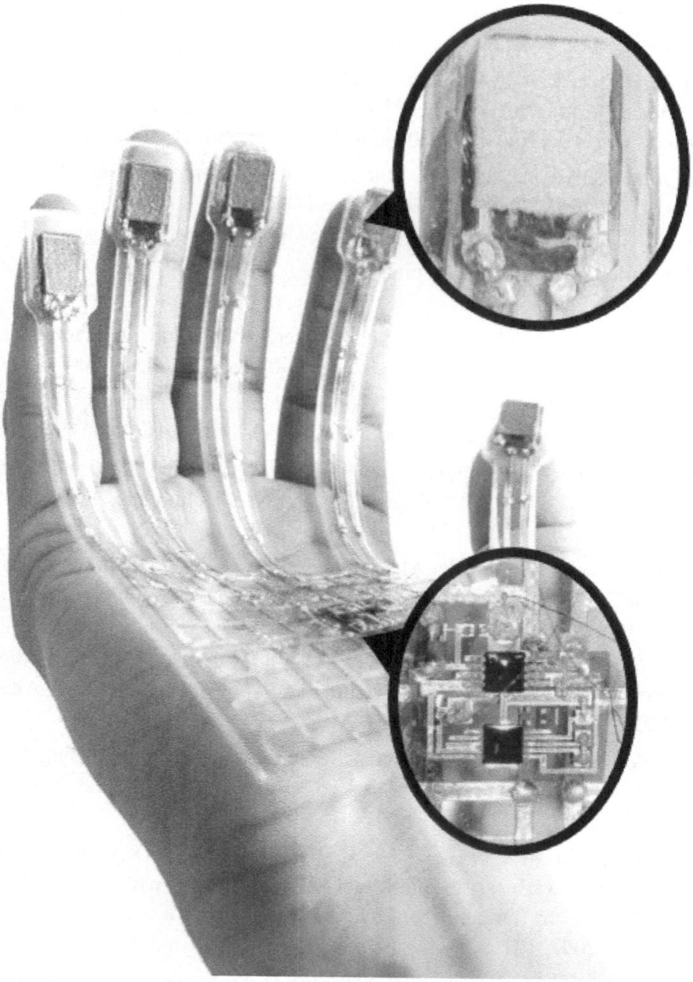

Figure 28. *Photograph of the stretchable multi-point, wireless pressure sensor prototype.*

Acknowledgments

Dr. Shi Cheng currently holds a Young Research Fellow position financed by the Swedish Research Council (VR) under the contract (Avtals-ID: C0486501). Also Dr. Jiantong Li acknowledges Göran Gustafsson Foundation for the financial support through the Young Researcher Prize (No. 1415 B).

REFERENCES

1. Rutherford J.J.Wearable Technology. Engineering in Medicine and Biology Magazine, IEEE. 2010;29(3):19-24. DOI: 10.1109/MEMB.2010.936550

2. Vertechy R., Accoto D. Wearable Robotics. Robotics & Automation Magazine, IEEE. 2014;21(4):19-110. DOI: 10.1109/MRA.2014.2364732

3. Honda W., Harada S., Arie T., Akita S., Takei K. Flexible Electronics: Wearable, Human-Interactive, Health-Monitoring, Wireless Devices Fabricated by Macroscale Printing Techniques. Advanced Functional Materials. 2014;24(22):3298. DOI: 10.1002/adfm.201470144

4. Rogers J.A., Someya T., Huang Y. Materials and Mechanics for Stretchable Electronics. Science. 2010;327:1603. DOI: 10.1126/science.1182383

5. Kim D.H., Ahn J.H., Choi W. M., Kim H.S., Kim T.H., Song J. Z., et al.. Stretchable and Foldable Silicon Integrated Circuits. Science. 2008;320(5875):507-511. DOI: 10.1126/science.1154367

6. Lacour S.P., Wagner S., Huang Z.Y., Suo Z. Stretchable Gold Conductors on Elastomeric Substrates. Applied Physics Letters. 2003;82(15):2404-2406. DOI: 10.1063/1.1565683

7. Lacour S. P., Jones J., Wagner S., Li T., Suo Z. Stretchable Interconnects for Elastic Electronic Surfaces. Proceedings of the IEEE. 2005;93(8):1459-1467. DOI: 10.1109/JPROC.2005.851502

8. Haopeng W., Debao Z., Jianguo C. Development of a Stretchable Conductor Array with Embedded Metal Nanowires. IEEE Transactions on Nanotechnology. 2013;12(4):561 -565. DOI: 10.1109/TNANO.2013.2258172

9. Brosteaux D., Axisa F., Gonzalez M., Vanfleteren J. Design and Fabrication of Elastic Interconnections for Stretchable Electronic Circuits. Electron Device Letters, IEEE. 2007;28(7):552 -554. DOI: 10.1109/LED.2007.897887

10. Siegel A. C., Bruzewicz D. A., Weibel D. B., Whitesides G. M. Microsolidics: Fabrication of Three-Dimensional Metallic Microstructures in Poly(dimethylsiloxane). Advanced Materials. 2007;19:727–733. DOI: 10.1002/adma.200601787

11. Kim H., Son C., Ziaie B. A Multiaxial Stretchable Interconnect Using Liquid-Alloy-Filled Elastomeric Microchannels. Applied Physics Letters. 2008;92(011904):1-3. DOI: 10.1063/1.2829595

12. Dickey M. D., Chiechi R. C., Larsen R. J., Weiss E. A., Weitz D. A., Whitesides G. M. Eutectic Gallium-Indium (EGaln): A Liquid Metal Alloy for the Formation of Stable Structures in Microchannels at Room Temperature. Advanced Functional Materials. 2008;18:1097-1104. DOI: 10.1002/adfm.200701216

13. Cheng S., Rydberg A., Hjort K., Wu Z. Liquid Metal Stretchable Unbalanced Loop Antenna. Applied Physics Letters. 2009; 94:144103. DOI: 10.1063/1.3114381

14. Cheng S., Wu Z., Hallbjörner P., Hjort K., Rydberg A. Foldable and Stretchable Liquid Metal Planar Inverted Cone Antenna. IEEE Transactions on Antennas and Propagation. 2009;57(12):3765-3771. DOI: 10.1109/TAP.2009.2024560

15. Kubo M., Li X., Kim C., Hashimoto M., Wiley B. J., Whitesides G.M. Stretchable Microfluidic Radiofrequency Antennas. Advanced Materials. 2010;22:2749-2752. DOI: 10.1002/adma.200904201

16. Cheng S., Wu Z. Microfluidic Stretchable RF Electronics. Lab on a Chip. 2010;10:3227-3234. DOI: 10.1039/c005159d

17. Cheng S., Wu Z. A Microfluidic, Reversibly Stretchable, Large-Area Wireless Strain Sensor. Advanced Functional Materials. 2011;21:2282–2290. DOI: 10.1002/adfm.201002508

18. Hayes G. J., So J. H., Qusba A., Dickey M. D., Lazzi G. Flexible Liquid Metal Alloy (EGaln) Microstrip Patch Antenna. IEEE Transactions on Antennas and Propagation. 2012;60(5):2151-2156. DOI: 10.1109/TAP.2012.2189698

19. Huang Y., Wang Y., Xiao L., Liu H., Dong W., Yin Z. Microfluidic Serpentine Antennas with Designed Mechanical Tunability. Lab on a Chip. 2014;14:4205-4212. DOI: 10.1039/c4lc00762j

20. Jeong S. H., Hagman A., Hjort K., Jobs M., Sundqvist J., Wu Z. Liquid Alloy Printing of Microfluidic Stretchable Electronics. Lab on a Chip. 2012;12:4657-4664. DOI: 10.1039/c2lc40628d

4 CHAPTER

One-Step Fabrication of Microchannels with Integrated Three Dimensional Features by Hot Intrusion Embossing

Mike Debono [1,2], Dan Voicu [1,2], Mohammad Pousti [3], Muhammad Safdar [3,4], Robert Young [2], Eugenia Kumacheva [1,2] and Jesse Greener [2,3]

[1] *Department of Chemistry, University of Toronto, Toronto, ON M5S 3H6, Canada*

[2] *FlowJEM Inc., Toronto, ON M5S 3H6, Canada*

[3] *Département de Chimie, Université Laval, Québec, QC G1V 0A6, Canada*

[4] *Department of Chemistry, University of Eastern Finland, Joensuu FI-80101, Finland*

ABSTRACT

We build on the concept of hot intrusion embossing to develop a one-step fabrication method for thermoplastic microfluidic channels containing integrated three-dimensional features. This was accomplished with simple, rapid-to-fabricate imprint templates containing microcavities that locally control the intrusion of heated thermoplastic based on their cross-sectional geometries. The use of circular, rectangular and triangular cavity geometries was demonstrated for the purposes of forming posts, multi-focal length microlense arrays, walls, steps, tapered features and three-dimensional serpentine microchannels. Process variables, such as temperature and pressure, controlled feature dimensions without affecting the overall microchannel geometry. The approach was demonstrated for polycarbonate, cycloolefin copolymer and polystyrene, but in principle is applicable to any thermoplastic. The approach is a step forward towards rapid fabrication of complex, robust, microfluidic platforms with integrated multi-functional elements.

Keywords: *microfluidics; microfabrication; thermoplastic; hot embossing; intrusion embossing*

1. INTRODUCTION

Thermoplastic-based microfluidic devices have a range of application-selective material properties and bonding methodologies [1,2,3]. Their physical robustness enables reliable fluidic and probe interfacing, embedded valves, as well as stable operating conditions, even under high pressures [3,4,5]. Cost still remains the bottleneck for higher penetration of these devices into the growing global market, especially for designs that require integration of functional and/or complex features. New microfabrication techniques that overcome this barrier will open a range of applications that benefit from integrated three-dimensional features. These include devices with tunable surface hydrophobicity [6]; optimized surface reactions [7]; ability to support surface acoustic waves [8]; resistance to biofouling, controlled cell growth, non-specific adsorption [9,10,11,12], and integrated optical elements [13,14]. Spot heating using laser pulses have been shown to generate pillars due to expansion of molten thermoplastic with some control of their size based on laser intensity, but suffers from large variances in feature morphology [15]. Grey-scale photolithography has been demonstrated in the creation of three-dimensional (3D) features, but is not widely utilized because of the need for specialized photomasks [13,14,16,17,18]. Material printing [18,19,20,21,22] and computer numerical controlled (CNC) micro machining [23] are becoming versatile options for rapid prototyping of 3D microfluidics. However, in the case of CNC machining, surface finish is typically rough. A recent explosion in 3D printing technologies is making impressive impact in consumer product fabrication markets. The microfluidic community is closely monitoring the advancements, [18,21] but for now, with the exception of two-photon polymerization methods, 3D printing feature resolution lateral directions is still roughly two orders of magnitude less than photolithography. In addition, both CNC machining and printing rely on a sequential approach to the formation of features, with resolution decreasing with fabrication speeds, making them more time-consuming and unfavourable for mass fabrication of microfluidic parts compared to templating methods, for the time being. In hot embossing and injection moulding, thermoplastics heated near to or above their glass transition temperatures can rapidly conform to template features with excellent fidelity. In these approaches, the hurdle becomes the fabrication of the template, which must be redone for each iteration during the prototyping process [24,25,26,27]. A variation on hot embossing called hot intrusion embossing (or partial embossing) offers a solution to this problem. Typically, with the use of a single 2D imprint template and proper calibration of embossing conditions, penetration of heated thermoplastic into cavities is varied, thus achieving control over feature height. However, after its demonstration as a means to produce templates for elastomeric microfluidic devices, there has been no significant development for this promising approach to be used for the direct fabrication of thermoplastic microfluidic devices [28].

Here we significantly expand on the concept of hot intrusion embossing by demonstrating a one-step method to fabricate thermoplastic microchannels containing integrated controllable microfeatures. We demonstrate that imprint templates containing microcavities with specific cross-section geometries locally control penetration velocity of heated polymer. The approach led to channels with integrated features such as posts, microlens arrays, semi-occluding walls, staircase patterns and features with 3D tapered edges. To demonstrate its versatility, the approach was implemented in different thermoplastics. As a proof-of-principle, we fabricated microchannels during one intrusion embossing step, containing two separate embedded functional regions: a 3D serpentine channel for mixing and a microlens array, featuring lenses with different focal lengths.

2. EXPERIMENTAL

2.1. Materials

Imprint templates were produced by FlowJEM Inc. (Toronto, ON, Canada). Thermoplastics used in this study included polycarbonate (PC, Lexan 9034-112, Sabic Polymershapes, Mississauga, ON Canada); cycloolefin polymer (COP, 1420, Zeon Chemicals L.P., Louisville, KY, USA); and polystyrene (PS, UVF Non-glare, Plaskolite Inc., Columbus, OH, USA). Glass transition temperatures were 145 °C, 138 °C, 100 °C, respectively.

2.2. Template Fabrication

Details regarding the experimental setup, materials used and characterization techniques are given in the Supplementary Materials. Imprint templates were fabricated based on a photolithographic approach using masks that contained opaque and transparent regions, corresponding to recessed and raised features, in the final imprint templates, respectively [29]. Each imprint template consisted of multiple raised, elongated features for microchannels with one inlet and one outlet. Within each raised channel structures on the imprint template, were a series of smaller cavities with different cross-sectional geometries: circular, rectangular and triangular. The height of the microchannel template feature was the same as the cavity feature depth, which was equal to the thickness of the photoresist layer in the imprint template.

2.3. Fabrication Parameters

Hot intrusion embossing (referred to simply as "embossing") was implemented inside a custom vacuum chamber located within a temperature controlled hydraulic press (Model 3851-C Carver Inc., Wabash, IN, USA). Temperature control was ±1 °C accuracy for top and bottom platens and included liquid leads for rapid cooling. The target thermoplastic was put in contact with the imprint template within the chamber which was then evacuated and the temperature was then raised to the embossing temperature (T_e). Following a 5 min temperature stabilization period, embossing pressure (P_e) was applied for $t_e = 2$ min. Cooling for approximately 30 s was applied until de-embossing temperature (T_d) was reached. Finally, the vacuum was broken and de-embossing was achieved with the aid of a blade to apply light prying action between the thermoplastic substrate and the imprint template. Afterwards, the thermoplastic substrate retained the exact inverse features of the imprint template everywhere, except in the microcavity region where the height was determined by the degree of polymer intrusion. Due to the reduced interaction between the intruding polymer and the microcavities, de-embossing was much easier than for fully embossed samples. In principle, we speculate that this should have the effect of extending the lifetime of the photoresist imprint template. Further details of the embossing process are given in the Supplementary Materials.

2.4. Imaging

Three-dimensional profiler measurements were acquired using an optical profilometer (NT1100, Veeco, Oyseter Bay, AZ, USA). Two-dimensional images of the embossed thermoplastic devices were acquired using an optical microscope (BX41, Olympus, Melville, NY, USA), with a CCD camera capture system (Evolution–VF, MediaCybernetics, Silver Spring, MD, USA) and image analysis of captured micrographs was conducted by software (Image-Pro Plus 5.0, Media Cybernetics, Silver Spring, MD, USA and ImageJ open-source program). Images of microfeatures were also acquired using a grazing angle camera system (32°) attached to a drop shape analyser (DSA 100, KRÜSS GmbH, Hamburg, Germany). Three-dimensional characterization of microlenses was done by confocal laser scanning microscopy (FV1200, Olympus, Richmond Hill, ON, Canada), using a 10× objective and by atomic force microscopy (Nanoscope III Multimode, Digital Instruments, Santa Barbara, CA, USA).

2.5. Numerical Simulations

Estimations of polymer melt flow velocity through the microcavities in the embossing templates were achieved by 3D finite element modeling (COMSOL Multiphysics, COMSOL Inc., Burlington, MA, USA). The simulation results were normalized so that relative differences in flow velocity within a single cavity could be made based on estimations of P_e, T_e and viscosity. Velocity profiles along the feature lengths were extracted for each model cavity shape. Microlens focal lengths were estimated by computer simulation software (COMSOL Multiphysics). This was accomplished by analyzing lens dimensions as measured by confocal laser scanning microscopy and atomic force microscopy. In order to reduce the computational cost, a 2D simulation was conducted on lens cross-sections. The refractive index for all lenses was chosen as $n = 1.58$ and $n = 1.33$ for a water filled channel. All optical simulations were operated at an incident light wavelength of 500 nm.

3. RESULTS

3.1. Hot Intrusion Using Cylindrical Microcavities

In the first experiment, the imprint template (IT_1) contained a 3 × 3 array of cylindrical microcavities in the main microchannel feature, which were 38.5 ± 0.2 μm deep. The cavities had hydraulic diameter d_h = 80 μm, which for cylindrical cavities is equal to its cross-section diameter. Profiler measurements for a subsection of this array are shown in Figure 1a,b. After embossing, the resulting cylindrical micropillars appeared as small circles from the top view (Figure 1c,d). A second imprint template (IT_2) also had cylindrical microcavities but was 44 ± 0.2 μm deep. See Supplementary Materials for more details.

For a particular embossing time, the height of the pillars was determined by the polymer melt flow velocity (v_f) into the cavities, whereas the overall channel depth was defined solely by the height of the channel features on the imprint template and not on embossing conditions. When the viscous flow dominates the capillary flow (see Supplementary Materials for calculation), the filling velocity (v_f) should increase with d_h and P_e and decreases with η_T [27,30]. In some cases the depth of the cavity could play a role due to trapped air, but experiments here were done under vacuum to avoid such complications [31]. In this proof of concept work, we rely on empirical relationships between process parameters and velocity of the heated polymer, because the system can only control processing variables (temperature, pressure) outside the home-built embossing chamber, which likely differs from those applied to the substrate and template (Supplementary Materials). For many polymers, including those used in this work, semi-empirical equations accurately predict the sharp decrease of η_T with increasing T_e [32,33].

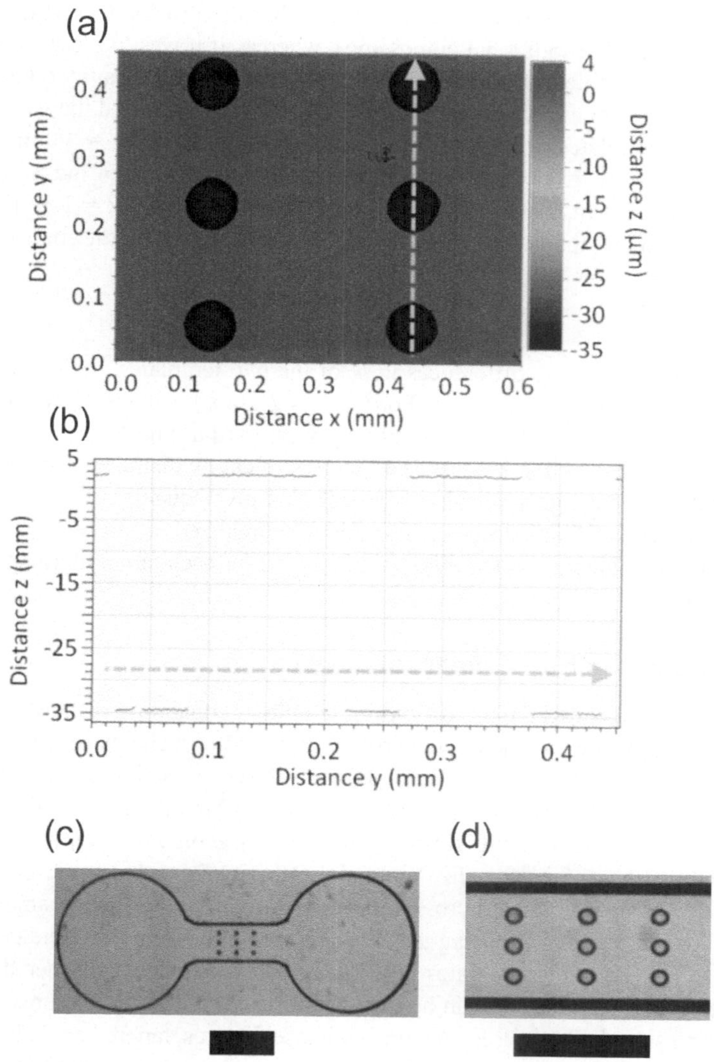

Figure 1. *(a) Three-dimensional profiler measurements of IT1 in the vicinity of the array of cylindrical cavities (dh = 80 µm). Channel width was 640 µm. The bottom of the cylindrical cavities are blue, 38.5 ± 0.2 µm below the top surface of the imprint template. (b) Two-dimensional height profile along the cross-section in (a). The direction indicated by the arrows in (a) and (b) are the same. Optical micrograph of the embossed PC device (Te = 175 °C, Pe = 1.7 MPa, t = 2 min) moulded from IT1 with imaging through a 1.25× objective (c) and a 4× objective (d). Scale bars in (c) and (d) are 1 mm and 500 µm, respectively.*

Figure 2a,b shows two different ways to achieve controlled intrusion of the heated polymer into cavities. The first is by controlling embossing conditions T_e or P_e. An increase to either results in increases to v_f within the cylindrical cavities (and corresponding increases to h of the final embossed features). The second approach to control intrusion is to change the d_h of the microcavities (Figure 2b). The first method was used as a calibration step to determine an appropriate combination of T_e and P_e for control over the polymer intrusion (Figure 2c). We note that if the T_e is too high, full intrusion was achieved into the cavity for all applied P_e (traditional embossing), whereas if it was too low, no value of P_e could result in appreciable intrusion. In the second approach where cavity diameters in the range 30 μm $\leq d_h \leq$ 100 μm were used controlled h. We demonstrated the generalizability with two imprint templates with different h_c values (h_{c1} = 38.5 μm, h_{c2} = 44 μm). The mask and embossed features from the two imprint templates are shown in the Supplementary Materials. The relationship between d_h and h in PS (T_e = 115 °C and P_e = 1.5 MPa) generated from both imprint templates, hown in Figure 2d, is represented by the empirical relation:

$$h = -0.0427d_h^2 + 0.496d_h - 108.99 \tag{1}$$

We note that despite the differences in h_c of the two templates, the h versus h_c trend is nearly the same for most values of h. For $h_c - h < 2$ μm h becomes nearly constant with d_h, due to the interaction of the heated polymer with the far side of the cavity. Equation (1) enabled prediction of pillar heights as a function of cavity diameter. This work demonstrates the potential for a generalizable approach that gives quantitative design rules that can predict the height of any structure. For this to be realized, a full multivariable fitting process relating h to all control variables (T_e, P_e, d_h, t_e) for each target thermoplastic material is required.

3.2. Hot Intrusion in Non-Cylindrical Cavities

Next, we demonstrated the use of imprint templates containing rectangular and triangular cavities to produce complex 3D microstructures in microchannels. Cavity shape and size resulted in different local d_h, permitting control over v_f within a single microcavity feature. Figure 3 is comprised of three subfigures which include the two-dimensional designs of the imprint templates with different microcavity geometries (Figure 3a), a two-dimensional simulation of the normalized intrusion velocity, $\tilde{v}_f$, within a microcavity feature and a plot of $\tilde{v}_f$ along an indicted cross-section (Figure 3b), and grazing angle images of the microfabricated parts in PS using a stamp (Figure 3c) bearing the same geometries as in Figure 3a. The latter demonstrates how selectively attenuated polymer flow through specific micro cavity geometries can be used to fabricate in-channel features such as partially occluding straight walls, free-standing staircase features, tapered occluding walls and free-standing tapered features. We note that the velocity maps are intended to qualitatively describe the range of polymer penetration velocities through a single microcavity based on the local geometry. This cannot necessarily be linked to the structure of the final embossed feature. For example in the case that the polymer flow front encounters the far side of the cavity, it will form a flattened shape and redistribution of polymer flow will ensue. This can be seen in embossing results for designs ii–iv, for example. Due to its relevance to biological sciences, we used PS for the remainder of the work.

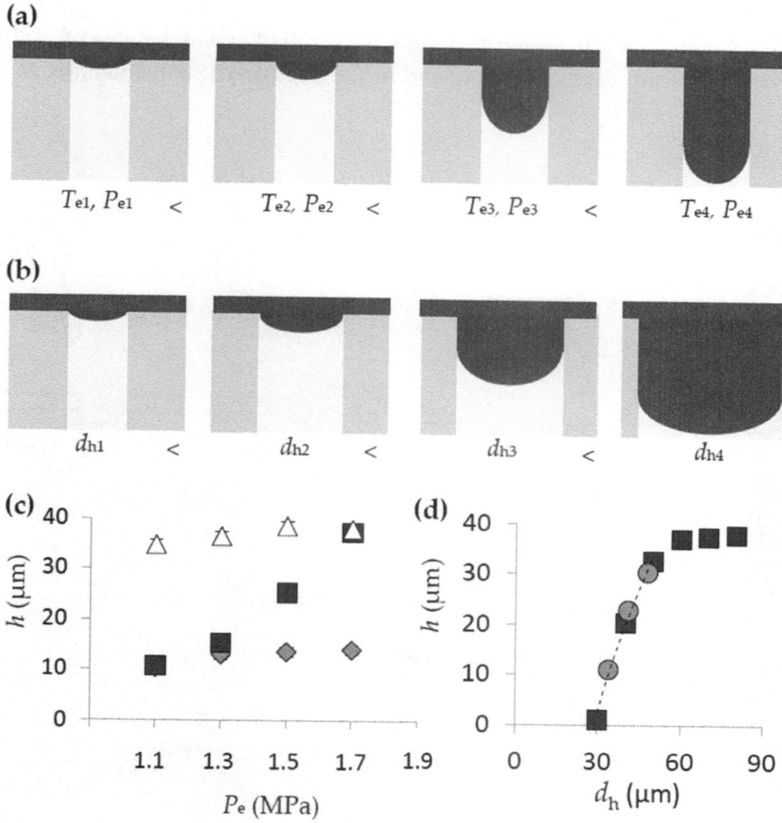

Figure 2. *(a) Cartoons depicting the progressive filling of one cylindrical cavity in the imprint template with the polymer melt (grey) as Te is increased from Te1 to Te4 and/or Pe is increased from Pe1 to Pe4. (b) Changes to the filling of a cavity with changes to hydraulic diameter from dh1 to dh4 for a particular Te, Pe set. (c) Variation in h posts versus Pe in the range 1.1 MPa to 1.7 MPa for Te = 165 °C (triangles), Te = 170 °C (squares), Te = 175 °C (diamonds) for devices fabricated in PC. (d) Plot of normalized h vs. dh for pillars fabricated using the imprint template IT1 (red squares) and IT2 (green circles) with embossing conditions were Te = 115 °C and Pe = 1.5 MPa in PS. The dashed trend line is the result of fitting the data to a second-order polynomial for points where hc − h > 1.5 µm, with R2 = 0.991. Data points were generated from the average from nine separate measurements. Error bars were calculated as the standard deviation of nine separate measurements, but were smaller than the data points.*

3.3. Walls and Stairs Using Rectangular Microcavities

Long, narrow trenches that cut perpendicularly across the raised microchannel feature in the imprint templates (h_c = 44 µm) resulted in partial polymer intrusion over long distances. Figure 3a(i) shows the schematic of an imprint template with two wall feature cavity widths (w_i = 20 and w_{ii} = 25 µm) intersecting the microchannel. Simulations predicted nearly constant v_f along the length of the trench, except near their intersection with the microchannel side-walls, where larger local d_h resulted in higher local v_f (Figure 3b(i)). Embossing PS with T_e = 110 °C and P_e = 1.5 MPa resulted in wall heights of 28.7 µm and 34.0 µm for the feature cavities with w_i and w_{ii}, respectively (Supplementary Materials).

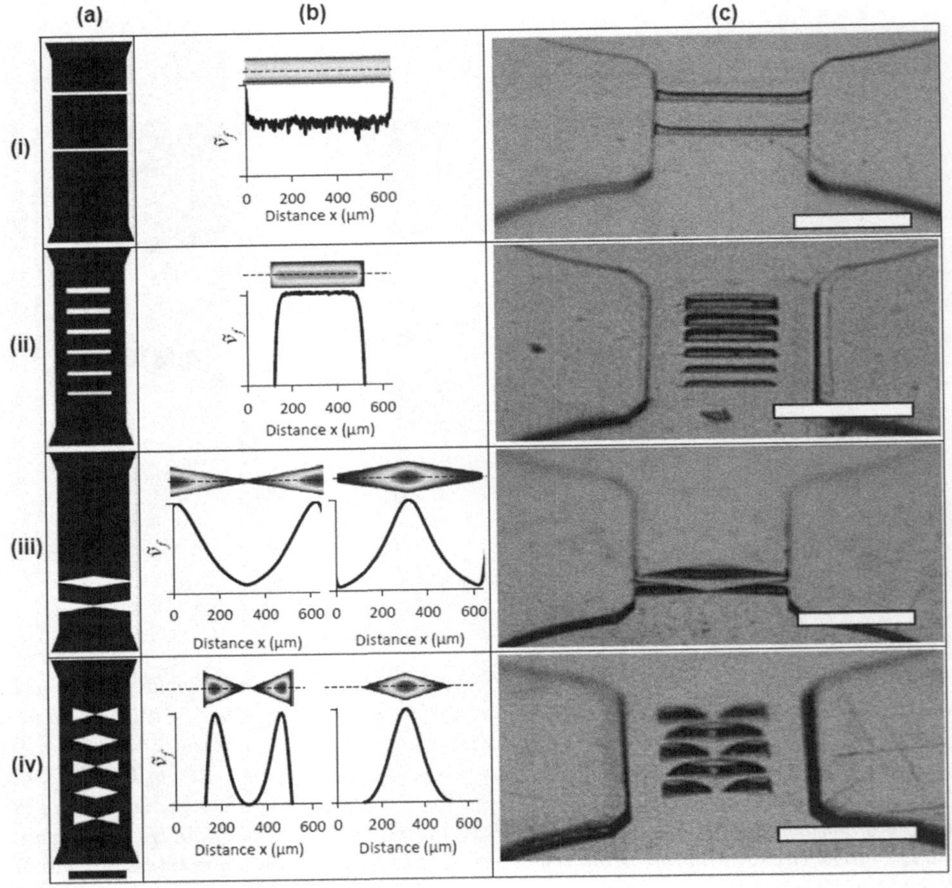

Figure 3. *Two-dimensional template design (a) and corresponding $\tilde{v}f$ simulation in one and two dimensions (b), and embossing results (c). Close up of imprint template schematics at the feature region for (i) partially occluding straight walls with widths 20 µm (bottom) and 25 µm (top); (ii) free-standing staircase features with widths (bottom to top) 25 µm, 30 µm, 35 µm, 40 µm, 45 µm and 50 µm; (iii) tapered occluding walls and (iv) free-standing tapered features. Results from numerical simulations of $\tilde{v}f$ in microcavities formed with masks in (a) are shown adjacent to the relevant design (i–iv) in (b). Colour scale ranges from $\tilde{v}f$ = 0 (dark blue) to $\tilde{v}f$ = 1 (red) are for qualitative comparison only. Velocity profiles were acquired along the length of the microcavity cross-section at its middle (marked with a dashed line). Grazing angle images of PS microchannels (c) that correspond to imprint template schematics in (a). Embossing conditions to produce images in (c) were: Pe = 1.5 MPa and Te = 110 °C (i), Pe = 1.5 MPa and Te = 105 °C (ii), Pe = 1.5 MPa and Te = 105 °C (iii) and Pe = 1.5 MPa and Te = 110 °C (iv). Common scale bar for (a) is 500 µm. All scale bars for images in (c) are 500 µm and all channels were 640 µm wide.*

Increased wall heights were observed at the intersection with the microchannel side wall as predicted by simulation. As seen in Supplementary Materials, reducing the temperature to $T_e = 105$ °C resulted in decreased wall heights (12.5 μm and 16.4 μm, respectively) due to increases to η_T.

Next, rectangular features with different w were demonstrated for the fabrication of a staircase pattern. Figure 3a(ii),b(ii) show the schematic of a microfluidic device with six rectangular cavities located within a microchannel and the simulated local $\tilde{v}_f$ within one of the microcavities, respectively. Each rectangular cavity was 400 μm long, with widths in the range of 25 μm ≤ w ≤ 50 μm at 5 μm intervals. The results from embossing PS (T_e = 105 °C and P_e = 1.5 MPa) show a series of free-standing rectangular features with heights between 8 μm and 44 μm (Figure 3c(ii)). Lower v_f near the edges of the features were predicted by simulation due to reduction in the local d_h.

3.4. Tapered Features Using Triangular Microcavities

Next we explored the use of triangular microcavities to impose continuous gradients in the local d_h along the feature length, which could systematically vary local v_f and feature h. Figure 3a(iii) shows a design with two elongated triangular features (θ_{apex} = 15°) that intersected the channel cross-section. Two variations were arranged, either with their most narrow part of the triangular microcavity (w = 13 μm) or the widest part (w = 100 μm) meeting in the middle of the microchannel. The gradual changes to $\tilde{v}_f$ along the length of the feature (Figure 3b(iii)) resulted in features with tapered edges in the embossed PS sheet, using P_e = 1.5 MPa and T_e = 105 °C (Figure 3c(iii)). The embossing was repeated for T_e = 110 °C and T_e = 115 °C, which produced features with larger portions of the cavity filled by thermoplastic, leaving an increasingly constructed orifice feature at the most narrowed regions. This opens the way for producing 3D orifice structures with tuned dimensions.

Finally, we used disconnected triangular cavities (Figure 3a(iv)) to create free-standing structures with tapered edges (Figure 3c(iv)). The features ranged in width from w = 100 μm to w = 0 μm (θ_{apex} = 34°). The d_h near triangle apexes resulted in zero local penetration, leading to strongly tapered wall edges. Cavities with tip to tip separation resulted in two separate PS features.

3.5. One-Step Fabrication of Integrated Element 1: Three-Dimensional Serpentine Channels

A microfluidic device with two functional elements was fabricated using the current method. The first element was a three-dimensional serpentine channel sequence that could be used for mixing. Fabrication of three-dimensional serpentine channels are typically far more complicated than for two-dimensional channels. This is because the former can require double sided wafer etching, aligned two layer photolithography and/or three-dimensional writing techniques [34,35,36,37]. Here, we demonstrated a design that uses localized intrusion embossing as a one-step fabrication method for three-dimensional switchback channels. Figure 4 shows the the design used to generate the appropriate imprint template and a three-dimensional rendering of the target channel volume. Embossing in PS was undertaken at T_e = 110 °C, P_e = 1.4 MPa with optical profiler results shown in Figure 4c. The approach also has the potential to form other passive mixing structures, such as chaotic mixers [38].

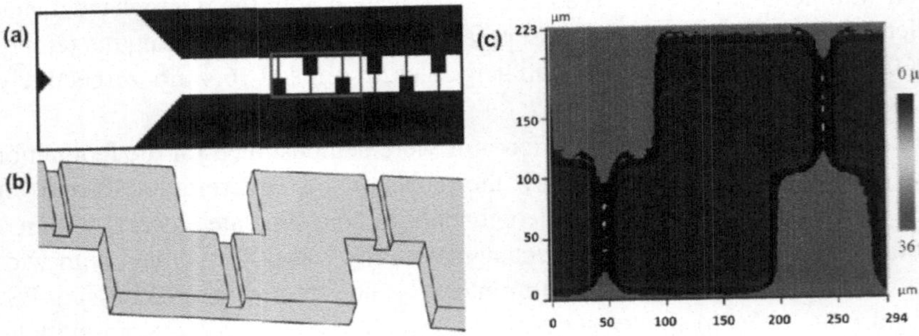

Figure 4. *(a) Portion of the design file used to create the mask for a three-dimensional mixer. Y-channel intersection at the left side of the image feeds into downstream mixing compartment. (b) A three-dimensional rendering of the expected results from intrusion embossing for the portion of the channel highlighted in red in (a). The channel interior volume (grey) shows 3D switchback in the x, y plane with channel constrictions in the z-direction. (c) Optical profilometry from embossed PS at Te = 110 °C, Pe = 1.4 MPa. The un-embossed surface (red) with depressions that formed the microfluidic channel (blue) and intra-channel constrictions in the z-direction (green).*

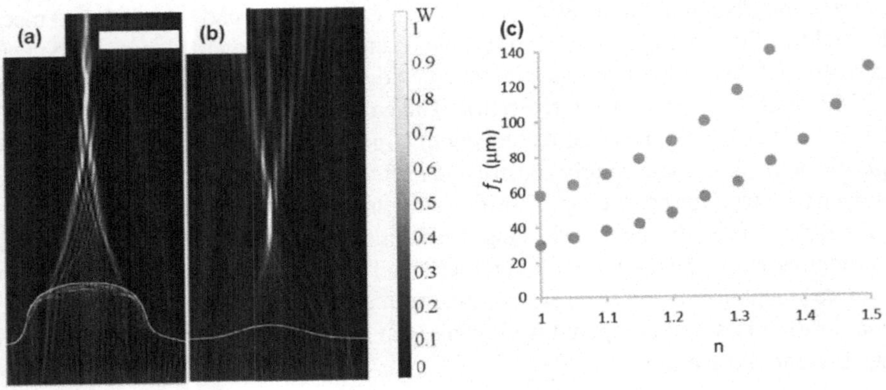

Figure 5. *The trace of two x, z lens profiles (white lines) acquired by CLSM (a) and AFM (b), produced by cylindrical cavities with dh = 37.5 µm and dh = 30 µm, respectively. Embossing conditions were Te = 110 °C, Pe = 1.4 MPa and t = 2 min in PS. Indices of refraction were nPS = 1.58 and nair = 1. Simulation results superimposed of the normalized electromagnetic power density, (W·m−2), resulting from a 500 nm light source placed at the bottom of the lens. Scale bar is 20 µm. The dependency of focal length on n is shown in (c) for lenses formed in cavities with dh = 37.5 µm (orange) and dh = 30 µm (blue).*

3.6. One-Step Fabrication of Integrated Element 2: Multi-Focal Length Microlenses

In a downstream segment of the same microfluidic device shown in the preceding section, hot intrusion embossing into an array of cylindrical cavities produced microlens arrays. The lens focal lengths (f_L) were varied based on the d_h of the cavities. In Figure 5a,b, two measured x, z profiles of lenses with d_h = 37.5 µm, d_h = 30 µm are shown. Using indices of refraction $n_{PS} = 1.58$ and $n_{air} = 1.33$ and the measured dimensions of the lens, a finite element simulation calculated an electromagnetic energy density map, giving f_L. Figure 5c shows the effect of channel filling with liquids with different values of n on f_L, for the two lenses in Figure 5a,b. See Supplementary Materials for corresponding optical images.

4. CONCLUSIONS

This work addresses the need for a cost-effective, one-step method to fabricate thermoplastic micro-channels containing complex micro features. We used a rapidly fabricated imprint template by photolithography with cavities that attenuated the intrusion of heated polymer during embossing based on their cross-sectional geometries. Fabrication was conducted in polycarbonate, cycloolefin copolymer and polystyrene, but in principle, can be extended to any thermoplastic. This approach enabled control over the in-channel micro feature height and geometry, while the channel height itself remained constant in all cases. The approach significantly advances the use of hot intrusion embossing by offering design rules for the fabrication of complex 3D structures. A numerical simulation provides a qualitative understanding of the filling velocities into different parts of the imprint template microcavities due to the local geometries. With a careful selection of microcavity geometries, fabrication of posts, multi-focal length microlense arrays, walls, steps, tapered features and three-dimensional serpentine microchannels was demonstrated. This work opens the way for applications requiring mixing, sorting, textured surfaces, cell manipulation and embedded optical sensors. In addition, applications requiring three-dimensional channel constrictions could be benefited, including burst valves for centrifugal microfluidics, blood constriction models. The technique can benefit from empirical studies and rigorous simulations to fine-tune cavity geometries and operating conditions to expand the range of applications [39]. Finally, we note that the approach is compatible with low-cost photoresist-based stamps.

Acknowledgments

The authors wish to thank Denis Boudreau for discussions on the micro-optics design parameters, NSERC and CMC Microsystems for funding. Funding from The Doctoral School of the University of Eastern Finland to MS is gratefully acknowledged.

Author Contributions

J.G. and E.K. conceived and designed the experiments; M.D., D.V. and R.Y. performed experiments; M.P. and M.S. conducted numerical simulations and image preperation and J.G. wrote the manuscript.

REFERENCES

1. Derzsi, L.; Jankowski, P.; Lisowski, W.; Garstecki, P. Hydrophilic polycarbonate for generation of oil in water emulsions in microfluidic devices. *Lab Chip* 2011, *11*, 1151–1156.
2. Greener, J.; Tumarkin, E.; Debono, M.; Dicks, A.P.; Kumacheva, E. Education: A microfluidic platform for university-level analytical chemistry laboratories. *Lab Chip* 2012, *12*, 696–701.

3. Chen, C.F.; Liu, J.; Hromada, L.P.; Tsao, C.W.; Chang, C.C.; DeVoe, D.L. High-pressure needle interface for thermoplastic microfluidics. *Lab Chip* 2009, *9*, 50–55.
4. Chen, C.-F.; Liu, J.; Chang, C.-C.; DeVoe, D.L. High-pressure on-chip mechanical valves for thermoplastic microfluidic devices. *Lab Chip* 2009, *9*, 3511–3516.
5. Greener, J.; Tumarkin, E.; Debono, M.; Kwan, C.-H.; Abolhasani, M.; Guenther, A.; Kumacheva, E. Development and applications of a microfluidic reactor with multiple analytical probes. *Analyst* 2012, *137*, 444–450.
6. Fang, G.; Li, W.; Wang, X.; Qiao, G. Droplet motion on designed microtextured superhydrophobic surfaces with tunable wettability. *Langmuir* 2008, *24*, 11651–11660.
7. Zhang, B.; Green, J.V.; Murthy, S.K.; Radisic, M. Label-free enrichment of functional cardiomyocytes using microfluidic deterministic lateral flow displacement. *PLoS ONE* 2012, *7*, e37619.
8. Ramakrishnan, N.; Nemade, H.B.; Palathinkal, R.P. Resonant frequency characteristics of a SAW device attached to resonating micropillars. *Sensors* 2012, *12*, 3789–3797.
9. Paul, D.; Saias, L.; Pedinotti, J.C.; Chabert, M.; Magnifico, S.; Pallandre, A.; De Lambert, B.; Houdayer, C.; Brugg, B.; Peyrin, J.M.; et al. A "dry and wet hybrid" lithography technique for multilevel replication templates: Applications to microfluidic neuron culture and two-phase global mixing. *Biomicrofluidics* 2011, *5*, 24102.
10. Andersson, M.; Berntsson, K.; Jonsson, P.; Gatenholm, P. Microtextured surfaces: Towards macrofouling resistant coatings. *Biofouling* 1999, *14*, 167–178.
11. Baker, D.W.; Liu, X.; Weng, H.; Luo, C.; Tang, L. Fibroblast/fibrocyte: Surface interaction dictates tissue reactions to micropillar Implants. *Biomacromolecules* 2011, *12*, 997–1005.
12. Dickinson, L.E.; Rand, D.R.; Tsao, J.; Eberle, W.; Gerecht, S. Endothelial cell responses to micropillar substrates of varying dimensions and stiffness. *J. Biomed. Mater. Res. Part A* 2012, *100A*, 1457–1466.
13. Galas, J.C.; Belier, B.; Aassime, A.; Palomo, J.; Bouville, D.; Aubert, J. Fabrication of three-dimensional microstructures using standard ultraviolet and electron-beam lithography. *J. Vac. Sci. Technol. B* 2004, *22*, 1160–1162.
14. Wu, M.H.; Park, C.; Whitesides, G.M. Fabrication of arrays of microlenses with controlled profiles using gray-scale microlens projection photolithography. *Langmuir* 2002, *18*, 9312–9318.
15. Nock, V.; Blaikie, R.J. Fabrication of optical grayscale masks for tapered microfluidic devices. *Microelectron. Eng.* 2008, *85*, 1077–1082.
16. Dun, A.; Ma, X.; Wei, J.; Gan, F. Laser-induced grayscale patterning in TeOx thin films. *Mater. Chem. Phys.* 2011, *131*, 406–412.
17. Fu, Y.; Ngoi, B.K.A. Virtual gray-scale mask for fabrication of micro-optical elements via focused ion beam direct writing. *Opt. Eng.* 2005, *44*, 128002–128005.
18. Bhattacharjee, N.; Urrios, A.; Kang, S.; Folch, A. The upcoming 3D-printing revolution in microfluidics. *Lab Chip* 2016, *16*, 1720–1742.
19. Ko, S.H.; Chung, J.; Hotz, N.; Nam, K.H.; Grigoropoulos, C.P. Metal nanoparticle direct inkjet printing for low-temperature 3D micro metal structure fabrication. *J. Micromech. Microeng.* 2010, *20*, 125010.
20. Kullmann, C.; Schirmer, N.C.; Lee, M.-T.; Ko, S.H.; Hotz, N.; Grigoropoulos, C.P.; Poulikakos, C.K. 3D micro-structures by piezoelectric inkjet printing of gold nanofluids. *J. Micromech. Microeng.* 2012, *22*, 55022.
21. Waheed, S.; Cabot, J.M.; Macdonald, N.P.; Lewis, T.; Guijt, R.M.; Paull, B.; Breadmore, M.C. 3D printed microfluidic devices: enablers and barriers. *Lab Chip* 2016, *16*, 1993–2013.

22. Tumbleston, J.R.; Shirvanyants, D.; Ermoshkin, N.; Janusziewicz, R.; Johnson, A.R.; Kelly, D.; Chen, K.; Pinschmidt, R.; Rolland, J.P.; Ermoshkin, A.; et al. Continuous liquid interface production of 3D objects. *Science* 2015, *347*, 1349–1352.
23. Lim, Y.C.; Altman, K.J.; Farson, D.F.; Flores, K.M. Micropillar fabrication on bovine cortical bone by direct-write femtosecond laser ablation. *J. Biomed. Opt.* 2010, *14*, 064021.
24. Tormen, M.; Carpentiero, A.; Vaccari, L.; Altissimo, M.; Ferrari, E.; Cojoc, D.; Di Fabrizio, E. Fabrication of three-dimensional stamps for embossing techniques by lithographically controlled isotropic wet etching. *J. Vac. Sci. Technol. B* 2005, *23*, 2920–2924.
25. Sharma, H.; Nguyen, D.; Chen, A.; Lew, V.; Khine, M. Unconventional low-cost fabrication and patterning techniques for point of care diagnostics. *Ann. Biomed. Eng.* 2011, *39*, 1313–1327.
26. Nguyen, D.; Taylor, D.; Qian, K.; Norouzi, N.; Rasmussen, J.; Botzet, S.; Lehmann, M.; Halverson, K.; Khine, M. Better shrinkage than Shrinky-Dinks. *Lab Chip* 2010, *10*, 1623–1626.
27. Worgull, M. *Hot Embossing: Theory and Technology of Microreplication*, 1st ed.; William Andrew: Oxford, UK, 2009.
28. Shiu, P.P.; Ostojic, M.; Knopf, G.K.; Nikumb, S.K. Rapid fabrication of polymethylmethacrylate micromold masters using a hot intrusion process. *J. Micro/Nanolit. MEMS MOEMS* 2008, *7*, 043012.
29. Greener, J.; Li, W.; Ren, J.; Voicu, D.; Pakharenko, V.; Tang, T.; Kumacheva, E. Rapid, cost-efficient fabrication of microfluidic reactors in thermoplastic polymers by combining photolithography and hot embossing. *Lab Chip* 2010, *10*, 522–524.
30. Fox, R.; McDonald, A.T. *Introduction to Fluid Mechanics*, 5th ed.; John Wiley & Sons: New York, NY, USA, 1998.
31. Taylor, H.; Lam, Y.C.; Boning, D. An investigation of the detrimental impact of trapped air in thermoplastic micro-embossing. *J. Micromech. Microeng.* 2010, *20*, 065014.
32. Vogel, H. The law of the relation between the viscosity of liquids and the temperature. *Phys. Z.* 1921, *22*, 645–646.
33. Liu, C.-Y.; He, J.; Keunings, R.; Bailly, C. New linearized relation for the universal viscosity–temperature behavior of polymer melts. *Macromolecules* 2006, *39*, 8867–8869.
34. Wang, J.; Wang, J.; Feng, L.; Lin, T. Fluid mixing in droplet-based microfluidics with a serpentine microchannel. *RSC Adv.* 2015, *5*, 104138–104144.
35. Liu, R.H.; Stremler, M.A.; Sharp, K.V.; Olsen, M.G.; Santiago, J.G.; Adrian, R.J.; Aref, H.; Beebe, D.J. Passive mixing in a three-dimensional serpentine microchannel. *J. Microelectromech. Syst.* 2000, *9*, 190–197.
36. Lin, C.-L.; Liu, Y.-J.; Lin, Z.-D.; Wu, B.-L.; Lee, Y.-H.; Shin, C.-S.; Baldeck, P.L. Laser direct writing 3D structures for microfluidic channels: Flow meter and mixer. In *Microfluid, BioMEMS, Medical Microsystems XIII, Proceedings of the SPIE 9320, San Francisco, CA, USA, 7 February 2015*; Gray, B.L., Becker, H., Eds.; SPIE Press: Bellingham, WA, USA, 2015.
37. Jo, B.H.; Van Lerberghe, L.M.; Motsegood, K.M.; Beebe, D.J. Three-dimensional micro-channel fabrication in polydimethylsiloxane (PDMS) elastomer. *J. Microelectromech. Syst.* 2000, *9*, 76–81.
38. Stroock, A.D.; Dertinger, S.K.W.; Ajdari, A.; Mezic, I.; Stone, H.A.; Whitesides, G.M. Chaotic mixer for microchannels. *Science* 2002, *295*, 647–651.
39. Taylor, H.; Lam, Y.C.; Boning, D. A computationally simple method for simulating the micro-embossing of thermoplastic layers. *J. Micromech. Microeng.* 2009, *19*, 075007.

5 CHAPTER

Bench-Top Fabrication of an All-PDMS Microfluidic Electrochemical Cell Sensor Integrating Micro/Nanostructured Electrodes

Sokunthearath Saem, Yujie Zhu, Helen Luu and Jose Moran-Mirabal

Department of Chemistry and Chemical Biology, McMaster University, 1280 Main Street West, Hamilton, ON L8S 4M1, Canada

ABSTRACT

In recent years, efforts in the development of lab-on-a-chip (LoC) devices for point-of-care (PoC) applications have increased to bring affordable, portable, and sensitive diagnostics to the patients' bedside. To reach this goal, research has shifted from using traditional microfabrication methods to more versatile, rapid, and low-cost options. This work focuses on the benchtop fabrication of a highly sensitive, fully transparent, and flexible poly (dimethylsiloxane) (PDMS) microfluidic (μF) electrochemical cell sensor. The μF device encapsulates 3D structured gold and platinum electrodes, fabricated using a shape-memory polymer shrinking method, which are used to set up an on-chip electrochemical cell. The PDMS to PDMS-structured electrode bonding protocol to fabricate the μF chip was optimized and found to have sufficient bond strength to withstand up to 100 mL/min flow rates. The sensing capabilities of the on-chip electrochemical cell were demonstrated by using cyclic voltammetry to monitor the adhesion of murine 3T3 fibroblasts in the presence of a redox reporter. The charge transfer across the working electrode was reduced upon cell adhesion, which was used as the detection mechanism, and allowed the detection of as few as 24 cells. The effective utilization of simple and low cost bench-top fabrication methods could accelerate the prototyping and development of LoC technologies and bring PoC diagnostics and personalized medicine to the patients' bedside.

Keywords: *fibroblast; shape memory polymer; flexible biosensor; cyclic voltammetry; cell sensor; xurography; stencil lift-off; on-chip electrochemical sensor*

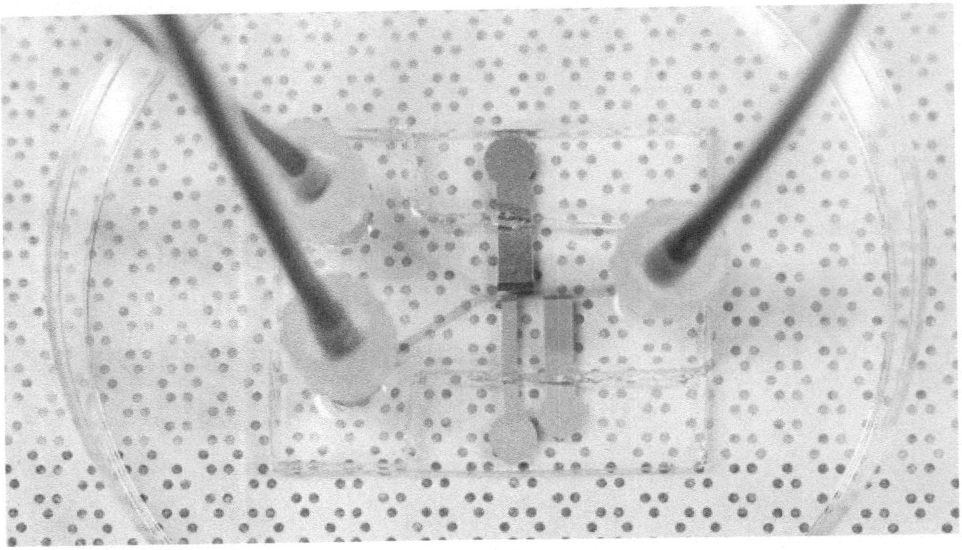

Graphical Abstract

1. INTRODUCTION

Sustainable global healthcare is a long sought-after idea [1]. Innovation in modern health care diagnostic techniques continues to improve patient outcomes, but the cost of bringing the technology to patients worldwide increases concomitantly. Ever since the inception of the portable glucose meter, scientists and engineers have been interested in lab-on-a-chip (LoC) devices for point-of-care (PoC) detection as low-cost and portable solutions to health screening, diagnostics, and personalized medicine [2,3]. To fabricate such devices on a mobile platform, LoC technology requires an all-in-one solution comprised of sensing and conducting elements within a microfluidic channel of sub-millimeter dimension [4].

Microfabricated on-chip microfluidic electrochemical biosensors have the advantage of being label-free, and offer high sensitivity and quantitative detection of analytes over a much broader concentration range than their fluorescence or colorimetric counterparts [4,5,6,7,8,9]. Most traditional microfabrication techniques are inherited from the semiconductor industry, where lithography [10,11,12], thin film deposition [12,13], and etching are routinely employed [12,14]. These techniques are very effective at producing high-resolution patterns at the micro- to nanoscale on a wide variety of materials. Yet, traditional microfabrication approaches can require expensive equipment, access to cleanroom facilities, multistep processes, and often limit themselves to rigid and non-transparent substrates. Some alternatives to making on-chip μF biosensors take advantage of soft silicone elastomers (e.g., poly (dimethylsiloxane)—PDMS) bonded to glass, silicon wafers, or polystyrene sheets through bonding techniques such as plasma oxidation and surface chemical grafting [13,15,16,17,18]. These methods offer a cost-effective benchtop alternative for the fabrication of μF devices without the use of expensive cleanroom facilities. Furthermore, μF devices for applications such as conformal pressure sensors, and flexible and transparent electronics, need to be implemented on soft and flexible materials, which sometimes cannot be processed through traditional methods. To overcome the limitations of cost, time, material properties, and processing, non-traditional techniques are being explored, including xurography [13,15,18], adhesive stencil lift off [15,18,19,20], shape-memory polymers (SMP) [13,15,20,21,22], ink jet printing [23,24], paper-based microfluidic devices [5,7,25,26,27], and 3D printing [28,29]. Despite the reduced resolution when compared to traditional microfabrication, these modern techniques offer lower costs and shorter turnaround time, two important factors for rapid prototype development and commercial scalability.

Recently, SMPs have gained significant traction in the microfabrication community for their ability to produce 3D micro/nanostructured surfaces on a variety of thin films (e.g., Au, Pt, CNT, SiO_2, and TiO_2) [13,20,30,31,32,33,34,35]. The structured surfaces are produced through a compressive stress applied by the SMP during shrinking once it is heated over its glass transition temperature [21,36]. This technique has been shown to successfully structure gold films for use as electrodes for electrochemical sensing [13,15,20,37] and as substrates for surface-enhanced Raman scattering [13,38,39,40]. The micro/nanostructures described in the literature showed increased electroactive surface area for working electrodes, ideal for sensing within μF devices. This simple bench-top fabrication approach presents an attractive route for the integration of highly sensitive electrochemical techniques into flexible LoC microfluidic devices.

This work presents a benchtop, low-cost, and rapid method for fabricating a highly sensitive, all-PDMS, μF electrochemical sensor, and demonstrates its sensing capabilities through the detection of the adhesion of murine 3T3 fibroblast cells—the most abundant cells in connective tissue and critical components in wound healing—onto the surface of the electrode. Commercial PDMS elastomer and pre-stressed polystyrene were used as the bulk substrates for the μF device and to fabricate the 3D micro/nanostructured electrodes, respectively. A salt-bridge-free three-electrode electrochemical μF device was fabricated to encapsulate the structured gold working, auxiliary, and platinum reference electrodes for an all-PDMS μF sensor. Cyclic voltammetry was employed to study the flow rate effects on charge transfer efficiency of the structured electrodes. Finally, fibroblasts were incubated and adhered onto the structured electrode and sensed using a redox reporter. This work demonstrates proof-of-concept for a novel rapid prototyping method to fabricate a transparent flexible sensor possessing high sensitivity and reproducibility. We anticipate that further surface functionalization of the structured electrodes will lead to inexpensive devices for the label-free detection of specific cell types, DNA-aptamer binding, and electrochemical or fluorescence sensing for applications in personalized precision medicine and PoC diagnostics.

2. MATERIALS AND METHODS

2.1. Electrode Fabrication

The structured electrodes used within the microfluidic device were fabricated using a vinyl stencil lift-off and pre-stressed polystyrene (PS) shrinking method previously described (Figure 1) [20]. Briefly, the PS sheets (Shrink Film, Graphix, Maple Heights, OH, USA) were cut to the desired shape using a blade cutter (ROBOPro CE5000-40-CRP, Graphtec

America Inc., Irvine, CA, USA) followed by a 3-step wash with isopropyl alcohol, ethanol, and 18.2 MΩ water under constant agitation at 60 rpm, and dried under a nitrogen (N_2) stream. The clean PS sheets were spin-coated with a positive-tone photoresist (PR, Shipley 1818, Marlborough, MA, USA) layer with a 1.8 μm nominal thickness. The PR-coated PS sheets were then baked on a hot plate at 90 °C for 3 min to remove residual solvent. The self-adhesive vinyl (FDC-4300, FDC graphics films, South Bend, IN, USA) shadow masks were also patterned using the blade cutter to the specific electrode dimensions for the working (WE) auxiliary (AE) and reference electrodes (RE), which were set to 10.8 mm × 1.6 mm × 4.8 mm, 10.8 mm × 3.0 mm × 4.8 mm, and 10.8 mm × 3.0 mm × 4.8 mm (length × width × pad diameter), respectively (Figure S1). The adhesive vinyl masks were then transferred onto the clean PS and served as stencils during the metal deposition process. Gold (99.999% purity, LTS Chemical Inc., Chestnut Ridge, NY, USA) and platinum were deposited using a Torr Compact Research Coater CRC-600 manual planar magnetron sputtering system (Torr International, New Windsor, NY, USA) at deposition rates of ~1 Å/s (100 nm) and ~0.1 Å/s (150 nm), respectively. Following this step, the vinyl stencils were lifted off and the PS sheets were placed in an oven heated at 160 °C, which shrunk the PS substrates to ~16% of their original size by area [15,20]. The structured metal films were lifted off from the PS by dissolving the PR in an acetone bath under constant agitation at 80 rpm for 30 min. Once the electrodes were lifted off from the PS, they were stored in acetone until further use.

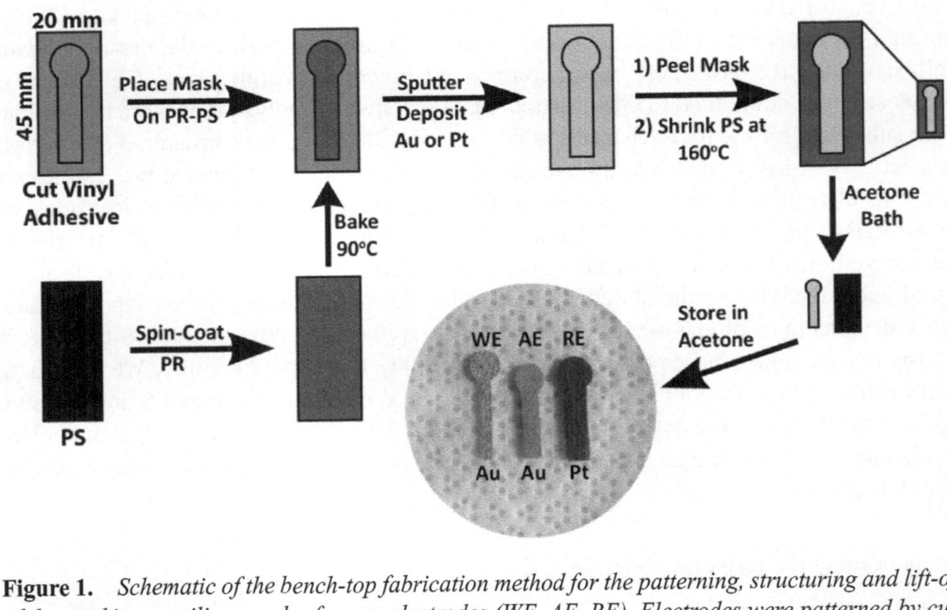

Figure 1. *Schematic of the bench-top fabrication method for the patterning, structuring and lift-off of the working, auxiliary, and reference electrodes (WE, AE, RE). Electrodes were patterned by cutting vinyl adhesive stencils (blue) to the desired shape and placing them on photoresist (PR, maroon) coated pre-stressed polystyrene (PS, black). Gold (100 nm) and platinum (150 nm) were sputtered onto the masked substrates followed by removal of the vinyl stencils. The sputtered flat electrodes were then subjected to heat at 160 °C to shrink the PS substrate down to 16% of its original area. The shrinking process resulted in Au and Pt micro/nanostructured electrodes, which were then lifted off by dissolving the PR in an acetone bath.*

2.2. All PDMS µF Device Fabrication

The µF channel mold was patterned using the blade cutter into a Bytac® PTFE surface protection laminate (Sigma-Aldrich, St. Louis, MO, USA) adhesive film with 125 µm nominal thickness. The patterned mold dimensions are shown in Figure 2a. 3D printed corrals defining the size of the PDMS µF layers (white, Figure 2a) were fabricated out of acrylonitrile butadiene styrene polymer using a Replicator 2X Experimental 3D Printer (MakerBot Industries, Brooklyn, NY, USA). The PTFE adhesive channel mold and 3D printed corral were placed on top of a 3-inch Si wafer and silicone tubing (Masterflex, Gelsenkirchen, Germany) was placed on top of the inlet and outlet reservoirs of the PTFE mold. A Sylgard-184 (Dow Corning, Midland, MI, USA) elastomer and hardener were mixed in a 10:1 ratio, degassed for 20 min, and poured into the 3D printed corrals. The top PDMS layer (2, Figure 2a) was placed first in an oven at 60 °C for 20 min followed by the bottom layer (1, Figure 2a) for the remaining 20 min before both halves were taken out of the oven and allowed to cool to room temperature. The structured Au [working (WE) and auxiliary (AE)] and Pt [reference (RE)] electrodes were placed on top of PDMS Layer 1. PDMS Layer 2 was removed from the 3D mold, and both layers were treated with air plasma for 30 s (30 sccm air inlet flow and 600 mTorr pressure) at a high-power setting (30 W) in a PDC expanded plasma cleaner (Harrick, Ithaca, NY, USA). The two layers were then bonded and left to fully cure at 60 °C for 1 h (Figure 2b).

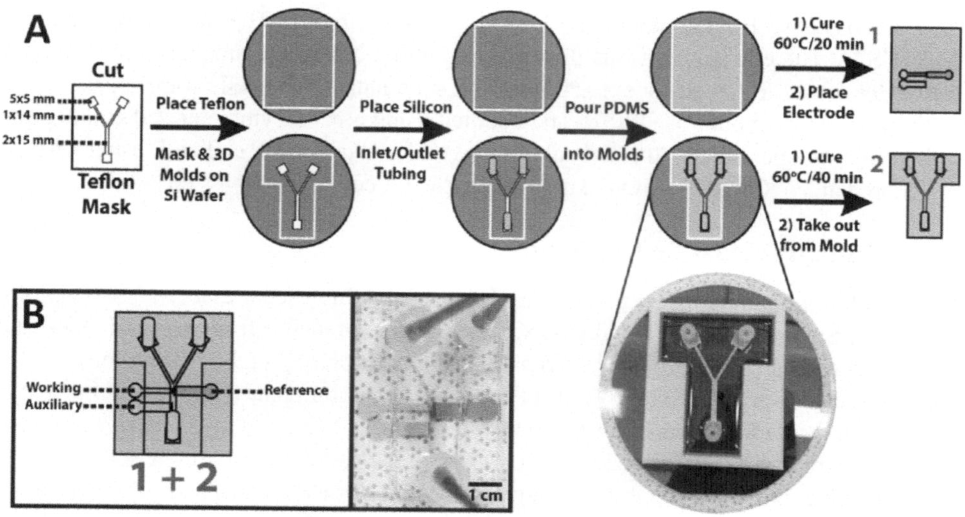

Figure 2. *Bench-top microfabrication process for making the all-PDMS on-chip µF electrochemical biosensors. (A) A Teflon adhesive µF channel mold was cut and placed inside a 3D printed device mold on top of a Si-wafer to ensure surface flatness. Silicon tubing was placed at the inlet/outlet positions, and PDMS was poured into the mold until full. The bottom layer (1) was partially cured at 60 °C for 20 min followed by placement of the three electrodes. The top layer (2) was cured at 60 °C for 40 min and removed from its mold. (B) PDMS Layers (1) and (2) were plasma treated for 30 s, bonded, and allowed to fully cure at 60 °C for 1 h.*

2.3. Dead-End Pressure Test of PDMS Bond Strength

All dead-end pressure tests were performed in triplicate for each bonding condition. A dead-end chamber with dimensions of 7 mm × 7 mm × 0.125 mm was made using patterned square PTFE molds as described above. The dead-end devices were attached to a N_2 (g) source with an inline pressure gauge and placed in a beaker filled with water. The N_2 (g) flow was increased until the water began to bubble, indicating that the device had burst through delamination or mechanical failure (Video S1 and Figure S2).

2.4. Murine 3T3 Fibroblast Cell Culture

Murine 3T3 fibroblast cells were prepared through standard cell culturing procedures. Briefly, Dulbecco's modified eagle's medium (DMEM, supplemented with 10% FBS, 1% L-glutamine, 1% penicillin/streptomycin (Invitrogen, Life Technologies, Burlington, ON, Canada)) was heated to 37 °C and 30 mL were pipetted into a 50 mL conical tube kept at 37 °C. A Frozen cryovial of murine 3T3 fibroblasts (ATCC, Manassas, VA, USA) was rapidly thawed by swirling the contents in a 37 °C water bath. To remove DMSO (Caledon Laboratory Chemicals, Georgetown, ON, Canada) from the cryovial contents, the contents were poured into the media kept at 37 °C and centrifuged at 500 g for 5 min. The supernatant was discarded, and the cell pellet was re-suspended in 15 mL of supplemented DMEM and incubated overnight in a T25 tissue culture flask (Sigma-Aldrich, Oakville, ON, Canada). Once the cells had grown to the desired confluency, they were detached from the culture flask by adding a trypsin solution (TrypLE, Thermo Fisher, Waltham, MA, USA) at 37 °C for 3–5 min. To neutralize the trypsin, an equal amount of DMEM was added (supplemented with 10% Fetal Bovine Serum and 1% Penicillin—both from Thermo Fisher, Waltham, MA, USA). The cell solution was then pipetted into a centrifuge tube and centrifuged at 500 g for 5 min followed by extraction of the supernatant. The cells were re-suspended with the desired amount of DMEM and counted using a hemocytometer cell counter and a light microscope. Finally, the cell solution was diluted to 2×10^6 cells/mL using DMEM and stored at 37 °C and 5% CO_2(g) until use in the μF cell incubation step.

2.5. CFSE Fibroblast Cell Staining

A solution containing 50 μM CFSE (5(6)-carboxyfluorescein diacetate N-succinimidyl ester, Sigma-Aldrich, Oakville, ON, Canada) was prepared from a 1 mM stock and 0.1% FBS in 1× PBS (phosphate buffer saline, 100 mM, pH = 7.4). Following this, 2×10^6 fibroblast cells were added to 200 μL of the CFSE solution, gently mixed and incubated at 37 °C in the dark for 1 h. After the incubation period, 200 μL of 100% FBS was added to the mixture and incubated at 37 °C for 10 min in the dark. The mixture was then centrifuged at 500 g for 5 min with the resulting pellet being washed 3 times with 10% FBS in 1× PBS solution. Finally, the pellet was re-suspended to the desired concentration in DMEM.

2.6. Cell Viability Assay

The viability assay for Au-PDMS substrates was carried out with PrestoBlue (Invitrogen, San Diego, CA, USA) in a 48-well plate. To start, 10^4 fibroblasts in 1 mL of DMEM were plated in each of the 12 wells and incubated at 37 °C/5% CO_2 for 24 h to ensure optimal cell adhesion. Among these, 4 wells contained Au-PDMS substrates, 4 were used to test 1 h 1× PBS incubation, and the last 4 were used for the DMEM control. For the viability assay, the media in all 12 wells was substituted with 180 μL of 1× PBS and 20 μL of 10x PrestoBlue solution. After 30 min of cell incubation at 37 °C and 5% CO_2, the change in the fluorescence of the samples was measured using a Cytation multi-well plate reader (Biotek Instruments Inc., Montreal, QC, Canada) with the excitation/emission wavelengths set at 535/615 nm.

2.7. Murine 3T3 Fibroblast Cell Sensing

The Au electrodes (WE and AE) in the PDMS μF device were preconditioned and baselined through cyclic voltammetry (CV). A 100 mM H_2SO_4 working solution and a 1× PBS reference solution were pumped through the μF device in separate laminar streams at 0.1 mL/min using a PHD ULTRA™ Syringe Pump (Harvard Apparatus, Holliston, MA, USA). The devices were preconditioned through 30 CV scans for the working electrode performed from 0.0 to 1.4 V at a scan rate of 0.10 V/s using a CHI600E electrochemical workstation (CH Instruments, Austin, TX, USA). Following the preconditioning step, the entire μF channel was washed with 1× PBS for 5 min at 0.10 mL/min. Prior to cell incubation, a baseline CV scan was performed on each μF device, yielding the total current for the clean WE. A sensing solution of 5 mM $K_4[Fe(CN)_6]$ in DMEM was continuously pumped through the device at 0.1 mL/min, and a 10 segment CV scan was performed from −0.4 to 0.2 V (with respect to the RE) at a scan rate of 0.10 V/s. Once again, the device was washed by flowing 1× PBS for 5 min at 0.1 mL/min. Finally, a fibroblast solution containing 2×10^6 cells/mL in DMEM was pumped through the μF stream flowing over the WE at 0.1 mL/min for 1 min and then stopped to allow for fibroblast attachment onto the WE for 1 h at 37 °C and 5% CO_2(g). After cell adhesion, the μF device was flushed with the sensing solution, and CV scans were performed as described above. Experiments were performed on three replicate devices to assess reproducibility and the statistical significance of the results.

2.8. Fluorescent Microscopy Image Acquisition

The murine 3T3 fibroblast cells inside the μF channel were imaged using a Nikon Eclipse LV100N POL epifluorescence microscope (Nikon Instruments, Mississauga, ON, Canada) equipped with excitation and emission filters for FITC dye, and a Nikon MRP50102 10×/0.25 NA Pol objective. Images were acquired with a Retiga 2000R cooled CCD camera (QImaging, Surrey, BC, Canada) and recorded with NIS-Elements AR software (Nikon Instruments, Mississauga, ON, Canada). The images were taken at 200 ms exposure time, 2 × 2 binning, and a hardware gain of 10.

3. RESULTS AND DISCUSSION

3.1. Optimization of PDMS Device Bond Strength for μF Device Fabrication

A viable flexible μF sensing platform must match or surpass current μF solutions in reproducibility, sensitivity, and reliability. The challenge in having a flexible all-PDMS μF sensor is that its intrinsic mechanical flexibility can introduce delamination at the bonding interface between the two PDMS layers of the μF device as it is being handled. In particular, the PDMS device should be strong enough to withstand the pressures generated from pressurized flow through the microfluidic channel. Therefore, the characterization and optimization of the bond strength between the two layers of a silicone-based μF device is crucial for the production of a reliable and flexible μF sensing platform. The bond strength of various bonding conditions was quantified using a dead-end chamber pressure test. To perform such a test, triplicate devices with dead-end chamber dimensions of 7 mm × 7 mm × 0.125 mm were made as described in the Materials and Methods section, and the bond strength was measured by pressurizing gas into the dead-end device until either the device delaminated or the PDMS layer/inlet tubing burst. The burst pressure was actively monitored using an inline pressure gauge. Performing the burst pressure test with the devices submerged in water provided an immediate indication of the device failure, as the escaping pressurized gas produced vigorous bubbling in the solution (Figure 3a, Video S1). Figure 3b shows a comparison of the burst pressure for different PDMS–PDMS bonding conditions benchmarked against the standard PDMS–glass interface, and a comparison of a device integrating a structured electrode 5 mm in width with and without PDMS sealing at the electrode–PDMS interface.

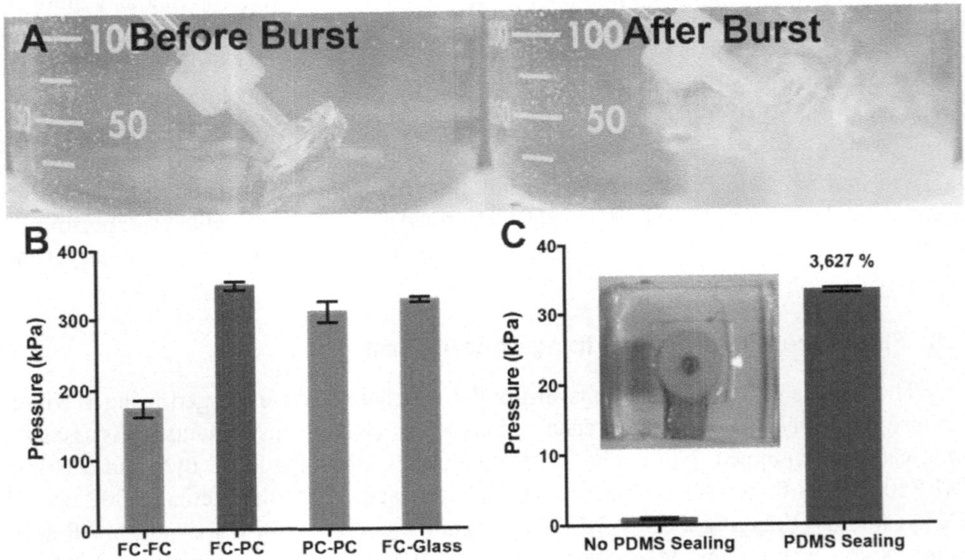

Figure 3. *PDMS bonding strength characterization All PDMS curing was done in a 60 °C oven with the following curing times as follows: fully cured (FC)—40 min; partially cured (PC)—20 min. (A) Image of the burst pressure test with the intact dead-end μF device (left) and the broken device (right). (B) Quantification of the burst pressure for different bonding combinations of FC–FC PDMS layers, PC–FC PDMS layers, PC–PC PDMS layers, and FC PDMS layer-glass. (C) Quantification of the burst pressure for a dead-end μF device with a 5 mm Au electrode in place. Graph compares devices without PDMS sealing around the device vs. with PDMS sealing after bonding. The inset shows a sample image of the dead-end burst pressure device with the Au electrode in place as would be seen in the μF device. All error bars represent the standard error of the mean, n = 3.*

The bonding conditions depicted in Figure 3b for PDMS devices without a structured electrode differ only in the combination of PDMS curing time (partially cured–PC vs. fully cured–FC), with every treatment subjected to 30 s air plasma treatment prior to device bonding. PC PDMS was made by subjecting the PDMS to 60 °C for 20 min and fully cured FC PDMS was subjected to 60 °C for 40 min prior to bonding followed by a final curing step at 60 °C for 1 h. The devices were allowed to form stable bonds at room temperature for 24 h after the initial bonding. FC–FC PDMS bonding exhibited an average burst pressure of 170±40170±40 kPa, which was ~50% lower than the FC–glass benchmark of 330±10330±10 kPa. The FC–PC and PC–PC combination produced statistically equivalent dead-end burst pressures to the benchmark, failing at 350±20350±20 and 310±50310±50 kPa, respectively. The increased bond strength observed in the FC–PC and PC–PC compared to the FC–FC devices can be attributed to having both hydrosilylation and dehydration reactions occurring at the interface. The hydrosilylation reaction is facilitated by a Pt catalyst found in the Sylgard-184 cross-linker and occurs when free silicon hydride (SiR_3-H) groups found in the Sylgard-184 elastomer base crosslinks with the vinyl-terminated polysiloxanes found in the Sylgard-184 cross-linker. Hydrosilylation can only occur with the partially cured PDMS mixture, where silicone chains are relatively mobile, while the condensation reaction occurs between two adjacent silanol (SiR_3–OH) groups, which will react to form Si–O–Si bonds and eliminate H_2O as a by-product. By exposing the two PDMS surfaces to plasma oxidation, we can generate surface SiR_3–OH, which will contribute to strengthening the PDMS–PDMS bonding interface over time. In addition, using the PC PDMS, the surface roughness effects on bonding are minimized, since the PC surface is still moldable and can conform to the complementary bonding layer. In view of the bonding strength test results for all-PDMS layers, we chose to use the FC–PC combination for the fabrication of devices incorporating structured electrodes, since it has a comparable failure pressure to the FC PDMS–glass benchmark, is relatively easy to handle, and has higher reproducibility than the FC–FC combination. Optimization of the bonding with layers containing structured electrodes was done using the same dead-end layout, but with the integration of a structured electrode that covered ~70% of one of the side walls (inset, Figure 3c). This setup was chosen to test the disruption in bonding strength with an extreme case of electrode-to-sidewall ratio. As anticipated, once a structured electrode was in place, the bonding strength was much weaker (0.9±0.10.9±0.1 kPa, Figure 3c) than that observed with the all-PDMS FC–PC combination. To overcome such weak bonding, we resorted to sealing the device edges with freshly mixed PDMS after the initial bonding, with the intent to reinforce the device bond at the site of the electrode. By sealing the device with the PDMS mix and then curing, the device burst pressure increased >36-fold (33.0±0.433.0±0.4 kPa, Figure 3c) when compared to the devices without sealing. This technique was then used on the µF device, and the burst flow rate was determined. With the PDMS sealing, the burst flow rate was determined to be >100 mL/min (maximum capability of the syringe pump). Thus, the bond strength with the optimized fabrication procedure is more than suitable for experimentation, where flow rates >4 mL/min have been shown to detach murine 3T3 fibroblast cells incubated within µF devices of similar dimensions without the use of fibronectin as an adhesion promoter [41].

Using FC–PC bonding and sealing procedure, all the subsequent PDMS µF devices were made in accordance to the process depicted in Figure 2a. The challenges in fabricating such devices with high reproducibility were in the careful placement of the three electrodes in the appropriate configuration, and the successful bonding over the structured electrode surfaces. It was found that following the 20 min curing step for the bottom half of the μF device, the three electrodes had to be immediately transferred from the acetone storage solution to the PDMS to take full advantage of the tackiness of the PC–PDMS for maximal electrode-to-PDMS adhesion. Once the electrodes had made contact with the PC–PDMS, the adhesion became too strong to remove the electrodes without destroying them. Since the transfer must be done successfully in one attempt, placing the electrodes one by one would require very high dexterity and precision to fabricate μF devices with acceptable reproducibility. To overcome this challenge, a Teflon filter membrane was used as a base support for the configuration and transfer of the electrodes in a single step. To ensure that the electrodes adhered to the Teflon membrane and that the membrane did not stick to the PC–PDMS, the membrane was pre-wet with a small amount of acetone (this also afforded transparency through the Teflon membrane for optimal electrode configuration). The electrodes were then removed from the storage solution using flat tweezers and placed on the membrane. At this stage, the electrodes could be easily positioned into the appropriate configuration without damaging them. Once in position, the electrodes were picked up by the Teflon membrane and placed on top of the PC–PDMS. Upon electrode to PC–PDMS contact, the membrane was then quickly removed. The final steps in the assembly and bonding of the μF devices were performed as described in the Materials and Methods section and depicted in Figure 2a.

3.2. Impact of Flow Rate on Electrochemical Sensing

Once the fabrication protocol for the μF devices was optimized, we turned our attention to the electrochemical sensing stability within the devices. Throughout this study, we utilized cyclic voltammetry (CV), a simple electrochemical technique that is commonly used to quantify redox processes and can be leveraged to implement cell-sensing strategies. CV is a diffusion-limited technique when the sensing of the redox process is performed in an unstirred solution, such that the analytes are not disturbed as they are undergoing reduction-oxidation cycles. Performing CV measurements in a flow cell, such as a μF channel, with variable flow rates could mimic a stirred cell condition resulting in limitations to charge transfer efficiencies across the working electrode, thus reducing the overall device sensitivity.

To perform on-chip electrochemical sensing and assess the impact of the flow rate on electrochemical sensing, a salt-bridge-free three electrode system was fabricated to incorporate Au WE and AE, and Pt RE in a Y-shaped bench-top fabricated PDMS μF device (cf. Figure 2b). The salt-bridge-free μF design exploits the laminar flow within the microfluidic channel to isolate the RE from the working sample solution, while still allowing the diffusion of the supporting electrolyte across the laminar flow interface. The separation between the working solution and the reference solution is critical in the stability of the potential output during sensing through CV. Given than the structured Pt film is used as a hydrogen pseudo-reference electrode, its half-cell potential is dependent on the concentration of hydronium ions (i.e., pH). To prevent fluctuations in pH, a buffer solution (1× PBS, pH = 7.4) was pumped through the reference inlet to be in constant contact with the RE, thus maintaining a constant reference electrode potential.

To assess the impact of the working flow rates used for sensing on charge transfer, we performed a series of CV measurements in the μF devices at flow rates from 0–0.5 mL/min. The CV for each flow rate was obtained in a 5 mM $K_4[Fe(CN)_6]$ in 1× PBS working solution and a 1× PBS reference solution utilizing the laminar flow inside the μF channel to separate the working and reference solutions (demonstrated in Video S2 using coloured solutions). Figure 4a shows an overlay of the cyclic voltammograms for the different flow rate conditions, with cathodic peaks shifting from −0.22 to −0.18 V with respect to stopped flow (blue curve, 0 mL/min). To perform the stopped flow sensing, the solution was pumped into the μF device, then the flow was stopped, and CV scans were then performed. The voltammogram shows a change in CV curve shape from a classic "duck" shape to a sigmoidal shape from low to high flow rates. The change in shape indicates that at higher flow rates we begin to reach a limit at which the convection introduced by the laminar flow is much greater than the redox kinetics of $K_4[Fe(CN)_6]$. The result is non-uniform anodic peak current to cathodic peak current and ultimately a change from a completely reversible process to a quasi-reversible process. Despite the signal shifts, the integration of the reduction peaks (Figure 4b) provides a quantification of the charge transfer, showing that there is an initial reduction in total charge to ~73% compared to the stopped flow. Between the flow rates of 0.1 and 0.5 mL/min, there is no statistical difference in charge transfer efficiencies. After confirming that the working flow rates (0.1–0.5 mL/min) did not negatively impact the electrochemical sensing capabilities of the PDMS μF device, we chose the lowest working flow rate tested of 0.10 mL/min for the murine 3T3 fibroblast cell sensing. This minimized the risk of cell detachment, while maintaining a continuous flow within the μF device. Minimizing cell detachment during the sensing period is essential during the acquisition of the voltammograms, since any such event would cause a significant increase in charge transfer efficiency at the working electrode.

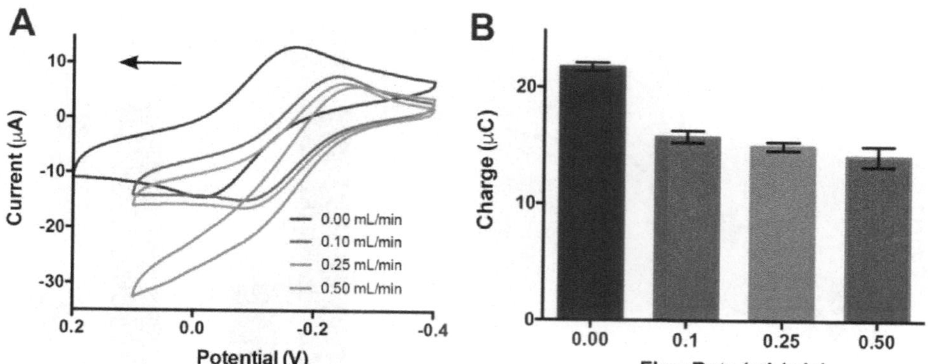

Figure 4. *Assessment of the impact of flow rate on charge transfer efficiency. The cyclic voltammetry experiments were performed using a 5 mM K4[Fe(CN)6] working solution at a 0.10 V/s scan rate. (A) Cyclic voltammograms showing changes in reduction and oxidation peaks and down-right shifting of the voltammograms as flow rate increased from 0 to 0.5 mL/min with a positive initial scan direction as indicated by the arrow. (B) Quantification of the total charge transfer in the cathodic peak for the different flow rates. Between flow rates of 0.1 and 0.5 mL/min, there is no statistical difference in charge transfer efficiency.*

3.3. On-Chip Cell Detection

To evaluate the impact of DMEM on electrochemical sensing over the incubation time, CV was performed immediately after being exposed to DMEM, and in DMEM after incubation for 1 h. We observed no statistical difference between the voltammograms obtained in DMEM immediately after exposure and those obtained after 1 h of incubation (Figure S4). Prior to on-chip cell detection, the murine 3T3 fibroblast cells were introduced into the μF channel by pumping through a solution containing 2×10^6 cells/mL in DMEM at 0.1 mL/min for 1 min, stopping the flow, and allowing the cells to attach to the WE by incubating them in the device for 1 h at 37 °C. The total volume of the device was ~2 μL, which translates into ~4000 fibroblast cells within the device at any given time. The experiment was performed on triplicate devices on the same day within a 4 h timeframe to minimize variability in cell adhesion and sensing due to ageing of the cells. After cell incubation, the μF was washed by pumping it through DMEM at 0.1 mL/min for 2 min to remove any weakly bound or unbound cells from the electrode surface. Using the minimal flow rate of 0.1 mL/min, the adhered murine 3T3 fibroblast cells were sensed in a DMEM working solution supplemented with 5 mM $K_4[Fe(CN)_6]$. During cell sensing, the redox-reporter was monitored through CV, where Figure 5a presents a typical voltammogram showing the cathodic peaks for a device before (blue) and after 3T3 cell incubation (red). The cells inside the μF device were monitored throughout the adhesion steps with fluorescence microscopy and a live CFSE cell stain, which can be seen in Figure S3. The figure shows images comparing the number of cells located on the Au–WE before incubation, after incubation (washed), and after sensing vs. control plain Au–WE. After sensing, the average number of cells counted on the WE (0.8 mm×1.60.8 mm×1.6 mm geometric area) was ~24 cells.

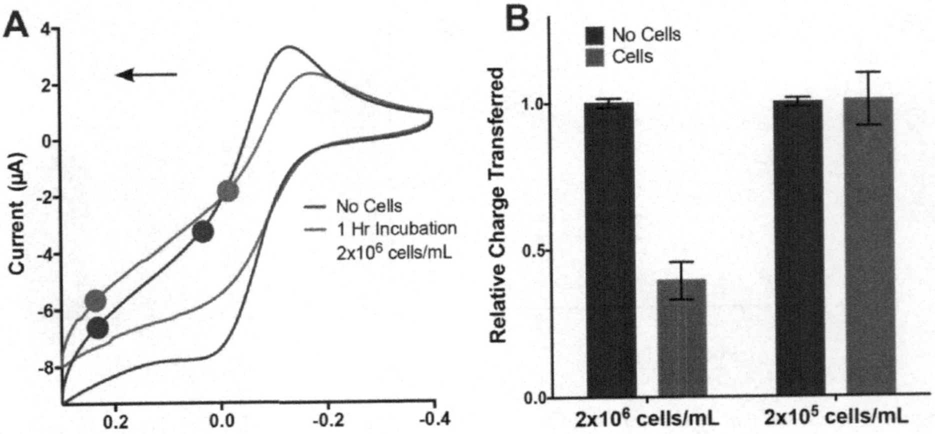

Figure 5. *Fibroblast sensing in 5 mM K4[Fe(CN)6] solution at a 0.1 mL/min flow rate. (A) Cyclic voltammogram of the detection of murine 3T3 fibroblast cells vs. control electrode. (B) Relative charge transferred by the working electrode during the detection of murine 3T3 fibroblast cell vs. the control electrode. All error bars represent the standard error of the mean, n = 3.*

To quantify the adhesion of cells to the electrode, the cathodic peaks of the CVs were integrated and the total charge transferred calculated. For the cathodic peak integration, the baseline correction was done through a linear regression using two points seen on each cyclic voltammogram (Figure 5a). To avoid any integration bias, the first point was placed at the initial leveling during the reduction sweep and the second point was placed at the base of the reduction peak. Upon integrating the cathodic peaks (Figure 5b), a 61% loss of charge transfer was recorded after incubation for the 2 × 1062 × 106 cell/mL solution. When the cell concentration was reduced by an order of magnitude, the electrochemical signal did not change. We attribute this to the low cell concentration in the seeding solution, where in the absence of adhesion proteins like fibronectin, we would expect one or two cells to adhere to the electrode area. This number of cells might be too low to produce a significant change in the electrochemical signal. The CV results show that the PDMS μF sensors were successful in detecting cell adhesion from murine 3T3 fibroblast cells at a relatively low number of cells (~24) and in a low sample volume. The sensitivity of the all-PDMS μF biosensor can be attributed to the use of structured working electrodes. Micro/nanostructuring of deposited thin metal films electrodes through the use of shape-memory-polymers has previously been shown to increase the electroactive surface area per geometric area by ~600% when compared to their flat counterparts [30]. Thus, the area blocked by any individual adhered cell is 6-fold higher than it would be for a traditional flat electrode. While we have demonstrated the capability of detecting murine 3T3 fibroblast cells, the described proof-of-concept detection approach is not specific. Our current efforts are directed towards the functionalization of the structured electrodes with capturing agents that selectively bind targeted cells. This will lead to a reduction in sensing time, improved sensitivity, and reproducibility.

4. CONCLUSIONS

This work demonstrates a novel, rapid, and cost-effective method for the integration of 3D structured high surface area electrodes into all-PDMS μF devices. The use of xurography on vinyl, PTFE, and PS and 3D printing offers a simple approach for the rapid and inexpensive prototyping and mold fabrication for μF devices. The optimized PDMS–PDMS bonding and sealing conditions offer bond strengths that allow the operation of the microfluidic devices under high flow rates and offer long-lasting encapsulation of the high surface area structured electrochemical sensors. In addition, the μF devices showed sustained charge transfer that was minimally disturbed by relatively low flow rates, as shown through cyclic voltammetry. The on-chip electrochemical sensing capabilities of bench-top fabricated μF devices were demonstrated by sensing the adhesion of model murine 3T3 fibroblast cells, where the signal decreased ~61% after the adhesion of an average of 24 cells over the working electrode. This highlights the excellent sensitivity of the structured Au electrodes. The bench-top fabrication method presented in this work offers a rapid, reproducible, and highly sensitive prototyping method that shows promise for the future prototyping of flexible devices for portable point-of-care diagnostics and personalized medicine.

Acknowledgments

This work was supported through the Natural Sciences and Engineering Research Council and a Canada Foundation for Innovation Leaders Opportunity Fund. Jose Moran-Mirabal is the recipient of an Early Researcher Award through the Ontario Ministry of Research and Innovation. We thank Fei Xu and Todd Hoare for kindly providing the murine 3T3 fibroblast cells used in this study.

Author Contributions

K.S. and J.M.-M. conceived and designed the experiments; K.S., Y.Z. and H.L. performed the experiments; K.S. and J.M.-M. analyzed the data; K.S. and J.M.-M. wrote the paper.

REFERENCES

1. Cordell, G.A. Sustainable medicines and global health care. *Planta Med.* 2011, *77*, 1129–1138.
2. Chin, C.D.; Linder, V.; Sia, S.K.; Daar, A.S.; Thorsteinsdottir, H.; Martin, D.K. Lab-on-a-chip devices for global health: Past studies and future opportunities. *Lab Chip* 2007, *7*, 41–57.
3. Yager, P.; Domingo, G.J.; Gerdes, J. Point-of-care diagnostics for global health. *Annu. Rev. Biomed. Eng.* 2008, *10*, 107–144.
4. Zhang, S.; Wright, G.; Yang, Y. Materials and techniques for electrochemical biosensor design and construction. *Biosens. Bioelectron.* 2000, *15*, 273–282.
5. Dungchai, W.; Chailapakul, O.; Henry, C.S. Use of multiple colorimetric indicators for paper-based microfluidic devices. *Anal. Chim. Acta* 2010, *674*, 227–233.
6. Chabinyc, M.L.; Chiu, D.T.; McDonald, J.C.; Stroock, A.D.; Christian, J.F.; Karger, A.M.; Whitesides, G.M. An integrated fluorescence detection system in poly (dimethylsiloxane) for microfluidic applications. *Anal. Chem.* 2001, *73*, 4491–4498.
7. Nie, Z.; Nijhuis, C.A.; Gong, J.; Chen, X.; Kumachev, A.; Martinez, A.W.; Narovlyansky, M.; Whitesides, G.M. Electrochemical sensing in paper-based microfluidic devices. *Lab Chip* 2010, *10*, 477–483.
8. Kwakye, S.; Goral, V.N.; Baeumner, A.J. Electrochemical microfluidic biosensor for nucleic acid detection with integrated minipotentiostat. *Biosens. Bioelectron.* 2006, *21*, 2217–2223.
9. Dhahi, T.H.S.; Bin Hashim, U.D.A.; Ahmed, N.M.; Mat Taib, A. A review on the electrochemical sensors and biosensors composed of nanogaps as sensing material. *J. Optoelectron. Adv. Mater.* 2010, *12*, 1857–1862.
10. Bhattacharyya, A.; Klapperich, C.M. Design and testing of a disposable microfluidic chemiluminescent immunoassay for disease biomarkers in human serum samples. *Biomed. Microdevices* 2007, *9*, 245–251.
11. Khine, M.; Ionescu-Zanetti, C.; Blatz, A.; Wang, L.-P.; Lee, L.P. Single-cell electroporation arrays with real-time monitoring and feedback control. *Lab Chip* 2007, *7*, 457–462.
12. Ziaie, B.; Baldi, A.; Lei, M.; Gu, Y.; Siegel, R.A. Hard and soft micromachining for BioMEMS: Review of techniques and examples of applications in microfluidics and drug delivery. *Adv. Drug Deliv. Rev.* 2004, *56*, 145–172.
13. Sharma, H.; Nguyen, D.; Chen, A.; Lew, V.; Khine, M. Unconventional low-cost fabrication and patterning techniques for point of care diagnostics. *Ann. Biomed. Eng.* 2011, *39*, 1313–1327.
14. Waits, C.M.; Morgan, B.; Kastantin, M.; Ghodssi, R. Microfabrication of 3D silicon MEMS structures using gray-scale lithography and deep reactive ion etching. *Sens. Actuators A Phys.* 2005, *119*, 245–253.

15. Sonney, S.; Shek, N.; Moran-Mirabal, J.M. Rapid bench-top fabrication of poly(dimethylsiloxane)/polystyrene microfluidic devices incorporating high-surface-area sensing electrodes. *Biomicrofluidics* 2015, *9*, 026501.
16. Dittrich, P.S.; Manz, A. Lab-on-a-chip: Microfluidics in drug discovery. *Nat. Rev. Drug Discov.* 2006, *5*, 210–218.
17. Eddings, M.A.; Johnson, M.A.; Gale, B.K. Determining the optimal PDMS–PDMS bonding technique for microfluidic devices. *J. Micromech. Microeng.* 2008, *18*, 67001.
18. Faustino, V.; Catarino, S.O.; Lima, R.; Minas, G. Biomedical microfluidic devices by using low-cost fabrication techniques: A review. *J. Biomech.* 2016, *49*, 2280–2292.
19. Pal, R.; Sung, K.E.; Burns, M.A. Microstencils for the patterning of nontraditional materials. *Langmuir* 2006, *22*, 5392–5397.
20. Zhu, Y.; Moran-Mirabal, J. Highly bendable and stretchable electrodes based on micro/nanostructured gold films for flexible sensors and electronics. *Adv. Electron. Mater.* 2016, *2*, 1500345.
21. Behl, M.; Lendlein, A. Shape-memory polymers. *Mater. Today* 2007, *10*, 20–28.
22. Liu, C.; Qin, H.; Mather, P.T. Review of progress in shape-memory polymers. *J. Mater. Chem.* 2007, *17*, 1543.
23. Cummins, G.; Desmulliez, M.P.Y. Inkjet printing of conductive materials: A review. *Circuit World* 2012, *38*, 193–213.
24. Derby, B. Inkjet Printing of Functional and Structural Materials: Fluid Property Requirements, Feature Stability, and Resolution. *Annu. Rev. Mater. Res.* 2010, *40*, 395–414.
25. Carrilho, E.; Martinez, A.W.; Whitesides, G.M. Understanding wax printing: A simple micropatterning process for paper-based microfluidics. *Anal. Chem.* 2009, *81*, 7091–7095.
26. Martinez, A.W.; Phillips, S.T.; Whitesides, G.M.; Carrilho, E. Diagnostics for the developing world: Microfluidic paper-based analytical devices. *Anal. Chem.* 2010, *82*, 3–10.
27. Dungchai, W.; Chailapakul, O.; Henry, C.S. Electrochemical detection for paper-based microfluidics. *Anal. Chem.* 2009, *81*, 5821–5826.
28. Kitson, P.J.; Rosnes, M.H.; Sans, V.; Dragone, V.; Cronin, L. Configurable 3D-Printed millifluidic and microfluidic "lab on a chip" reactionware devices. *Lab Chip* 2012, *12*, 3267.
29. Gross, B.C.; Erkal, J.L.; Lockwood, S.Y.; Chen, C.; Spence, D.M. Evaluation of 3D printing and its potential impact on biotechnology and the chemical sciences. *Anal. Chem.* 2014, *86*, 3240–3253.
30. Gabardo, C.M.; Zhu, Y.; Soleymani, L.; Moran-Mirabal, J.M. Bench-top fabrication of hierarchically structured high-surface-area electrodes. *Adv. Funct. Mater.* 2013, *23*, 3030–3039.
31. Fu, C.C.; Grimes, A.; Long, M.; Ferri, C.G.L.; Rich, B.D.; Ghosh, S.; Ghosh, S.; Lee, L.P.; Gopinathan, A.; Khine, M. Tunable nanowrinkles on shape memory polymer sheets. *Adv. Mater.* 2009, *21*, 4472–4476.
32. Pegan, J.D.; Ho, A.Y.; Bachman, M.; Khine, M. Flexible shrink-induced high surface area electrodes for electrochemiluminescent sensing. *Lab Chip* 2013, *13*, 4205–4209.
33. Gabardo, C.M.; Yang, J.; Smith, N.J.; Adams-McGavin, R.C.; Soleymani, L. Programmable Wrinkling of Self-Assembled Nanoparticle Films on Shape Memory Polymers. *ACS Nano* 2016, *10*, 8829–8836.
34. Bowden, N.; Brittain, S.; Evans, A.G.; Hutchinson, J.W.; Whitesides, G.M. Spontaneous formation of ordered structures in thin films of metals supported on an elastomeric polymer. *Nature* 1998, *393*, 146–149.

35. Lacour, S.P.; Wagner, S.; Huang, Z.; Suo, Z. Stretchable gold conductors on elastomeric substrates. *Appl. Phys. Lett.* 2003, *82*, 2404–2406.
36. Meng, Q.; Hu, J. A review of shape memory polymer composites and blends. *Compos. Part A Appl. Sci. Manuf.* 2009, *40*, 1661–1672.
37. Woo, S.M.; Gabardo, C.M.; Soleymani, L. Prototyping of wrinkled nano-/microstructured electrodes for electrochemical DNA detection. *Anal. Chem.* 2014, *86*, 12341–12347.
38. Gautam, M.; Jayatissa, A.H. Detection of organic vapors by graphene films functionalized with metallic nanoparticles. *J. Appl. Phys.* 2012, *112*.
39. Steuwe, C.; Kaminski, C.F.; Baumberg, J.J.; Mahajan, S. Surface enhanced coherent anti-stokes Raman scattering on nanostructured gold surfaces. *Nano Lett.* 2011, *11*, 5339–5343.
40. Zhang, L.; Lang, X.; Hirata, A.; Chen, M. Wrinkled nanoporous gold films with ultrahigh surface-enhanced raman scattering enhancement. *ACS Nano* 2011, *5*, 4407–4413.
41. Lu, H.; Koo, L.Y.; Wang, W.M.; Lauffenburger, D.A.; Griffith, L.G.; Jensen, K.F. Microfluidic Shear Device for Quantitative Analysis of Cell Adhesion. *Anal. Chem.* 2004, *76*, 5257–5264.

6 CHAPTER

Elastomer Application in Microsystem and Microfluidics

Shuang (Jake) Yang and Kunqiang Jiang

Department of Pathology, Johns Hopkins University, Baltimore, USA

Department of Mechanical Engineering, University of Maryland, USA

Department of Chemistry and Biochemistry, University of Maryland, USA

1. INTRODUCTION

Elastomer has been widely used in academic and industry since it was invented in the nineteen century [1]. By definition, elastomer is a polymer which is viscoelastic and able to regain its shape after deformation. Compared to other materials, elastomer has notably low Young's modulus and high yield strain. In general, amorphous elastomer is made of carbon, hydrogen, oxygen and/or silicon whose glass transition temperature is well below the application temperature, e.g., room temperature. As a result, it has excellent properties in elasticity [2], transparency [3], permeability [4], and insulation [5].

One of the most commonly used elastomers is silicone elastomer such as poly(dimethylsiloxane) (PDMS), which has been widely used in electronics and microfluidics due to its elasticity, optical transparency, UV transmission, permeability, biocompatibility, and availability. PDMS generally consists of repeat unit dimethylsiloxane. The repeat methyl is modifiable using functional groups such as urea or bis-urea for development of high mechanical strength of elastomer. The modified PDMS presents both a high mechanical strength and elasticity at room temperature due to functionality of the cross-linking domains and intrinsic reversibility of the supramolecular chains. PDMS is optically transparent and is usually inert, non-toxic and non-flammable [6]. PDMS is viscoelastic and is normally measured using dynamic mechanical analysis (DMA), in which a specialized instrument is used to determine the material's flow characteristics over a wide range of temperature, flow rates, and deformations. In general, the shear modulus of PDMS is in the range of 100 kPa to 3 MPa, varied with the preparation conditions.

Characterization of elastomer by dynamic-mechanical testing includes temperature sweep, frequency sweep, strain sweep, damping and friction properties. The dynamic mechanical measurement is particularly useful to characterize conductive particles in elastomer for understanding the particles effect on overall elastomer mechanical properties. It is very common to measure damping factor (tan) change with temperature for understanding the elastomer degree of crosslinking [7]. The dynamic mechanical testing provides information on both uncrosslinked and crosslinked phases of the elastomer, as well as viscoelasticity. To this end, nanoindentation plays a significant role on viscoelastic characterization of elastomers for their creep and relaxation properties [8, 9].

Elastomer is an excellent matrix which can mix with other materials such as conductive particles and thermoplastics to enhance mechanical properties. For example, silver particles are added into silicone elastomer matrix to formation of conductive elastomer for high-density electrical contacts in microelectronics. Polypropylene and ethylene-propylene copolymer elastomer significantly improves the dynamic mechanical, tensile and impact properties and even changes the modes of failure [10]. The size, shape and spatial packing of elastomer inclusions are the most significant factors in controlling the mechanical behavior of blends. Since PDMS can be cured with different ratio of dimethylsiloxane and curing reagent, it is very useful to fabricate microfluidic devices for a variety of applications, including cell culture [11], ELISA assay [12], capillary electrophoresis [13], DNA sequencing [14].

In this chapter, we briefly discussed elastomer elastic and viscoelastic properties, as well as permeability. Elastomer is widely used in microelectronics and microfluidics due to all or part of these characteristics. Then we discussed several applications in microelectronics and microfluidics using elastomer as substrate. Finally, we introduce potential development of elastomer.

2. ELASTOMER PROPERTY AND ITS APPLICATIONS

Elasticity is the ability of a material to return to its original shape and size after being stretched, compressed, twisted or bent [15]. Elastic deformation (change of shape or size) lasts only as long as a deforming force is applied to the object, and disappears once the applied force is removed. The elasticity of elastomer makes it extremely useful in microelectronics and microfluidics. For example, the Intel CPU uses land grid array sockets (LGA), in which the socket has over thousands of elastomer contacts that consist of silver particles in elastomer matrix [16, 17]. When socket, the CPU and printed circuit board (PCB) are assembled by compressing force, the silver-embedded elastomer is compressed to increase contact area between socket and the CPU, and between socket and the PCB. The excellent elastic deformation of elastomer can keep the electrically tight contact for minimum contact resistance. Elasticity of PDMS minimizes its stress relaxation and creep effect, which is severe in other polymer with low elasticity. The excellent elasticity of PDMS allows to release from microfluidic masters without damaging PDMS device [18, 19] and significant deflection to occur when a pressure difference develops [20]. Therefore, it is important to characterize PDMS elasticity and understand how to improve such property for various applications.

The most common elasticity experiment for many decades involved mechanically blending a reinforcing filler such as carbon black into natural rubber, crosslinking the elastomer with sulfur, and characterizing it network structure by stress-strain measurements in simple extension, or by extents of equilibrium swelling [21]. However, in most instances, such as thin-film elastomer, land grid array elastomer contact, and PDMS microfluidics, these methods may not be applicable due to effects of supporting materials. To characterize elastomer in those applications, nanoindentation has been an emerging technology for elastomer characterization and is an indispensable tool in many instances where other techniques are inapplicable. Nanoindentation is such tool for characterization of polymer on its viscoelasticity. Indentation technique was first introduced as advent of indentation fracture mechanics [22] in 1975 and ever since it has been developed for characterization of elastic-plastic materials using the Oliver-Pharr's method [23]. No valid method was reported for polymer nanoindentation analysis based on stress-strain experiments until the viscoelastic model was proposed in 2004 and other emerging studies that followed [8, 24, 25]. Because of viscoelasticity, elastic-plastic model cannot be directly used to derive modulus and hardness. Also these measurements are dependent on the experimental parameters, such as loading rate, maximum load, holding duration, unloading rate, and minimum unloading force [26]. To minimize the effects by those experimental parameters during indentation, the experimental conditions were proposed in such a way that fast loading was applied for minimizing viscoelastic deformation, long dwell time for full viscoelastic deformation, and quick unloading rate. The fast loading and unloading showed that only elastic-plastic deformation was observed, while viscoelastic deformation was observed in the dwell duration. Also, it was observed that the viscoelastic deformation can be measured by keeping a minimum force on the indenter so that the instrument can record the deformation progress with time. The modulus and viscosity coefficient of elastomer can be derived using flat-ended punch-tip [8], the elastic-viscoelastic-viscous deformation is described by following equation,

$$h = \frac{P_0 h_{in}}{E_0 A_0} + \sum_{1}^{n} \frac{P_0 h_{in}}{E_i A_0} \left(1 - e^{-E_i t / \eta_i}\right) + \frac{h_{in}}{\eta_o} t \qquad \text{E1}$$

where P represents load, h is deformation and subscript e is elastic, coefficient of viscosity, E modulus, t time. Elastic modulus is thus derived by $P_0 h_{in} / A_0 h_e$. This model provides a means for characterization of elastomer, such as investigation of effect of crosslinking, work of adhesion and fluid environment on elastic modulus [24]. Nanoindentation on PDMS indicated that it is capable of measuring several order of magnitudes of elastic modulus of PDMS, with expected increase of the elastic modulus at higher crosslinked PDMS (2.80 MPa at 10:1 ratio, and 0.88 MPa at 30:1 ratio; monomer:crosslinker, wt%) [24, 27].

Elastomer is permeable to solvents, gases, moisture or even some small molecules [4, 28-30]. Permeability has been employed to fabricate non-contact pump for the manipulation of aqueous solutions within PDMS microfluidic devices [20] or supply oxygen for large-scale culture of hepatocytes [31]. On the other hand, PDMS permeability limits its application to microfluidic technology, in which coating such as glass-like layer using sol-gel chemistry is developed on PDMS to minimize permeability [32]. The permeability of a PDMS film or membrane to a penetrant (i.e., gas or solvent) is defined as [28, 33],

$$P_A = \frac{N_A l}{p_{2A} - p_{1A}} \qquad \text{E2}$$

Where P_A is the permeability coefficient, N_A is the steady-state penetrant flux through the film, p_{1A} and p_{2A} are the downstream and upstream partial pressures of component A, and l is the membrane or film thickness. In general, PDMS permeability depends strongly on the feeding gas/solvent composition and temperature: an increase in vapor concentration or liquid pressure leads to increased chain mobility at constant feeding pressure and temperature, resulting in high diffusivity and permeability. The increase of permeability with pressure is attributed to the swelling of PDMS.

Due to the permeability nature of PDMS, many researchers synthesized co-polymer which contained PDMS as a permeable medium for gas/solvent. For example, the 'smart' tri-component amphiphilic membranes consisting of poly(ethylene glycol) (PEG), PDMS and polypentamethylcyclopentasiloxane (PD5) domains were synthesized for biological applications, including immunoisolation of cells [29, 30]. The high oxygen permeability of PDMS makes it feasible for the application in microdevices in cell culture [34]. Compared to large-scale system / platform, microfluidics provides suitable environments for sample interaction, such as cell culture, with the significant increase of surface-to-volume ratio while the fluidic behavior is similar to the environments in vivo [29]. Integration of PDMS into microfluidic platform can implement studies on cell-to-cell interactions and understanding of cell behaviors in vitro, emulation of situations observed in vivo. Several studies have demonstrated that microfluidic devices consisted of two PDMS layers were capable for cell culture [35]. They further fabricated a complex microfluidics which consisted of stacking ten PDMS layers, in which four cell culture chambers and one-oxygen chamber inserts between.

3. ELASTOMER AS ELECTRICAL CONTACT MATRIX IN MICROELECTRONICS

Insulation elastomer in microelectronics plays a significant role to prevent a short-circuit [36]. Electrical insulating polymers have been introduced nearly one hundred years ago and proved to be excellent long-term physical and electrical properties, including weatherability, moisture-sealing, corrosion resistant and durability. The polymer insulating materials are used in a wide range of applications such as surge arresters, insulators, insulation enhancement, and bushings [5]. The initial polymers used in insulating materials were mainly on radiation-crosslinked and semicrystalline polyolefin copolymers. With development of manufacturing and materials technologies, elastomer has been increasingly designed for insulating applications [5, 37].

The elastomer formulations have been optimized so that an exceptional electrical and weathering performance is achieved, which are equal or above that of polyolefin copolymer materials. The silicone elastomer has many advantages such as hydrophobicity, hydrophobic recovery, weathering resistance, processability, and elastomeric mechanical properties. These advantages make silicone elastomer suitable for outdoor insulating applications. For example, PDMS has been used for high-voltage (HV) outdoor insulation in the real world as a replacement over conventional porcelain and glass insulation. As one silicone elastomer, PDMS as HV insulating materials is light weight, has superior vandal resistance, and contains better contamination performance. However, many research works have been conducted on molecular modification in order to improve PDMS properties, including degradation under exposure to discharge and arcing, mechanical strength etc. It has been demonstrated that hydrophobicity of PDMS can be rendered by exposure to discharges, which can be recovered when exposure ceases. Their studies showed that the lowly crosslinked PDMS can readily reorient between hydrophilic and hydrophobic states, therefore improving its weathering capabilities [38].

Recently, PDMS has been micropatterned using a photoresist lift-off technique for selective electrical insulation in microelectrode array applications [39]. The micropatterning technique is able to manufacture PDMS patterns with feature as small as 15 μm and various thicknesses down to 6 μm. With micropatterned PDMS insulation layer, the electrical resistance between adjacent electrodes is within the specification. Additionally, the micropatterned PDMS insulation can be applied to biosensing microdevices such as in an extensive neuronal network. Other researchers used PDMS to develop a replaceable insulator for a single-use planar microelectrode array (MEA) in the study of electrogenic tissues [40]. They demonstrated applications using microstencils for rejuvenation of an old MEA and the fabrication of a single-use MEA.

Conductive materials are used to produce shielding gaskets for military, aerospace, electronics, and communications. The conductive elastomers are designed to balance requirements for electrical conductivity, thermal management, and cost performance [41]. Conductive elastomer materials are ideal for customer applications requiring both excellent electromagnetic interference shielding and environmental sealing across a wide range of temperature. The conductive elastomers could be blended polyacetylene with different types of thermoplastic elastomers: styrene-butadiene-styrene, styrene-isoprene-styrene, and styrene-ethylenebutylene-styrene tri-block copolymers [42]. The polyacetylene is a highly conjugated polymer which contains high concentration of unsaturated sites. Those sites are easily attacked by ozone / UV light and reacted with elastomer, formation of conductive elastomer composites. Or insulation elastomers are coated with compliant electrode material on both sides of the elastomer film [43]. The dielectric elastomers were actuated by means of electrostatic forces applied via compliant electrodes. The performance of conductive elastomers is dependent on many factors, e.g., elastomer crosslinking method and type of monomer. In an electrically conductive heterogeneous binary polymer blends, consisting of ethylene-propylene-diene-monomer (EPDM) and polyaniline (PAni), the performance was significantly affected by the crosslinking method [44]. In the blend, PAni underwent doping upon exposure to a protonic acid and became electrically conductive. Several methods have been studied for monomer crosslinking, such as the use of a phenolic resin for unsaturated rubbers and high service temperature products, inter-chain C-C bonds formation after the reticulation reaction, electron beam induced crosslinking for 3-D network. Results indicated that electron beam irradiation crosslinking method was not affected by the presence of the acid necessary for doping the conductive elastomers.

Insulation elastomer has been used as a matrix to hold conductive particles in land grid array socket since it significantly improves contact resistance and increases interconnect density. The need for high density electronics has led to the development of semiconductor array packages, such as ball grid array (BGA) and land grid array (LGA) packages. However, soldering a package to a printed circuit board (PCB) is challenging for input/output (I/O) counts greater than a thousand, and solder joint defects can induce reliability problems in the application. These problems can be solved by using an LGA socket, which eliminates the soldering process, provides a separable interconnect between the component and the PCB, and also enables easy product rework, repair, and upgrade. The conductive particles in elastomer matrix has been introduced into electronics in 1980's [16, 45]. One such conductive elastomer composite is called Z-direction anisotropically conductive materials which are made of the magnetic alignment of conductive particles in elastomers [46]. The composites contain many vertically aligned but laterally isolated chains of ferromagnetic metal spheres, and the ends of which protrude from the surfaces for better electrical contact. These materials exhibit Z-direction only electrical conduction, in combination with the compliant nature of the materials, can be exploited advantageously for a variety of electronic applications, including fine-pitch, area-array, circuit interconnects, circuit-testing, heat sink interfacing, and sensor devices [47, 48]. The schematic diagram of this design structure is shown in Figure 1. This design has been used in a tactile shear sensor for applications such as robotic skins or grippers, touch sensitive actuators, and finger operated controls [49].

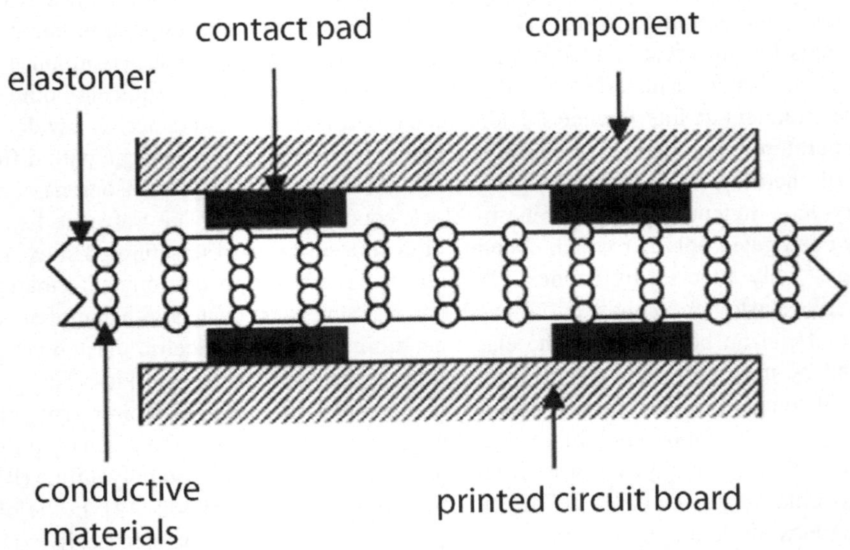

Figure 1. *Schematic diagram of a Z-direction only electrical conduction, in which the component is electrical connected to the printed circuit board through vertically aligned and laterally isolated conductive materials. The contact is made by external contact force. Reprint and permission from [16]*

An LGA socket assembly consists of component, socket, and board as shown in Figure 2. The contact force is applied through a fixture [16, 45]. The socket design can vary in terms of target component, contact, and housing. One class of LGA sockets incorporates conductive particles in elastomer (polymer) as electrical interconnects for LGA. This class of socket has comparatively low cost, ease of assembly, and short signal path; the elastomer has a sealing effect. However, elastomer-based interconnection is susceptible to failure due to creep and stress relaxation, whenever a force or deformation is provided. A decrease in the contact force due to stress relaxation may lead to the degradation of contact resistance. Increase of elasticity can effectively mitigate creep and stress relaxation, in that crosslinked elastomer has instantaneous recovery after removal of stress.

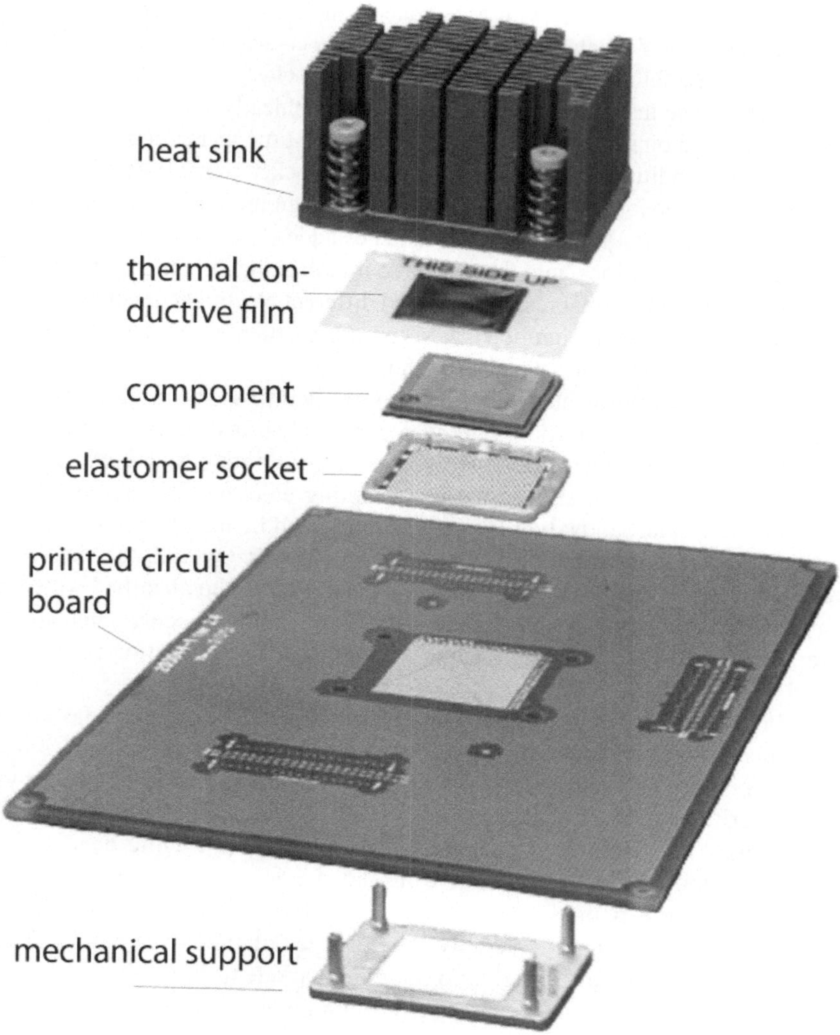

Figure 2. *Schematic diagram of elastomer land grid socket assembly. The interconnection part, elastomer socket, can be replaced with different type of sockets, including metalized particle interconnect (MPI), PariPoser film, and conductive rubber sockets (CRS). Reprint and permission from [16]*

Several mechanisms can lead to the elastomer failure, such as electrochemical migration [50-52], stress relaxation and creep [45, 53] and corrosion [16]. This section will extensively emphasize these mechanisms and use respective models to predict the life-cycle of the components. The metal electrochemical migration was first identified in the short-circuit failure of Sun Microsystems Sun Fire 6800. Investigation on the failed parts did not observe any dendrite to bridge adjacent components and/or contacts. In general, dendrites were considered to short the electrical circuit. This finding indicated that even without dendrite formation a short-circuit can happen. The design of the experiment, which mimicked the device environment with accelerated conditions, was conducted to identify the failure mechanism and electrochemical migration processes based on the failure products [50-52]. The new failure mechanism in the metal-in-elastomer socket was proposed on the facts that surface insulation resistance has catastrophically decreased and unknown materials were formed on surface after a certain period of time incubation in the temperature-humidity-bias conditions. The metal-in-elastomer socket consists of silver particles in elastomer matrix. Previous studies showed that silver can migrate in the humidity-temperature-bias conditions [54]. However, those studies added the external conductive ions, e.g., contaminants or silver ions, into the system to initialize the migration process [55]. In recent studies, several substantial results exploited different electrochemical migration, including silver electrochemical migration (ECM) processes, migration products, and failure criterion on surface insulation resistance (10^5 ohm).

The silver ECM could occur when bias was applied to silver-in-elastomer socket in the humidity-temperature. Initially, the water layer absorbed on the surface is electrolyzed to hydrogen and hydroxide ions, which further migrate to cathode and anode respectively. Silver oxides to silver ions by accepting electrons through electrolyzed water solution under bias, and silver ions are accumulated in absorbed water layer. The silver ions are also migrated to cathode through water layer. Overall, the surface insulation resistance (SIR) decreases when ions are generated in the absorbed water layer, or formation of conductive electrolyte. The short-circuit failure occurs when silver ion concentration increases to the specific value when the surface insulation resistance decreases to 10^5 ohm.

Both film and dendrites were observed on the film after permanent failure when the surface insulation resistance of the silver-in-elastomer was less than 1000 ohm. Studies on this regard showed that the film growth was developed in deionized water, while the dendrite growth was developed in high conductivity water. It was found that the film grew on the surface without distinctive direction while dendrites grew from cathode to anode. The analysis of film and dendrites from electron scanning spectrometry and X-ray photoelectron microscopy showed that the film consists of silver oxide and dendrites are silver only. The process of the silver electrochemical migration is progressed as follow: silver oxidation at anode to formation of silver ion; silver ion migration from anode to cathode through absorbed water layer (6-7 water monolayers); partial silver ion reduction to silver oxide with hydroxide from water electrolysis (film); other silver ion reduction to formation of silver at the cathode (dendrites) [50-52].

New failure criterion on surface insulation resistance was proposed. In the IPC standard (IPC-TR-476A), the failure was defined as the surface insulation resistance below 10^6 ohm. New experimental findings showed that the component didn't fail when resistance below the resistance, no electrochemical migration related products detected either [50-52]. In addition, the surface insulation resistance immediately decreased to the order of 10^6 ohm when the metal-in-elastomer was conditioned in the highly accelerated stress testing (130C/85% RH/bias). Until SIR was below 10^5 ohm, the silver dendrites were observed. The resistance ultimately was within hundreds of ohms when the dendrites bridged the adjacent contacts.

The new ECM process was proposed as: presence of moisture, anodic metal dissolution or ion generation and ion migration to cathode (electrochemical reaction), ion accumulation, and metal dendritic growth. The ion accumulation was found to be the rate controlling step in which the surface insulation resistance degraded significantly when the ion concentration accumulated to a specific value. This added step explained the failure occurrence without dendritic formation/bridging and perfected the process. Time-to-failure (TTF) model considered these factors to estimate the failure time, including water adsorption, ion generation, ion migration, and ion accumulation. The dendritic growth was excluded from the model based on the experimental findings. Furthermore, incorporation of dendritic growth model with TTF model can be used to estimate time-to-permanent component failure, since the dendritic bridge creates a permanent conductive path between conductors.

The ion-accumulation based model assumed that the silver discharged at the anode was directly proportional to the amount of electrical charge that passed through the electrode, or Faraday's Law, the total mass of metal dissolution equaled the total mass of metal ions discharged into the electrolyte. Unlike the bulk resistance of conductive particles in elastomer [56], by use of Ohm's law and Brunauer-Emmett-Teller (BET) model [57], the time-to-failure model was given by,

$$TTF = nF \times \frac{m_0}{M} \times \beta \times \frac{1}{V} \times \frac{(1-RH)[1+(c-1)RH]}{cRH} \times e^{(\frac{E_a}{RT})} \quad \text{E3}$$

Where $_s$ is the surface insulation conductivity, R is the resistance, m_0 is the critical mass loss required to dissolve to reach the critical ion concentration when resistance decreases to value of failure criterion, V is potential, RH is relative humidity, n is chemical valence, M is the molecular weight, F and R are constant, , c and E are coefficient relevant to relative humidity, temperature and materials properties. Using this model, the time-to-failure is estimated for elastomer sockets. For silver in elastomer socket, the *TTF* can be simplified as,

$$TTF = 1.11 \times 10^{-5} \times \frac{1}{V} \times \frac{(1-RH)}{RH} \times e^{(\frac{7846}{T})} \quad \text{E4}$$

Based on this model, the silver-in-elastomer socket fails in 150 hours after incubation in 90C/90%/20 V conditions; while it can resist up to 27000 h in service at conditions of 25C/50%/110 V.

Another failure mechanism is called creep and stress relaxation due to viscoelastic elastomer as matrix in socket. As shown in Figure 2, the mechanical support fixed elastomer deformation as constant. As a result, the stress applied to the socket is expected to decrease gradually due to the stress relaxation of elastomer. To understand failure mechanism and its effect on contact resistance, the dynamic mechanical analyzer (DMA) was used to characterize elastomer stress relaxation. The chamber of DMA can control temperature by air gas and liquid nitrogen. Creep is performed by constant stress and stress relaxation by constant strain. The creep process is often categorized into three stages: primary creep (a stage of decreasing creep rate), secondary creep (a stage of constant creep rate), and tertiary creep (a stage of increasing creep rate). When the load is initially applied, there is an instantaneous elongation, a primary stage of a transient nature during which slip and work hardening occur in the most favorably oriented grains. Then there is a secondary stage of steady-state creep during which the deformation continues at an approximately constant rate. The third stage (tertiary stage) takes place when the stress is high enough that the creep rate accelerates until fracture occurs [58]. The results for the creep of the silicone elastomer are shown in Figure 3 (a). Both primary and secondary creeps are observed. The results show that the creep deformation increases with temperature. The constant creep rate during the secondary stage was also a function of temperature.

Stress relaxation refers to the time-dependent contact force decrease under a constant strain. Figure 3 (b) plotted behavior of stress relaxation of elastomer under various temperatures. The significance of stress relaxation on elastomer socket is that contact resistance increases at a decreased contact force. It is thus important to derive the relationship time-dependent contact resistance due to stress relaxation of elastomer. Combination of stress relaxation and contact resistance, the time-dependent contact resistance model is derived [45],

$$R_c = \frac{\rho}{2}\sqrt{\frac{\pi H}{F_0[1 - Bln\left(\frac{t}{C}+1\right)]}} + \frac{\sigma_f H}{F_0[1 - Bln\left(\frac{t}{C}+1\right)]} \quad \text{E5}$$

Where the R_c is contact resistance, is electrical resistivity (ohm.m), H is contact hardness, F_0 is initial contact force, t is time, and B, C are temperature dependent parameters, given by,

$$B = 1.20 \times 10^{-4}T, \text{ and } C = 6.62 \times 10^{-7}Te^{\frac{1130}{T}} \text{ (for silicon elastomer)}$$

The $Rc - t$ can be used for estimation of contact resistance degradation of metal-in-elastomer socket for scheduled socket replacement.

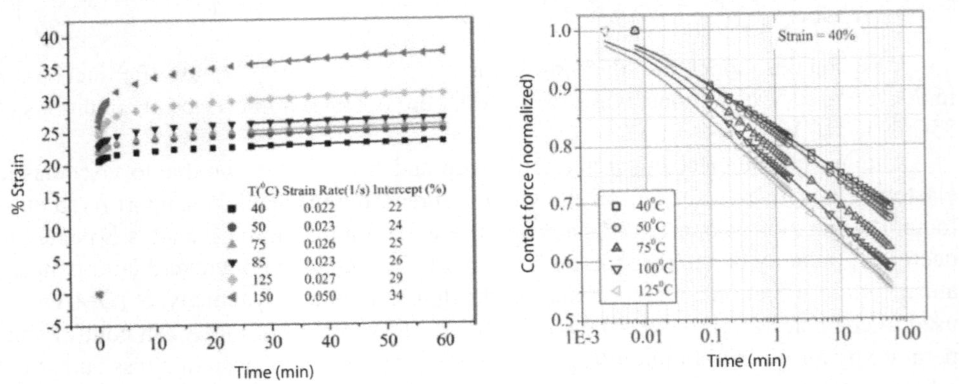

(a) elastomer creep (b) elastomer stress relaxation

Figure 3. *Elastomer creep (a) and stress relaxation (b) behavior at various temperatures. Dots: experimental data; solid line: fitted curve. Reprint and permission from 45*

4. ELASTOMER AS SUBSTRATE IN MICROFLUIDICS

Elastomer is also one of popular materials for microfluidics application as substrate, micro-valves, pumps [59-61]. Elastomer can be easily patterned by curing on a micromachining mold as an alternative microfabrication technique. The soft lithography technique, which generates micropatterns by contact printing and microstructures by embossing and replica molding, has been used to manufacture blazed grating optics, stamps for chemical patterning, and microfluidics devices [19, 62, 63]. Different from photolithography technique [64], the soft lithography is not subject to the limitations set by optical diffraction and optical transparency, with feature size down to 10-100 nm. Soft lithography techniques is termed as 'multilayer soft lithography' and developed by Quake et al. [59] in 2000. Briefly, multilayer structures were formed by bonding PDMS layers from separated cast in a micromachining mold. Since each layer has excess of one of the components, the reactive molecules remained at the interface form permanent bonding after further curing. They used such technique for development of microfluidic large-scale integration chips that contain plumbing networks with thousands of micromechanical PDMS valves and hundreds of individually addressable chambers. Their chips can be used to construct the microfluidic analog of a comparator array and a microfluidic memory storage device similar to the behavior of random-access memory.

PDMS microfluidics have great potential to be biomedical applications which only require small amounts of sample, routine operation by untrained personnel and low cost [65]. One of such examples was developed on a simple, inexpensive PDMS microfluidic diagnostic device, which performs sandwich immunoassays for medicine and biological studies [12, 63]. Screws are used in the device as virtual valves to on and off fluids. The low-cost on soft lithography microfabrication and raw materials makes this device great potential for portable healthcare delivery and monitoring. PDMS microfluidics can be used for reagent mixing in microscale channel [66]. The challenge for flow mixing in a small dimension is mainly due to the difficult liquid turbulence inside chip. By use of PDMS as microfluidics substrate, a topological structure can be fabricated to exploit the laminarity of the flow to repeatedly fold the flow and double the lateral concentration gradient in a compact chip. Two different solutions are contained in a T-junction, and fluid flow is repeatedly split, rotated, and merged [67]. The chip design was obtained by a rapid-prototyping master mold, and results showed that effective mixing can be achieved on short length scales.

PDMS microfluidics has applied to a variety of applications such as cell sorting, DNA sizing and sorting. DNA sequence is successfully performed in PDMS devices including sample preparation and electrophoresis analysis [14]. PDMS has substantial advantages over glass or silicon dioxide substrates, since these latter materials were adhesive for the cells of interest. In addition, Chou et al. [68] developed a PDMS device for the sizing and sorting of restriction fragments of DNA based on single DNA molecule detection. The device sizes DNA on the basis of fluorescence intensity in which the DNA was labeled with fluorescence dye.

PDMS device can be easily fabricated by non-covalent bonding or plasma treated permanent bonding after replica by molding. The PDMS microfluidics for capillary electrophoresis was developed by Effenhauser et al. [69] in 1997. They used replica molding to obtain PDMS device and bonded reversibly the device with a flat piece of PDMS. The performance of PDMS CE is not ideal on unmodified surface, especially for proteins/peptides analysis, due to hydrophobicity of the unmodified PDMS surface. Later, Whitesides et al. developed PDMS microfluidics device by a rapid prototyping and irreversible bonding through plasma oxidation [19]. Different PDMS devices were developed to demonstrate separation performance of fluorescently labeled amino acids and proteins using capillary zone electrophoresis (CZE) and gel electrophoresis (GE).

One-dimensional sodium dodecyl sulfate (SDS) capillary gel electrophoresis (CGE) can be also performed in PDMS microfluidics due to its versatile surface chemistry. PDMS surface is treated by UV-Ozone so that it could immobilize other hydrophilic molecules to prevent electroosmotic flow and protein/peptide non-specific binding. For complex cell lysate analysis, isoelectric focusing (IEF) can be interfaced with SDS-CGE as the first dimension separation. What makes PDMS special for IEF-SDS/CGE is because of conventional procedures for fabricating 3-D microfluidic channels [70], and the facility with which PDMS-based devices can be readily assembled and disassembled. For example, after IEF separation, the PDMS was disassembled and connected with a 3-D channel for SDS/CGE. Recent development on 2-D IEF-SDS/CGE showed that other thermoplastic would be better substrate of choice for integrated 2-D complex cell lysate analysis [71, 72].

The design of elastomer microfluidics can be made based on these studies [71-73]. The electrokinetic injection of defined sample plugs within single dimension microfluidic systems has been extensively studied, with the most widely used injector configurations including the cross, double-T, and triple-T topologies [74, 75], and extended configurations employing continuous sample injection leveraging flow switching techniques. Floating injection and pinched-injection methods [76, 77] have been employed in simple cross-injectors for the controlled definition of small sample plugs, while double- and triple-T designs are generally employed for the introduction of larger sample volumes. Sample leakage [78] was observed in a 2-D device during sample injection which was the central issue for separation performance as shown in Figure 4. After the desired sample plug has been transferred to the separation channel, additional samples enter the injection region due to diffusion and fringing of the electrical field during the electrokinetic transfer process. The excess sample tailing degrades separation performance. To solve this issue, backbiasing is a commonly used to eliminate sample leakage. Practically, bias voltages can be applied to sample inlet and waste reservoirs, electrokinetically pulling back excess sample from the injection zone. The effective of this approach depended on the accurate selection of backbiasing bias. This requires characterization of device geometry, medium in device, and buffer resistivity. A few groups [49, 72, 79] have proposed 2-D chip designs consisting of an array of second dimension separation channels aligned with an identical number of injection channels on the opposite side of the first dimension microchannel, a more efficient design for spatially multiplexed 2-D separation chips is shown in Figure 4 (a). The design consists of a single first dimension separation channel intersected by multiple second dimension separation channels on one side, and sample injection channels on the opposite side, with the injection channels staggered with respect to the separation channels to ensure complete and simultaneous sampling of species separated in the first dimension channel. In this staggered design, the parallel second dimension microchannels may be regarded as an array of double-T injectors operating in parallel. However, unlike the single-channel case, the double-T injectors are not electrically isolated in the multidimensional system. This interconnected design can result in significant variations in performance between the different injectors, as depicted in Figure 4 (b). In this image, the sample within the first dimension channel is being electrokinetically transferred into the second dimension channel array by applying a uniform bias voltage in the injection channel reservoirs while grounding the second dimension reservoirs. Three main features are evident in this injection process. The substantial tailing of sample occurs at the head of each second dimension channel, resulting in sample dispersion during transfer; in all but the center second dimension channel, the injection is highly asymmetric, reflecting a non-uniform electric field distribution within the first dimension channel; sample from the outermost regions of the first dimension channel continually leaks into the second dimension array due to a combination of diffusion and electric field fringing, leading to additional tailing which can continue long after sample from the center region of the first dimension channel has been fully transferred separation devices. The solution for unsymmetrical injection and tailing was proposed by adding backbias channel s parallel to the separation channels as shown in Figure 4 (b) [70]. Analytical calculation indicated that two-fold of resistance in the backbias channel generated uniform current and potential distribution in the 2-D channel network, not only in 1-D channel, but also injection and separation channels. Additionally, the angled 1-D channel can level the electric field distribution in the local region of T intersection. As a result, the angled 1-D channel substantially reduced sample tailing with negligible effect on the first dimension separation. These design perspectives have been used for complex biological sample analysis.

Use of PDMS for liquid chromatography has been limited largely due to its low bonding strength [80] and high gas/liquid permeability [81]. It was reported to fabricate column in a device using PDMS [82-84]. Since PDMS is inexpensive and thus it can be fabricated as a disposable column, it can be used as a stationary phase in liquid chromatography columns. The siloxane monomer was added into a molding to formation of desired features. The featured PDMS was reversibly sealed with cover PDMS by placing the molded chip in contact with cover PDMS, or irreversibly sealed by plasma oxidation. The PDMS chip was treated with silanes to minimize sample non-specific absorption. In general, PDMS is not the good materials of choice for high performance liquid chromatography. In such instance, other thermoplastics such as polyimide (PI) [85], fuse silica capillary [75], cyclic olefin copolymer (CoC) [86], are recommended.

Droplets formed by mixing one fluid in another or emulsions are very useful in a wide range of applications including personal care products, foods, microbiology, cell biology, and drug delivery vehicles [87, 88]. It has been recognized that a successful droplet device for applications is able to control over size and distribution in microscale and nanoscale emulsions [88]. Droplet formation can be obtained by either turbulence to break apart an immiscible mixture or continuous flow. PDMS devices for micro-droplet formation are fabricated by soft lithography in a molding. The native PDMS surface is usually hydrophobic and excellent surface for water-in-oil emulsions. The hydrophobic PDMS surface ensures that the oil droplets encapsulation of water phase inside without being contacted the PDMS channel walls [89, 90]; while hydrophilic PDMS surface can be used for oil-in-water droplets and biological sample analysis [91, 92]. Several surface modifications have been reported by the sequential layer-by-layer deposition of polyelectrolytes yielding hydrophilic microchannels. Others used oxygen-plasma and oxygen-C2F6 to modify PDMS surface in order to enhance hydrophilicity and reduce channel electroosmotic flow (EOF). Alternatively, use of chemical modification, e.g., 2-hydroxyethyl methacrylate (HEMA), can permanently enhance PDMS hydrophilicity [91]. In layer-by-layer (LbL) surface modification approach, segments of sodium chloride, poly(allylamine hydrochloride), and poly(sodium 4-styrenesulfonate) solutions are separated by air plugs and loaded into a piece of tubing and sequentially flushed through the channel at a constant flow rate. This LbL modification provides long storage time up to 5 months without noticeable behavior change of droplet formation. The LbL modified PDMS device can be used for water-oil-water double emulsion system for applications such as pharmaceutical compound delivery.

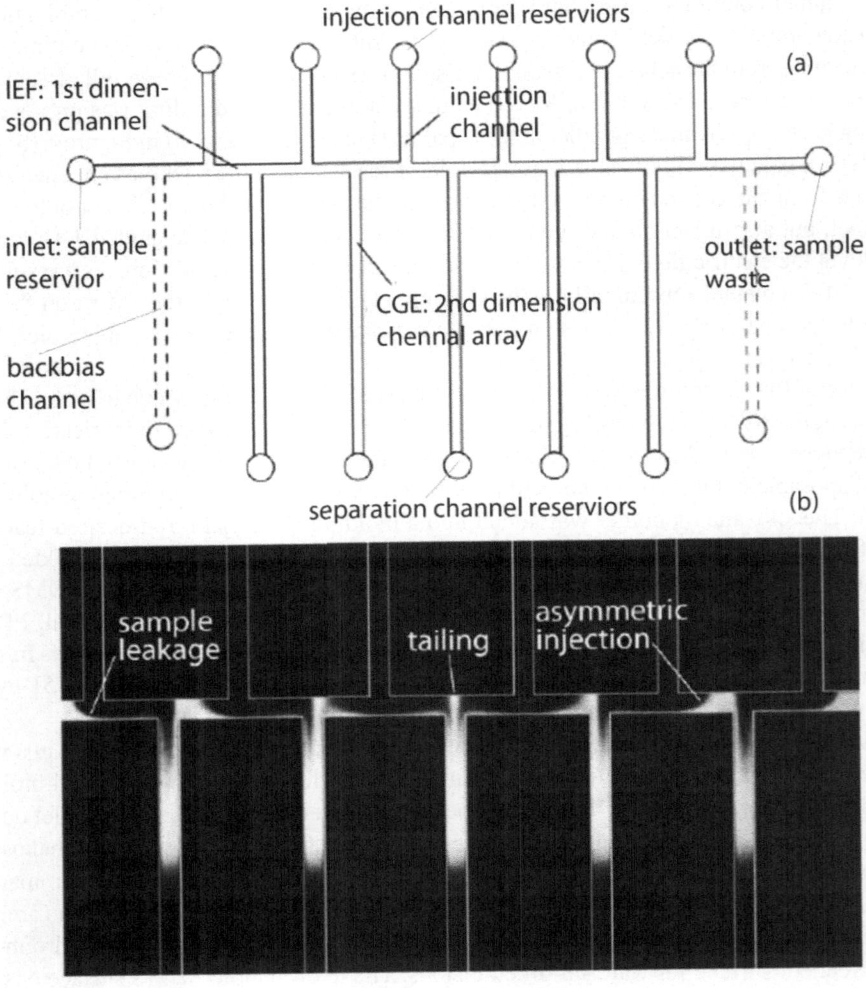

Figure 4. *Schematic diagram of two-dimensional capillary electrophoresis microfluidics design with (a) simplified device of a spatially-multiplexed separation platform with five second dimension microchannels, and (b) image during the sample transfer process using a fabricated chip without backbias channels. Reprint and permission from 71*

An interesting PDMS device for static microdroplet arrays is fabricated to droplet trapping, incubation, and release of enzymatic and cell-based assays [93]. They simply used a single-layered PDMS microfluidic structure. The aqueous droplets are trapped en masse and optically monitored for extended periods of time. The array droplet approach is used to characterize droplet shrinkage, aggregation of encapsulated biological cells and enzymatic reactions. On the other hand, to optimize the droplet formation, a design which use feedback control loop is developed [88]. They successfully demonstrated that the closed loop control system with negative feedback can produce the required droplet size with minimal noise and no systematic steady sate error. This close-loop design is utilized for a number of applications, including micro-chemistry and emulsion production.

5. OUTLOOK OF ELASTOMER APPLICATION

The application of elastomer is being continuously exploited. The future of elastomer is closely related to the specific requirement on applications. One of the hot topics is development of nano- and microfluidics for assisted reproductive technologies [94]. For example, the studies have demonstrated that thickness of PDMS layers at 100 μm in chips produced significantly fewer embryos due to drastically shift in osmolality of embryo-culture media. It could be the solution by development of a hybrid membrane consisting of PDMS-parylene-PDMS for maintaining a stable cell growth environment. Therefore, investigation of new type of PDMS materials or hybrid PDMS can provide means for such applications.

Due to the low bonding strength, PDMS has seldom used for high performance liquid chromatography application (HPLC). The HPLC can be directly interfaced with detection system such as electrospray ionization or mass spectrometry. As a result, it becomes research of interest to synthesize PDMS which has strong bonding strength (via covalent bond) and low permeability, while PDMS optical transparency and elasticity are largely remained. In summary, PDMS is becoming more and more popular as the materials of choice for biological analysis due to its good physical, mechanical, and chemical properties, as well as its excellent biocompatibility.

A thermoplastic rubber material has recently been developed that can completely mend itself when the fracture interfaces are rejoined and left to heal for a moderate time [95]. This “smart” rubber is easy to synthesize and displays excellent mechanical properties. This material could be used for microelectronics, microfluidics and biology system.

REFERENCES

1. Legge, N.R. Thermoplastic elastomers. 60, 83 (1987).
2. Eisenbach, C., Baumgartner, M. & Gunter, C. Advances in Elastomers and Rubber Elasticity. 51 (1986).
3. Markham, R.L. & Mueller, W.J. (Google Patents, 1986).
4. Massey, L.K. Permeability Properties of Plastics and Elastomers: A Guide to Packaging and Barrier Materials, Edn. 2nd. (Plastics Design Library / William Andrew Publishing, Norwich, NY; 2002).
5. Simmons, S., Shah, M., Mackevich, J. & Chang, R.J. Polymer outdoo insulating materials Part Ill-silicone elastomer considerations. 13, 8 (19978).
6. Lin, Y.H., Kang, S.W. & Wu, T.Y. Fabrication of polydimethylsiloxane (PDMS) pulsating heat pipe. 29, 573-580 (2009).

7. Jafari, S.H. & Gupta, A.K. Impact strength and dynamic mechanical properties correlation in elastomer-modified polypropylene. J Appl Polym Sci 78, 962-971 (2000).
8. Yang, S., Zhang, Y.W. & Zeng, K.Y. Analysis of nanoindentation creep for polymeric materials. J Appl Phys 95, 3655-3666 (2004).
9. White, C.C., Vanlandingham, M.R., Drzal, P.L., Chang, N.K. & Chang, S.H. Viscoelastic characterization of polymers using instrumented indentation. II. Dynamic testing. J Polym Sci Pol Phys 43, 1812-1824 (2005).
10. Jancar, J., Dianselmo, A., Dibenedetto, A.T. & Kucera, J. Failure Mechanics in Elastomer Toughened Polypropylene. Polymer 34, 1684-1694 (1993).
11. Toepke, M.W. & Beebe, D.J. PDMS absorption of small molecules and consequences in microfluidic applications. 6, 1484-1486 (2006).
12. Eteshola, E. & Leckband, D. Development and characterization of an ELISA assay in PDMS microfluidic channels. 72, 129-133 (2001).
13. Martin, R.S., Gawron, A.J., Lunte, S.M. & Henry, C.S. Dual-electrode electrochemical detection for poly (dimethylsiloxane)-fabricated capillary electrophoresis microchips. 72, 3196-3202 (2000).
14. Paegel, B.M., Blazej, R.G. & Mathies, R.A. Microfluidic devices for DNA sequencing: sample preparation and electrophoretic analysis. 14, 42-50 (2003).
15. Scott, N. An area modulus of elasticity: definition and properties. 58, 269-275 (2000).
16. Yang, S., Wu, J. & Pecht, M.G. Reliability Assessment of Land Grid Array Sockets Subjected to Mixed Flowing Gas Environment. Ieee T Reliab 58, 634-640 (2009).
17. Liu, W. & Pecht, M. IC component sockets. (Wiley-Blackwell, 2004).
18. Lee, J.N., Park, C. & Whitesides, G.M. Solvent compatibility of poly (dimethylsiloxane)-based microfluidic devices. 75, 6544-6554 (2003).
19. McDonald, J.C. et al. Fabrication of microfluidic systems in poly(dimethylsiloxane).Electrophoresis 21, 27-40 (2000).
20. Eddings, M.A. & Gale, B.K. A PDMS-based gas permeation pump for on-chip fluid handling in microfluidic devices. 16, 2396 (2006).
21. Mark, J.E. Some unusual elastomers and experiments on rubberlike elasticity. Prog Polym Sci 28, 1205-1221 (2003).
22. Lawn, B. & Wilshaw, R. Indentation Fracture - Principles and Applications. J Mater Sci 10, 1049-1081 (1975).
23. Oliver, W.C. & Pharr, G.M. An Improved Technique for Determining Hardness and Elastic-Modulus Using Load and Displacement Sensing Indentation Experiments. J Mater Res 7, 1564-1583 (1992).
24. Carrillo, F. et al. Nanoindentation of polydimethylsiloxane elastomers: Effect of crosslinking, work of adhesion, and fluid environment on elastic modulus. J Mater Res 20, 2820-2830 (2005).
25. Bao, T., Morrison, P.W. & Woyczynski, W. AFM nanoindentation as a method to determine microhardness of hard thin films. Mater Res Soc Symp P 517, 395-400 (1998).
26. Oyen, M.L. Sensitivity of polymer nanoindentation creep measurements to experimental variables. 55, 3633-3639 (2007).
27. Mata, A., Fleischman, A.J. & Roy, S. Characterization of polydimethylsiloxane (PDMS) properties for biomedical micro/nanosystems. 7, 281-293 (2005).

28. Singh, A., Freeman, B.D. & Pinnau, I. Pure and mixed gas acetone/nitrogen permeation properties of polydimethylsiloxane [PDMS]. J Polym Sci Pol Phys 36, 289-301 (1998).

29. Kurian, P. et al. Synthesis, permeability and biocompatibility of tricomponent membranes containing polyethylene glycol, polydimethylsiloxane and polypentamethylcyclopentasiloxane domains. Biomaterials 24, 3493-3503 (2003).

30. Pinnau, I. & He, Z.J. Pure- and mixed-gas permeation properties of polydimethylsiloxane for hydrocarbon/methane and hydrocarbon/hydrogen separation. J Membrane Sci 244, 227-233 (2004).

31. Leclerc, E., Sakai, Y. & Fujii, T. Microfluidic PDMS (polydimethylsiloxane) bioreactor for large-scale culture of hepatocytes. 20, 750-755 (2004).

32. Abate, A.R., Lee, D., Do, T., Holtze, C. & Weitz, D.A. Glass coating for PDMS microfluidic channels by sol–gel methods. 8, 516-518 (2008).

33. Ghosal, K. & Freeman, B.D. Gas separation using polymer membranes: an overview. 5, 673-697 (1994).

34. Shiku, H. et al. Oxygen permeability of surface-modified poly(dimethylsiloxane) characterized by scanning electrochemical microscopy. Chem Lett 35, 234-235 (2006).

35. Leclerc, E., Sakai, Y. & Fujii, T. Cell culture in 3-dimensional microfluidic structure of PDMS (polydimethylsiloxane). Biomed Microdevices 5, 109-114 (2003).

36. Tummala, R.R., Rymaszewski, E.J. & Klopfenstein, A.G. Microelectronics Packaging Handbook. (Kluwer Academic Publishers Group, Norwell, MA; 1997).

37. Kim, J., Chaudhury, M.K. & Owen, M.J. Hydrophobicity loss and recovery of silicone HV insulation. Ieee T Dielect El In 6, 695-702 (1999).

38. Liu, H.P., Cash, G., Birtwhistle, D. & George, G. Characterization of a severely degraded silicone elastomer HV insulator - an aid to development of lifetime assessment techniques. Ieee T Dielect El In 12, 478-486 (2005).

39. Park, J., Kim, H.S. & Han, A. Micropatterning of poly(dimethylsiloxane) using a photoresist lift-off technique for selective electrical insulation of microelectrode arrays. J Micromech Microeng 19 (2009).

40. Nam, Y., Musick, K. & Wheeler, B.C. Application of a PDMS microstencil as a replaceable insulator toward a single-use planar microelectrode array. Biomed Microdevices 8, 375-381 (2006).

41. Tong, X.C. Advanced Materials for Thermal Management of Electronic Packaging, Edn. 1st. (Springer Science & Business Media, LLC, New York; 2010).

42. Lee, K.I. & Jopson, H. Electrically Conductive Thermoplastic Elastomer Polyacetylene Blends. Polym Bull 10, 105-108 (1983).

43. Pelrine, R., Kornbluh, R., Pei, Q.B. & Joseph, J. High-speed electrically actuated elastomers with strain greater than 100%. Science 287, 836-839 (2000).

44. Faez, R., Schuster, R.H. & De Paoli, M.A. A conductive elastomer based on EPDM and polyaniline. II. Effect of the crosslinking method. Eur Polym J 38, 2459-2463 (2002).

45. Yang, S., Wu, J., Tsai, D. & Pecht, M.G. Contact resistance estimation for time- dependent silicone elastomer matrix of land grid array socket. Ieee T Compon Pack T 30, 81-85 (2007).

46. Jin, S. et al. New, Z-Direction Anisotropically Conductive Composites. J Appl Phys 64, 6008-6010 (1988).

47. Jin, S., Tiefel, T.H., Wolfe, R., Sherwood, R.C. & Mottine, J.J. Optically Transparent, Electrically Conductive Composite Medium. Science 255, 446-448 (1992).

48. Ausanio, G. et al. Giant resistivity change induced by strain in a composite of conducting particles in an elastomer matrix. Sensor Actuat a-Phys 127, 56-62 (2006).

49. Chen, L.H., Jin, S. & Tiefel, T.H. Tactile Shear Sensing Using Anisotropically Conductive Polymer. Appl Phys Lett 62, 2440-2442 (1993).

50. Yang, S., Wu, J. & Pecht, M. Electrochemical migration of land grid array sockets under highly accelerated stress conditions. 238-244 (2005).

51. Yang, S., Wu, J. & Christou, A. Initial stage of silver electrochemical migration degradation. Microelectron Reliab 46, 1915-1921 (2006).

52. Yang, S.A. & Christou, A. Failure model for silver electrochemical migration. Ieee T Device Mat Re 7, 188-196 (2007).

53. Liu, W.F., Pecht, M.G. & Xie, J.S. Fundamental reliability issues associated with a commercial particle-in-elastomer interconnection system. Ieee T Compon Pack T 24, 520- 525 (2001).

54. Krumbein, S.J. Metallic Electromigration Phenomena. Ieee T Compon Hybr 11, 5-15 (1988).

55. Zamanzadeh, M., Meilink, S.L., Warren, G.W., Wynblatt, P. & Yan, B. Electrochemical Examination of Dendritic Growth on Electronic Devices in Hcl Electrolytes. Corrosion 46, 665-671 (1990).

56. Ruschau, G.R., Yoshikawa, S. & Newnham, R.E. Resistivities of Conductive Composites. J Appl Phys 72, 953-959 (1992).

57. Brunauer, S., Emmett, P.H. & Teller, E. Adsorption of gases in multimolecular layers. J Am Chem Soc 60, 309-319 (1938).

58. Findley, W.N., Lai, J.S. & Onaran, K. Creep and Relaxation of Nonlinear Viscoelastic Materials. (Dover, New York; 1975).

59. Unger, M.A., Chou, H.P., Thorsen, T., Scherer, A. & Quake, S.R. Monolithic microfabricated valves and pumps by multilayer soft lithography. Science 288, 113-116 (2000).

60. Melin, J. & Quake, S.R. Microfluidic large-scale integration: The evolution of design rules for biological automation. Annu Rev Bioph Biom 36, 213-231 (2007).

61. Erickson, D. & Li, D.Q. Integrated microfluidic devices. Anal Chim Acta 507, 11-26 (2004).

62. de Mello, A. On-chip chromatography: the last twenty years. Lab Chip 2, 48n-54n (2002).

63. Xia, Y.N. & Whitesides, G.M. Soft lithography. Angew Chem Int Edit 37, 551-575 (1998).

64. Levenson, M.D., Viswanathan, N. & Simpson, R.A. Improving resolution in photolithography with a phase-shifting mask. 29, 1828-1836 (1982).

65. Paguirigan, A.L. & Beebe, D.J. Microfluidics meet cell biology: bridging the gap by validation and application of microscale techniques for cell biological assays. 30, 811- 821 (2008).

66. Song, H. & Ismagilov, R.F. Millisecond kinetics on a microfluidic chip using nanoliters of reagents. 125, 14613-14619 (2003).

67. Chen, H. & Meiners, J.C. Topologic mixing on a microfluidic chip. 84, 2193 (2004).

68. Chou, H.P., Spence, C., Scherer, A. & Quake, S. A microfabricated device for sizing and sorting DNA molecules. 96, 11 (1999).

69. Effenhauser, C.S., Bruin, G.J.M., Paulus, A. & Ehrat, M. Integrated capillary electrophoresis on flexible silicone microdevices: analysis of DNA restriction fragments and detection of single DNA molecules on microchips. 69, 3451-3457 (1997).

70. Jo, B.H., Van Lerberghe, L.M., Motsegood, K.M. & Beebe, D.J. Three-dimensional microchannel fabrication in polydimethylsiloxane (PDMS) elastomer. 9, 76-81 (2000).

71. Yang, S., Liu, J. & DeVoe, D.L. Optimization of sample transfer in two-dimensional microfluidic separation systems. Lab Chip 8, 1145-1152 (2008).

72. Yang, S., Liu, J.K., Lee, C.S. & Devoe, D.L. Microfluidic 2-D PAGE using multifunctional in situ polyacrylamide gels and discontinuous buffers. Lab Chip 9, 592-599 (2009).
73. Liu, J.K., Yang, S., Lee, C.S. & Devoe, D.L. Polyacrylamide gel plugs enabling 2-D microfluidic protein separations via isoelectric focusing and multiplexed sodium dodecyl sulfate gel electrophoresis. Electrophoresis 29, 2241-2250 (2008).
74. Fu, L.M., Yang, R.J., Lee, G.B. & Liu, H.H. Electrokinetic injection techniques in microfluidic chips. 74, 5084-5091 (2002).
75. Blas, M., Delaunay, N. & Rocca, J.L. Electrokinetic-based injection modes for separative microsystems. 29, 20-32 (2008).
76. Hirokawa, T., Takayama, Y., Arai, A. & Xu, Z. Study of a novel sample injection method (floating electrokinetic supercharging) for high-performance microchip electrophoresis of DNA fragments. 29, 1829-1835 (2008).
77. Zhang, L., Yin, X. & Fang, Z. Negative pressure pinched sample injection for microchip-based electrophoresis. 6, 258-264 (2006).
78. Duffy, D.C., McDonald, J.C., Schueller, O.J.A. & Whitesides, G.M. Rapid prototyping of microfluidic systems in poly (dimethylsiloxane). 70, 4974-4984 (1998).
79. Li, Y., Buch, J.S., Rosenberger, F., DeVoe, D.L. & Lee, C.S. Integration of isoelectric focusing with parallel sodium dodecyl sulfate gel electrophoresis for multidimensional protein separations in a plastic microfludic network. 76, 742-748 (2004).
80. Gent, A. & Tobias, R. Effect of interfacial bonding on the strength of adhesion of elastomers. III. Interlinking by molecular entanglements. 22, 1483-1490 (1984).
81. Kuncová-Kallio, J. & Kallio, P.J. 2486-2489 (IEEE, 2006).
82. Weng, X., Chon, C.H., Jiang, H. & Li, D. Rapid detection of formaldehyde concentration in food on a polydimethylsiloxane (PDMS) microfluidic chip. 114, 1079-1082 (2009).
83. Slentz, B.E., Penner, N.A., Lugowska, E. & Regnier, F. Nanoliter capillary electrochromatography columns based on collocated monolithic support structures molded in poly (dimethyl siloxane). 22, 3736-3743 (2001).
84. Ocvirk, G. et al. Electrokinetic control of fluid flow in native poly (dimethylsiloxane) capillary electrophoresis devices. 21, 107-115 (2000).
85. Levkin, P.A. et al. Monolithic porous polymer stationary phases in polyimide chips for the fast high-performance liquid chromatography separation of proteins and peptides. 1200, 55-61 (2008).
86. Mair, D.A., Geiger, E., Pisano, A.P., Fréchet, J.M.J. & Svec, F. Injection molded microfluidic chips featuring integrated interconnects. 6, 1346-1354 (2006).
87. Anna, S.L., Bontoux, N. & Stone, H.A. Formation of dispersions using "flow focusing" in microchannels. 82, 364 (2003).
88. Miller, E., Rotea, M. & Rothstein, J.P. Microfluidic device incorporating closed loop feedback control for uniform and tunable production of micro-droplets. 10, 1293-1301 (2010).
89. Link, D.R. et al. Electric control of droplets in microfluidic devices. 45, 2556-2560 (2006).
90. Fujii, T. PDMS-based microfluidic devices for biomedical applications. 61, 907-914 (2002).
91. Bodas, D. & Khan-Malek, C. Formation of more stable hydrophilic surfaces of PDMS by plasma and chemical treatments. 83, 1277-1279 (2006).
92. Bauer, W.A.C., Fischlechner, M., Abell, C. & Huck, W.T.S. Hydrophilic PDMS microchannels for high-throughput formation of oil-in-water microdroplets and water- in-oil-in-water double emulsions. 10, 1814-1819 (2010).

93. Huebner, A. et al. Static microdroplet arrays: a microfluidic device for droplet trapping, incubation and release for enzymatic and cell-based assays. 9, 692-698 (2009).

94. Beebe, D., Wheeler, M., Zeringue, H., Walters, E. & Raty, S. Microfluidic technology for assisted reproduction. 57, 125-135 (2002).

95. Wietor, J.L. & Sijbesma, R.P. A Self-Healing Elastomer. 47, 8161-8163 (2008).

7 CHAPTER

Microfluidic Devices Fabrication for Bioelectrokinetic System Applications

Nurul Amziah Md Yunus
Universiti Putra Malaysia, Malaysia

1. INTRODUCTION

The research and development of microfluidic technologies has increased tremendously over the past twenty years. The technology allows designers to create small, portable, robust, low cost and easy-to-use diagnostic instruments that offer high levels of capability and versatility. Microfluidic systems will decrease reagent consumption and reduce cost per analysis. It also reduces analysis time and provides better controllable process parameters in chemical reactions.

Microfluidic is also known as miniaturised total (chemical) analysis system, (μTAS). It is a generic term for a small system or device (microfluidic) designed to perform one or more chemical/biochemical processes. The early stage of microfluidics has been dominated by the development of microflow sensors, micropumps and microvalves in late 1980s and the beginning of 1990s and the concept has been introduced by Manz and co-workers from Imperial College London [1]. Since then, chip-based analytical systems have been rapidly applied to a variety of fields such as separation science, chemical production, DNA analysis, medical diagnostics and environmental analysis [2, 3]. Examples of microfluidic devices are shown in Fig. 1 (a)-(b).

Surprisingly, many fabrication methods used for making a microfluidic device are based on silicon microelectronics fabrication industry, with one of the first devices on silicon by Terry [a gas chromatographic air analyser in the late 70s, but little study was done about producing one again at that time. Interest in this type of technology was revived in the 90s with current microfluidic devices fabricated not only from silicon based but to a range of polymers, glass wafers and foils. This technology is used to build miniature devices as shown in Fig. 1(c) with packaging that could perform as a conventional "big bench-top" such as shown in Fig. 1 (d).

Thus, the challenge has been to create such as medical, chemical, biological devices that are capable of doing a fast and accurate analysis of the reagent and has the capacity to be green technology products in the future. With the recent advances in the synthesis and the characterization of size-selected particles in colloids (submicron and nanometer range) such as blood itself, an investigation on their physical and chemical properties has been made possible [5]. In microfluidic, a fabrication of it devices with integrated microchannels and microelectrodes of dimensions are made comparable to biological cells or particles size. Additionally, if there is a capability of producing small-scale devices, it will allow the development of entirely novel experiments, which currently is rapidly under study.

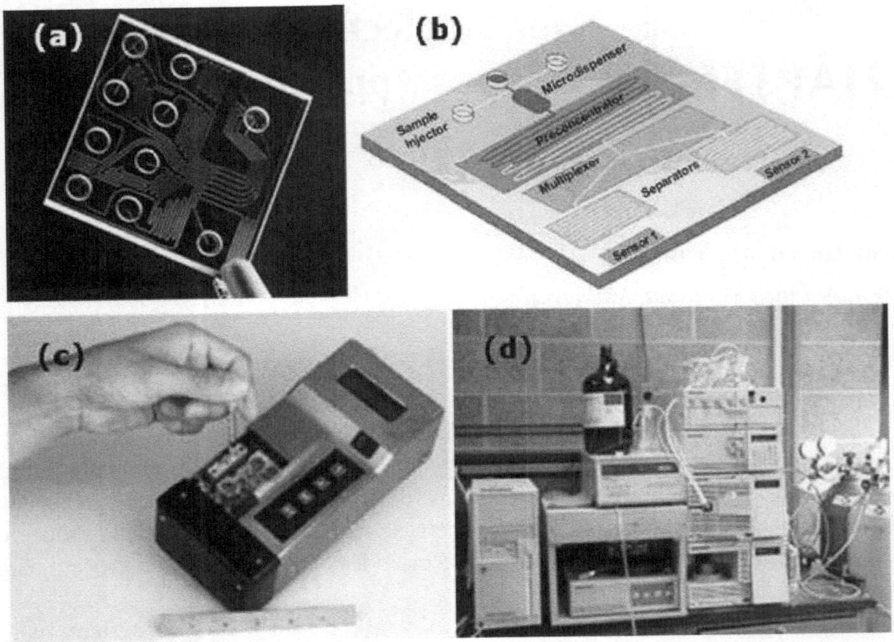

Figure 1. *a) Microfluidics Device –(image is taken from Agilent Technologies website) (b) Possible analysis on the microfluidic device with integrated microfluidic with sample injector, microdispenser, preconcentrator, multiplexor, separator and sensor (c) Handheld microfluidic in the proper packaging (d) Example of a big benchtop system –the conventional version of chemical/biological analytical technique.*

Microchannels are very important features for microfluidic systems. There are many ways to fabricate the microchannel, such as by using the silicon based organic polymer, polydimethylsiloxane (PDMS) or the femtosecond laser micromachining and inscription or using deep reactive ion etching (DRIE). The method emphasized in this Chapter will be the fabrication of microfluidic microchannel using the laminate sheet, dry film resist (DFR), for example, the SY series from Elga Europe [7]. This laminate sheet is a biocompatible material and it adheres easily on the co-planar substrate, i.e glass or silicon substrate. It is cheap, simpler and easier to use.

The work is also concerned on the application of electrokinetic technique to bioelectrokinetic monitoring system. Electrokinetic is defined as the study of fluid or particle motion in electric fields. It generally works in direct current, DC and alternating current, AC.

Electrokinetics include electrophoresis (EP), dielectrophoresis (DEP), electro-osmosis (EO), electrorotation (ROT) and electroorientation [6]. These electrokinetic processes are able to manipulate, concentrate, collect, mix and separate different types of bioparticles, biomolecules, organic and inorganic materials.

2. THE PHYSICS OF MICROFLUIDIC SYSTEMS

In this section, an overview of the physics of microfluidic will be highlighted because at the microscale, there are many different forces which become significant over forces experienced in everyday life. It is begun with the motion of fluid described by Navier-Stokes equation named after Claude-Louis Navier (1785-1836), a French engineer and physicist and George Gabriel Stokes (1819-1903), a mathematician and also a physicist. The equation was derived from conservation of momentum arguments [8]. This equation arises from applying Newton's second law to fluid motion, together with the assumption that the fluid stress is the sum of a constant viscous term (proportional to the gradient of velocity), plus a pressure term [9]. For an incompressible Newtonian fluid (named after Isaac Newton a mathematician and physicist (1642-1727) [10]) the equation is [11, 12]:

$$\rho_m \frac{\partial \mathbf{u}}{\partial t} + \rho_m \left(\mathbf{u} \bullet \nabla\right)\mathbf{u} = -\nabla p + \eta \nabla^2 \mathbf{u} + \mathbf{f} \quad \text{E1}$$

where ρ is the mass density, uis the velocity of fluid, pis the pressure, η is the viscosity and f is the total applied body force. The ratio of the inertial term to the viscous term can be determined from the coefficients in equation which gives a factor referred to as the Reynolds' number (named after Osborne Reynolds (1842–1912)) [13-15] shown in equation (2):

$$\mathrm{Re} = \frac{\rho_m u_o l_o}{\eta} \quad \text{E2}$$

'Re' stands for Reynolds' number, uo is the magnitude of a typical velocity and lo is a length scale. It is a dimensionless number and can be estimated as follows: Assuming that typical working fluid is water with typical velocities uo of *uo* of 10^{-6} to 10^{-2} m/s Typical channel radii of 10^{-4} m, $\rho m = 10^3$ kgm^{-3} and $\eta = 10^{-3}$ kgm$^{-1}s^{-1}$. For fluids in microfluidic devices, which have small Reynolds number (Re <<1), their behaviour is dominated by viscous forces over inertial forces and the resulting flows are linear and the fluid is considered to move in laminar sheets. If inertial forces are much bigger, Re>>1, turbulent flow will occur. The low values of the Reynolds number in the microfluidic microsystem indicate that, if the fluid is moving under an applied pressure or force, which is suddenly removed, the fluid flow will stop immediately.

2.1. Laminar flow

In microfluidic microsystem, the microchannel is fabricated to guide the fluid flow through the device. This channel can be in any size in the range of less than 5μm to 100s μm and the flow is always laminar. Laminar by definition is a layer. Laminar flow is a condition whereby the velocity of a particle in a fluid stream is not a random function of time thus the fluid flow follows streamlines [16]. One of the effects of laminar flow is that, two or more streams flowing in contact with each other will not be mixed except by diffusion [17].

The fluid in steady state flow is defined as Poiseuille flow (named after Jean Louis Marie Poiseuille 1797-1869) a French physician and physiologist) [18]. The flow establishes a parabolic flow profile after some distance from the entry point of the fluid into the channel. The velocity is zero at the walls and reaches maximum at the centre of the channel.

2.2. Viscous drag force

When there is particle in fluid, it will experience a viscous drag force due to the action of the fluid on the particles. Viscous drag is the force that resists the movement of a particle through the fluid. The force is proportional to but opposite the relative velocity of the particles. For a sphere particle of radius a in a fluid of viscosityη, viscous drag force is given by Stokes's law [11] in equation(3).

$$\mathbf{F}_{Viscous} = 6\pi\eta a\mathbf{v} \quad \text{E3}$$

vis the velocity of the particle. For a constant applied force the particle will eventually reach a terminal velocity, which makes it, does not accelerating. If the fluid is at rest, the terminal velocity is simply proportional to the applied force. If the fluid is moving, the terminal velocity experienced by the particle depends on the velocity of the fluid.

For a sphere particle, the friction factor is given by equation (4). The constant *f,* which refers to the friction factor, depends on a range of particle parameter such as size, shape and surface characteristics.

$$f = 6\pi\eta a \quad \text{E4}$$

3. FABRICATION OF MICROFLUIDIC PROCESS

When the microfluidic is introduced in the early 1990s, a glass substrate is used as a material for the substrate for most microfluidic devices [19-21].The advantages of glass in relation to the other materials are it is chemically inert to most liquid and gases, hydrophilic and optically 100% clear. There are three types of glass that are commonly used in the laboratory. There are consisting of borosilicate, soda lime and fused silica glass. In this work, the borosilicate glass or pyrex glass, which is the type of the microscopic glass slide, is used as the substrate. It can resist strong acids, saline solutions, chlorine, bromine, iodine, and strong oxidizing and corrosive chemicals. This glass can be produced in a mass product since it is not expensive.

The other reason glass slide is used as a substrate, is that glass is hydrophilic, in which it attracts and holds moisture. Most plastics, in comparison, are hydrophobic and need treatment to become hydrophilic. Glass has material properties that are stable in time and it is thus preferable for applications in which devices are used extensively, such as in high throughput screening. It is a reliable shelf life. It is non-porous, implying that small molecules will not be able to diffuse into it.

3.1. Glass as a substrate

To start utilize glass for microfluidic device, we could cut it into pieces, of dimension for example 2 cm by 1.5 cm, and we call it as fragmented chip glasses. These glasses will go through the cleaning process. They were placed in the glass holder filled with 95% pure water and 5% of soap. Then this holder is placed in the ultrasonic bath (Ultrawave Ltd) for washing away all the fouls with the aid of the beaker filled with DI water for 15 minutes. After cleaning, the glasses are thoroughly rinsed with DI water followed by acetone, methanol and iso-propanol (IPA), the RGB solvents, and blow-dried with nitrogen.

The glasses are placed in the oven for dehydration purposes at the temperature of 200°C for 2 hours minimum after the cleaning process is completed. This step is done to make sure that the glasses used are dried totally so that attachment of laminate sheet mentioned earlier in the introduction on these glasses will be excellent.

3.2. The device

Manz and co-workers introduced photolithographic techniques to microfabricated electrophoretic separation channels [19]. Lithography uses photoresist materials to cover areas on the wafer that will not be subjected to material deposition or removal.

There are two types of photoresist materials, namely, negative and positive photoresists. Negative photoresists are those that will become less soluble in the developer solution when they are exposed to light, forming negative images of the mask patterns on the wafer (the substrate) while positive photoresists are those that will become more soluble in the developer when they are put under the same exposure of light, forming positive images of the mask patterns on the wafer.

Advances in micro and nano fabrication techniques allow small microelectrodes of order of 1μm in size and smaller with selective materials to be manufactured with relative ease. Therefore microfluidic device has been designed and fabricated by taking into consideration the material used for making the microelectrode so that only small amount of voltage needed to gain high current. Thus, this microelectrode structures can provide sufficient force to manipulate particles without requiring high voltage signal generators. This will reduce joule heating in the microfluidic device. Although, it is found that if the microelectrode gap and chamber height is made smaller and lower respectively can impinge on the joule heating problem when higher conductivity medium is employed especially by biologist in biology application, but with selective material for microelectrode i.e biocompatible, it will overcome that problem.

The microelectrodes can be fabricated on approximately 10 cm diameter, (4 inch), 1 mm thick glass wafers or substrates. The total chip that can fit the 4 inch wafer is about 14 chips. This is depending on the size of the device/chip designed by the engineer for a particular function. The microelectrode will be deposited on the glass wafer by using the electron beam lithography. This technique will ensure the layer of microelectrode is thinner and the electrode width and gap are precisely small.

The microelectrodes are fabricated using gold, Au. The gold layer is essential because of its low resistance and its biocompatible nature, but unfortunately it does not adhere well to the glass surface. Thus titanium, Ti on the other hand, adheres well on the glass substrate and was used to improve gold adhesion. The palladium, Pd layer acts as a diffusion barrier between the titanium, Ti and the gold, Au. The other metal that can promote adhesion to gold, Au and platinum, Pt is chromium, Cr. However Cr is known to diffuse into the overlying gold, Au over time more quickly than titanium, Ti.

Microelectrodes on the glass were patterned using photolithography and ion beam milling over titanium, Ti, seed layer. Pt or Au microelectrodes are chosen because they are highly conductive, chemically unreactive, and able to resist tarnishing and corrosion, which then would allow the maximum current flow through it and hence, greater force or energy could be produced. They are also biocompatible because of its quality of not having toxic or injurious effects on biological systems. Some evaporated metal layers for microelectrode are 10 nm titanium, Ti, 10 nm palladium, Pd, and 100 nm gold, Au. For example, in this work, the width of the wire of the microelectrode is 20 µm and it is made of three layers of metal consisting 30 nm Ti, 100 nm Au and 30 nm Ti. The top layer of Ti is to reduce the effects of corrosion at low frequency and high potentials [22]. The electron-beam lithography technique is used to fabricate smaller size of microelectrode ranging 1-5 µm and gap sizes ranging from 1-10µm and its reproducibly.

3.3. Laminate

The next step is to perform the channel on these glasses. The fresh laminate sheet, dry film resist (DFR) of 50 µm is taken out from the fridge. The purpose of keeping the laminate in cold place is to lengthen the freshness/lifetime/dryness/unwanted crosslink. It is folded in black plastic cover to prevent exposure to or penetration of the UV light. A layer at one side of the laminate is the protective film, the polyethylene (PE) layer, is then peeled off. The photoresist is then applied on the substrate, which is placed on the cardboard as shown in Fig. 1. The substrates are put through the hot-roll laminator with speed 1 m/min and at temperature 100°C -110°C. The protective polyester (PET) layer should remain on top of the laminate when the exposure process is performed. This layer is to be removed once the processes of developing and rinsing are followed.

3.4. Adhesive bonding

Our procedure for bonding the fragmented chips uses a technique called adhesive bonding. There are two appliances that can be used for adhesive bonding process, the hot press bonder and the oven. For the hot press bonder, the temperature used must be ramped up by 0.83°C/min, from 150°C to 200°C. The length of time of the first temperature, 150°C is set for 30 minutes while the second temperature 200°C is set for 1 hour. After the course, the temperature is ramped down slowly, with 1°C/min until it settled to the room temperature. This process will take approximately 6 hours to be completed.

For bonding in the oven, the chip will be pressed between several standard size microscope glass slides (rigid plates) with additional big black paper clamps to be used for clamping as shown in Fig. 2. There is other presser that we can use i.e the ceramic type presser. This cheap procedure follows the ideas from Pan et al. [23]. The process is started by setting the temperature ramped up 0.83°C/min to 150°C for 30 minutes, and then it is ramped down 1°C/min to the room temperature. The step is followed by setting up back the temperature 0.83°C/min to 200°C for two hours and then the temperature is ramped down 0.83°C/min to the room temperature. This process will take longer time as compared to the hot press bonder [7].

(a) (b)

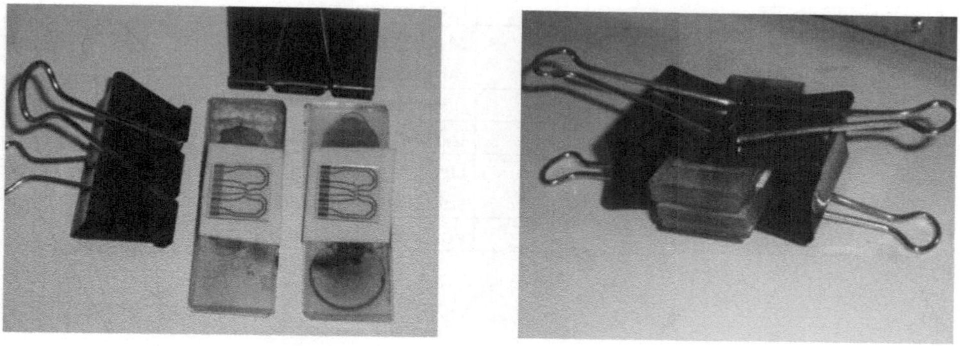

Figure 2. *a) Picture of the substrates that are ready to be placed in between the set of glass slides with Teflon sheets as the absorber/suspension to give room for the expansion and contraction when the temperature rise or fall and (b) they will be clamped with two big black paper clamps which are used for clamping the glasses at its place.*

Subsequently, the channel made will be drilled at specific inlets and outlets (marked) to form the holes for the fluid to flow into and out from the channel. The test fluids are flowed through the channel to check for any leakage and clogging before the device get fully utilised [24]. Sometimes the bonded channel are tested using fluorescent dyes to confirm a good seal [25] otherwise the UV glue will be used if the channel has leakage. This UV glue is used to cover the high probability of leak area at the side of the two bonded substrates. The glue will flow by capillary force in between the substrates. And, when the glue has covered the whole area around the channel wall, the chip will then be exposed under the UV light for about 40s. As a result, the close channel will be formed successfully.

3.5. Channel fabrication

The mask design can be created using L-Edit/CleWin or any CAD drawing tools and this design can be transferred in a high-resolution transparency/acetate mask using a high-resolution printer suggested to be from 3386 dpi (dots per inch) to 128000 dpi. Fig. 3 below shows the ready made acetate mask of the channel. The kind of this mask is preferred and the advantages of it are stated in Table 1.

Illuminated areas will be remained on the glass during development process because we will be using negative photoresist. The exposure time and the developing time are then recorded. This process, of course, requires a very clean substrate (wafer) so that no airborne debris or dust imbedded onto mask when the mask makes contact with the wafer. The imbedded particle could cause permanent damage to the mask and result in defects on the glass wafer with each succeeding exposure.

Table 1. The advantages of using pattern acetate mask-substrate transfer.

Parameter	Mask
Pattern	1:1 mask-substrate
Critical dimension	Easy to pattern micron dimensions on mask and substrate
Exposure Field	Entire substrate
Mask technology	Mask has same dimensions as substrate -a rapid prototype
Throughput	Potentially higher
Die (chip/device) alignment and focus	Individual die alignment and focus
Defect density	Defects are not repeated multiple times on a substrate
Surface flatness	For overall global focus and alignment

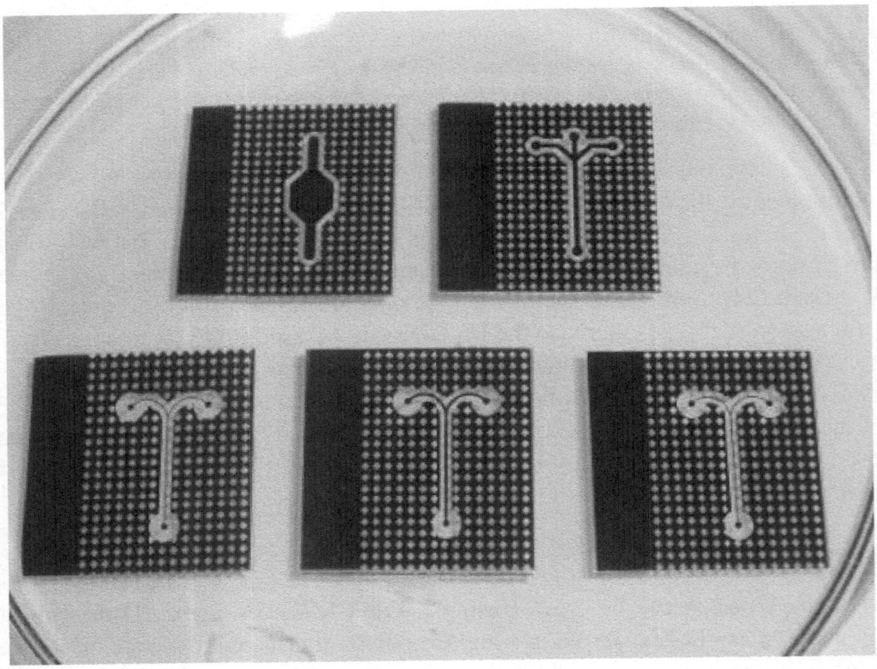

Figure 3. *Acetate mask of the channel, with white area is the part where the photoresist is to be remained on the substrate and black area is the part where the photoresist is to be removed after the patterning process.*

3.6. Bonding

Prior to bonding, the devices with microfluidic channel are first treated with plasma asher, (Oxford Instruments Plasmalab 80 Plus System) where it is used for ashing, etching and cleaning the surface of polymer, the microchannel. It will remove the surface contamination and prevent any contamination from interfering the adhesion especially during bonding process. It makes the surface more hydrophilic and thus enhances the adhesive transfer. In other words, plasma treatment will improve polymer analysis, wettability problems, painting/coating of plastics, gluing results and clean surfaces from carbon, grease and oil. The interactions of plasma with a polymer surface can be divided into four general categories as shown in block diagram in Fig. 4.

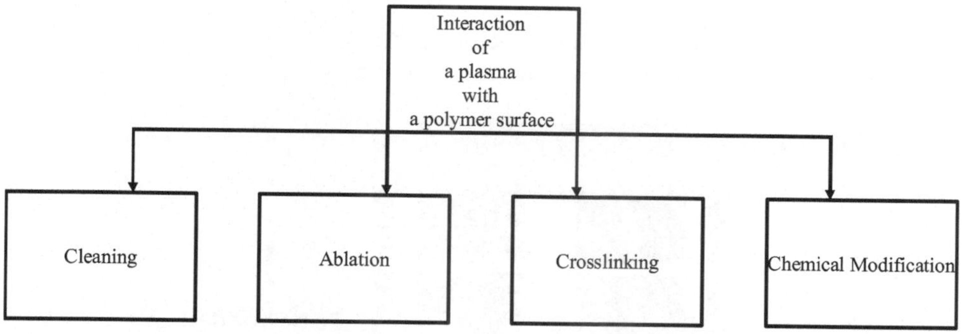

Figure 4. *Four main interactions of plasma with a polymer surface. Cleaning, Ablation, Crosslinking and Chemical Modification.*

A surface cleaning is the removal of organic contamination. Ablation is to remove material by micro-etching mainly to increase surface area and it is also used to remove a weak boundary layer. A cross-linking or branching is used to strengthen the surface cohesively, and surface chemistry modification is to improve chemical and physical interactions at the bonding inter-phase. The setting of the equipment is shown in Table 2.

Table 2. Sample using plasma asher for 2 minutes with its settings and conditions.

Setting	**Condition**
O_2	10 sccm (process gas out)
Pressure	50 (set)
Forward Pressure	100 (RF)
Chiller	20 degree

These microchannels are sandwiched between two glass substrates to form a closed microchannel for microfluidic systems applications such as a micropump, micromixer and separation systems. For example, the microchannel size is 500 μm wide and the step by step of forming the channel has been discussed in [7]. First, the microchannel is placed at one side of the glass substrate with microelectrodes array on it. The other side of the glass substrate will only with another microchannel on it without microelectrode as shown in Fig. 4. The disadvantage of this configuration with one side of microelectrodes array and two layer of laminates is that the electrokinetic force is weaker due to small gradient of electric field strength generated and it will be difficult to be used to control the bioparticle motion [26].

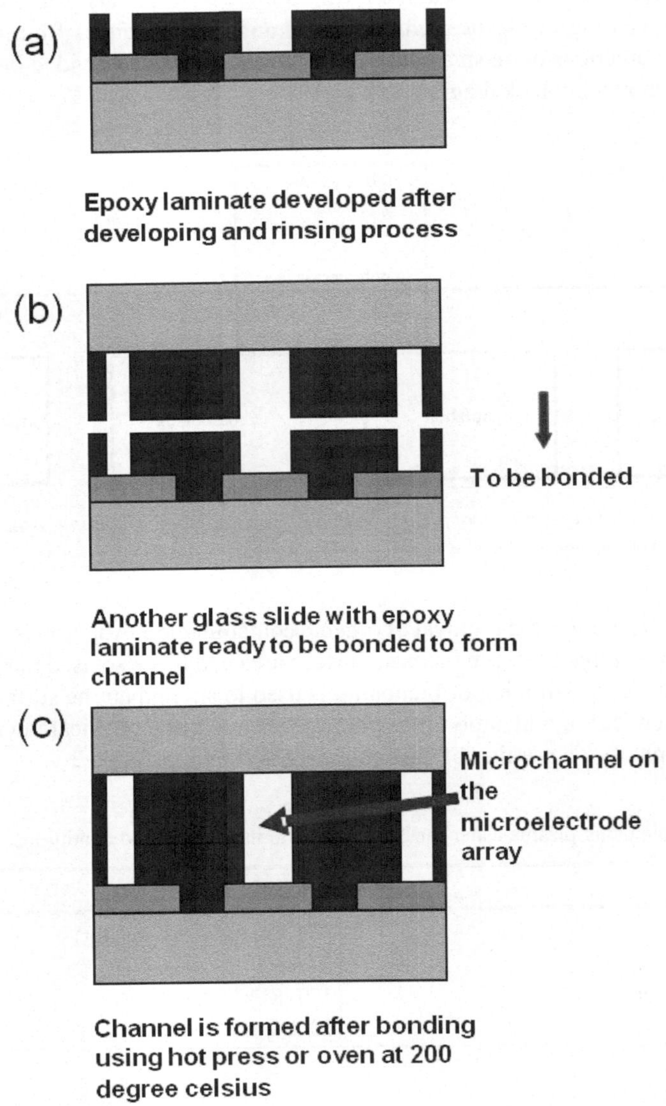

Figure 5. *a) The developing and rinsing process of laminate sheet deposited on the substrate with microelectrodes array (b) The two channels from two glass substrates are aligned and then bonded and (c) hence form the close channel after placing them in the hot press or oven at 200°C.*

The electric field strength of the microfluidic device is further increased and this is done by placing the microelectrodes array on the top and the bottom of the glass substrate and with only one layer of laminate as the microchannel. The microfluidic device for bioelectrokinetic working much better with this arrangement, because the gradient of the electric field generated becomes stronger [27]. The thickness or the height of the channel is reduced from 50 μm to approximately 35-37 μm since only one laminate sheet is used [28]. The channel height is reduced compromising the desired electrokinetic operation. For example, we can use it as a separation device to separate a mixture of submicron particles.

Nevertheless, there are times where the bonding process has not been successful. If more than one device is intended to be bonded at the same time using the hot press bonder as shown in Fig. 6, mainly when they are collected to be bonded together, the thickness of the device must really be considered. Otherwise the results shown will be a part of the microfluidic channel area bonded and a part will not be bonded. Fig. 7 shows the result after using the hot press bonder with collection of devices having different thicknesses to be bonded and assembled in one go.

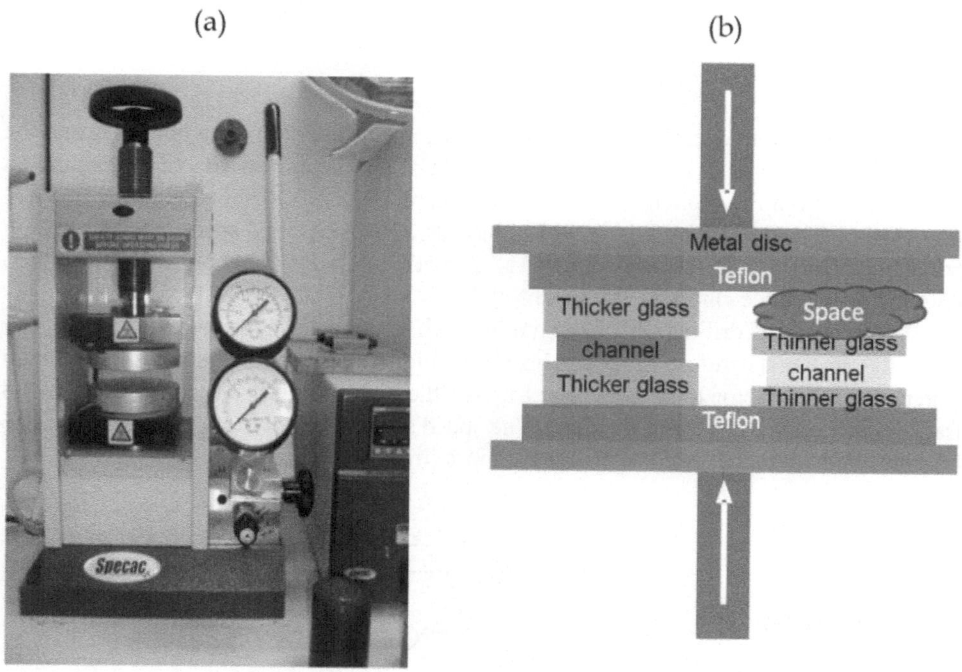

Figure 6. *a) The hot press bonder (b) Illustration of two uneven sized/height devices bonded together in the hot press which will cause the non-bonded area.*

Furthermore, high internal pressures could be developed in the hot press bonder as the fabricated channels were compressed between two contact glasses (wafers), and this might create non-bonding forces as well [29]. These occurrences are not preferred because, the channel will not be formed properly and if the aperture is drilled, there will be a leakage. The optimization of the bonding process for fragmented device is still under investigation.

Figure 7. *a) The non-bonded area of the device approximately 2 mm thick was shown at the bottom part of the device with light shaded area (b) the non-bonded area of the device approximately 1.4 mm thick is shown at the right hand top corner near the right most top inlet.*

3.7. Drilling

It is important to exercise a great caution when handling or drilling apertures into the device since the device is small and made from glass. The machine used to drill aperture on the device is from ProxxonGmbH i.e the main adapter NG 5/E with bench drill TBM 220 (220-240V). The drill bit used is a solid carbide drill, DCSPADEG, spade type, 1 mm, 60° point from Drill Service (Horley) Ltd. It is specially made for glass substrate. The sharp point helps to reduce chipping at breakthrough.

The process of drilling will generate heat when the drill bit is spinning and it may break the device. Therefore to reduce heat, the chip is placed in the petri dish filled with water. The other factor to be considered is that the drill bit must be very sharp to avoid fractal crack on the surface of the glass. The speed and the height of the drill bit must also be controlled. Here the speed of drilling is approximately 600rpm.

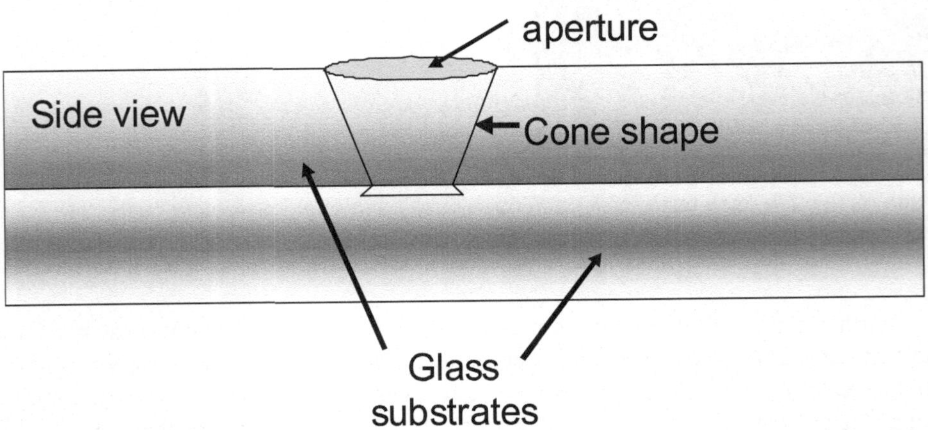

Figure 8. *The chip is tilted (side view) to see the shape of the cone has been formed on the both side near the drilled aperture.*

The process of drilling starts after the correct place of aperture is marked. In order to make sure that the aperture has already been drilled, the chip is tilted to the side under the light to check whether the shape of the cone has been formed on both glass sides near the drilled aperture as shown in Fig.8. The geometry of the aperture depends on the shape of the drill bit. The device could crack at apertures as shown in Fig. 9(a) and (b). The chip might also break into pieces if the apertures are not drilled properly as shown in Fig. 9(c)-(e) and to overcome this occurence, a temporary tape could be placed on top of the device to prevent leakage as shown in Fig. 9(f).

Since the apertures are drilled using a machine and we are the one who controlled the speed, we might get a different aperture diameter in the end. This will result some differ-

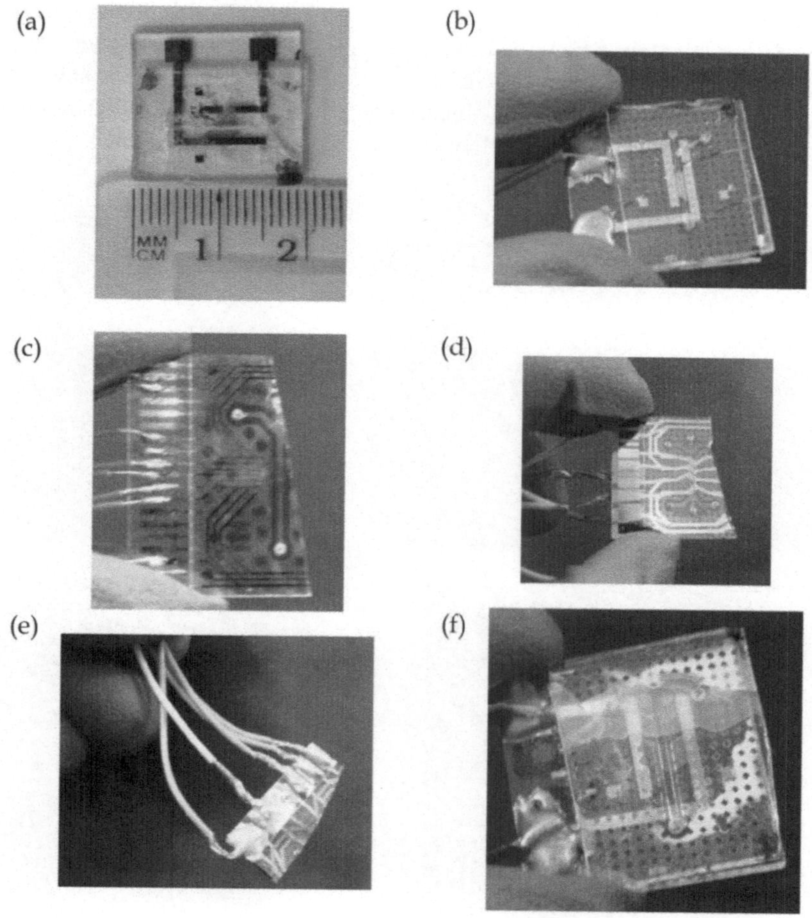

Figure 9. *a) and (b) show the devices crack because of the drilling process. (c), (d) and (e) show the devices break into pieces when there is severe crack occur during drilling process. This breakage will cause trouble during fitting the chip into the holder or handling/preparation of the chip before the experiment. (f) The device is plastered using the tape to cover the other side of the substrate when the drilling process has accidentally passed through the other side of the substrate making the hole throughout the substrates.*

ences on the input or output resistance. Once this occurred, there will be a problem if one is working to balance the sheath flow or to sort the particles at the output. In other word, a robotic and controllable driller for microfluidic device should be designed to avoid differences in input and output resistance of the device.

3.8. Channel formation

Finally, the channel is produced after laminating and drilling apertures are completed. This close channel is shown in Fig. 10. Fig. 10(a) shows the earlier device tested however has residues in it. Fig. 10(b) shows the successful bonded device under the temperature of 200°C with the length of heating is 2 hours. Fig. 10(c) shows the successful bonded device under temperature 200°C with length of heating of 4 hours. Notice that the image of the device is darker in colour when it is heated longer. Fig. 10(d) shows a device that can be examined by running dyed water into it. This is done to observe if there is any leakage by using the syringe connected with the silicon tube. The bonded device is then tested with chemical resistance by continuously flow the chemical through the channel.

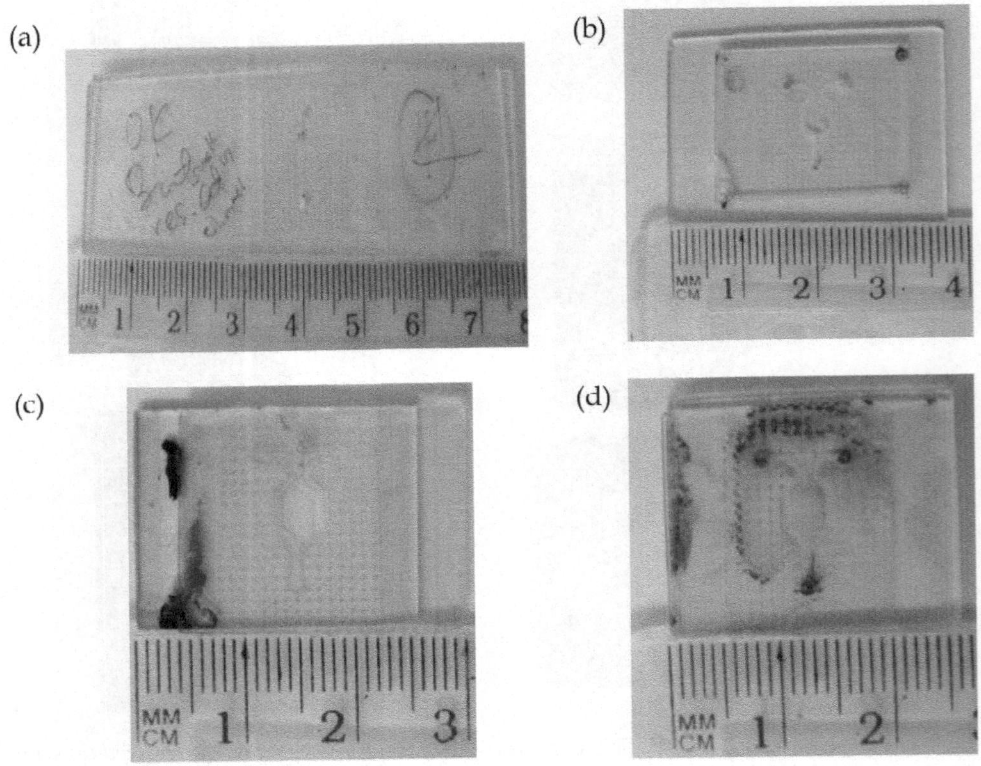

Figure 10. *Some tested device using the microscope slide bonded in the oven. (a) a device is successfully bonded but has residues in it and (b) a successful bonded device with in 2 hours. (c) Successful bonded device in the oven 4 hours and (d) bonded device tested using dyes for leakage discovery.*

3.9. The device

Fig. 11 shows one of the bonded microelectrodes array device that is ready for the experiment. The connecting wires is used to connect the device to the controller. The gold, Au and chromium, Cr layers are easily leached (dissolved) into the solder therefore soldering iron is not suitable to be used on gold, Au [30]. These wire connections are glued to gold, Au bonding pads, the large connection areas which were included in the microelectrode designs. The glue used is the silver paint or conductive epoxy glue from Chemtronics. It is cured at room temperature for about 4 hours. Table 3 shows the hours in the making of the epoxy conductive glue to cure under the room temperature (RT). This glue is suitable to be used when microelectrodes array is made from gold, Au or platinum, Pt which coated with titanium, Ti.

Table 3. Hours for the epoxy conductive glue to be cured in ambient air/room temperature (RT)

Hours on Bonding pad	Epoxy Conductive Glue
1 hour	Not cured
2 hour	Not cured
3 hour	Not cured
4 hour	Cured
5 hour	Cured

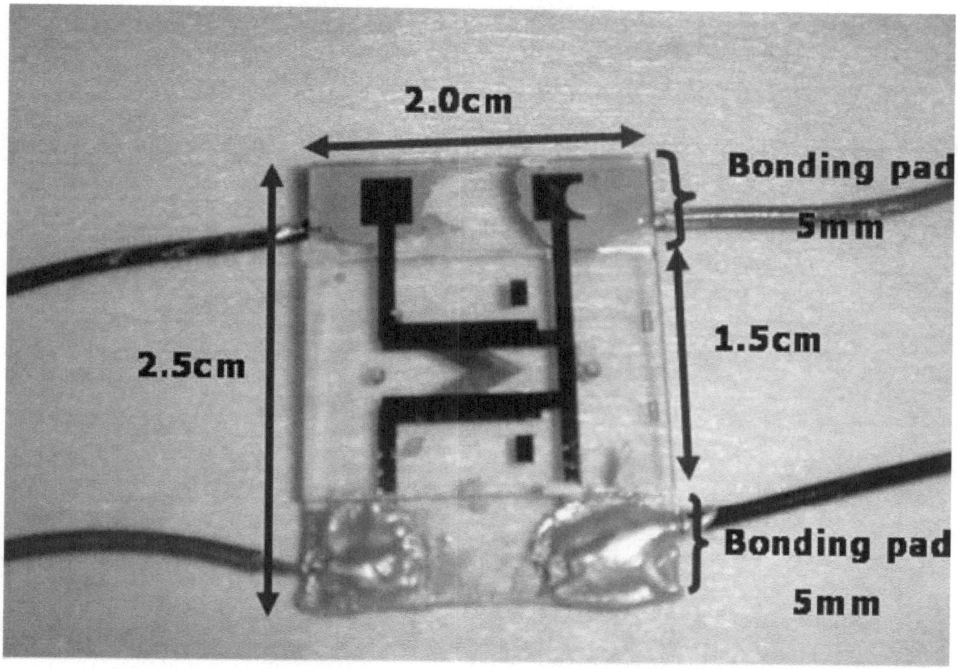

Figure 11. *A bonded microelectrodes array with dimensions. The width of the device is 2.0 cm and length is 2.5 cm. The 5 mm length each side of the microfluidic device is made especially for bonding pad of both sided top and bottom microelectrodes [7].*

3.10. Device holder

The holder is designed custom made for the device of size 20 mm x 20 mm shown in Fig. 12. The holder base is made from the PEEK material. The lid is made from brass material and it should be firmed enough to hold the device tightly and strongly. The gasket, which is made from polytetrafluoroethylene (PTFE), is a sheet or rubber type ring made of Teflon or a type of sealing material that can fill the space between two objects, the device and the holder, preventing leakage and breaking between the two objects, under compression/pressure.

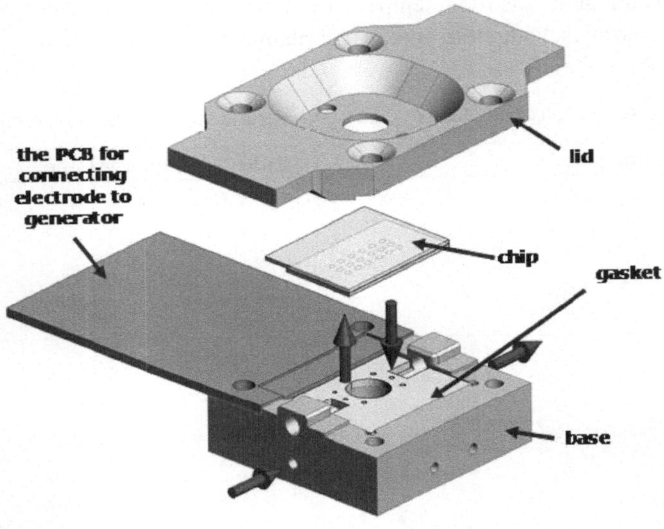

Figure 12. *An illustration of a device/chip holder in 3D showing the lid, possible chip, gasket, base and the PCB connector to the electrode on the chip [illustrated by Rupert Thomas].*

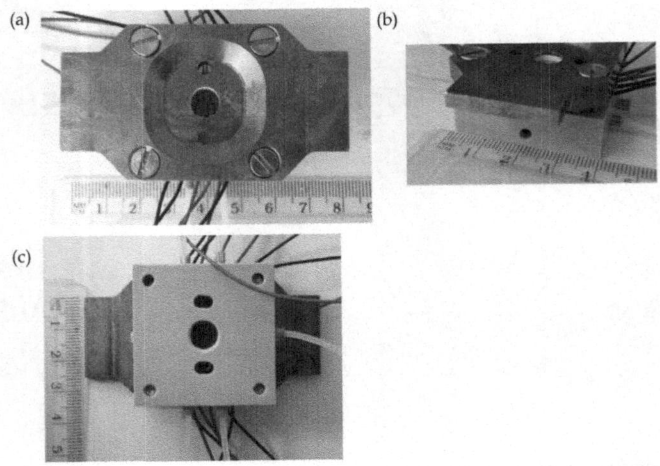

Figure 13. *a) The device in the holder ready to be experimented. (b) The dimension of the base of square shape is 4 mm x 4 mm (c) The top holder is turned upside down, showing the base of the holder.*

The device is clamped in the holder tightly as shown in Fig. 13. The holder can clamp the device of at least made of two substrates of each 700 μm thick and above.

The holder could be the same but the connectors to the device can be modified depending on the device connectors design. The connectors can be made using the cable of 26 pin and the working pin on the electrode could be selected. Uniquely with this holder, it can be used for viewing the channel of the device on either side of the holder as shown in Fig. 14 (a)-(b) as long as the microscope objective used can meet its working distance. The thickness of the base of the holder is 1 cm and the microscope objective used for this holder is X 10. Fig. 14 (c) and (d) show the dimension of the connectors, made from printed circuit board (PCB).

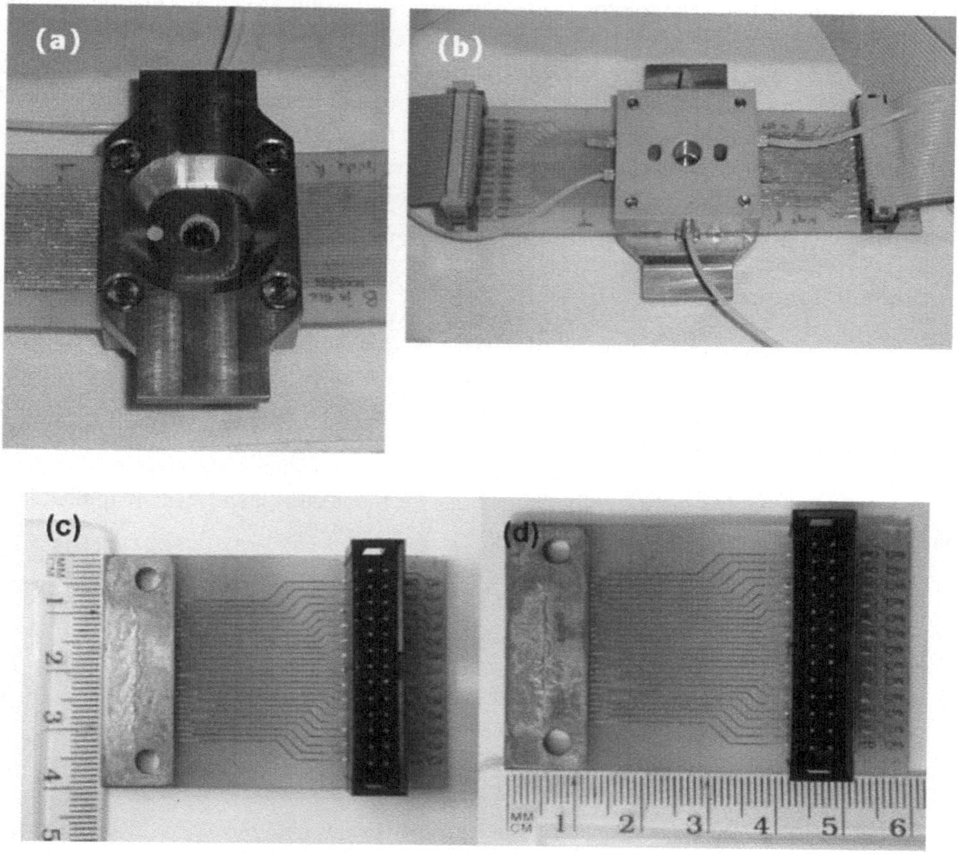

Figure 14. *a) The holder (b) The back of the holder with the connectors (c) and (d) show the dimension of the PCB connectors.*

4. DISCUSSION

The microfluidic device (from fragmented chip glasses) is successfully fabricated with an array of microelectrode in it and the microchannel made from the laminate sheet, dry film resists (DFR). The bonding process for the microchannel is seemed excellent to be done in the oven. The thermosetting material for this microchannel makes it can be cured in high temperature and also it cannot be dissolved easily after cured. This is a good indicator for a microfluidic device to be used for longer times in the experiment and during analysis. Hence the complete fabricated microfluidic devices depending on the microelectrode and microchannel designs, they can be used as a particle separator, mixer, focusing device and so on which definitely be functional in the bioelectrokinetic system.

5. CONCLUSION

In this chapter, the microfluidic device fabrication for bioelectrokinetic system applications is described. The microelectrode is fabricated using a standard microelectronic fabrication technique including the microchannel made of laminate sheet.

In order to make the system efficient, the steps of fabrication and cautions have had to be taken into account. All steps would require thorough precautions and might only need slight changes in time duration and it will depend on the size of the substrate, the thickness of the microelectrode and microchannel. In fabrication of microfluidic the microelectrode and microchannel dimensions are important for getting a good electrokinetic force to do manipulation of fluid and particle in bioelectrokinetic system [26].

Acknowledgement

The author would like to thank Prof. Hywel Morgan and Reader Dr. Nicolas G. Green for ideas, support and laboratory facilities during her research study at Bionanoelectronic Laboratory, Nano Group, School of Electronic and Computer Science (ECS), University of Southampton. A huge appreciation to Dr. Shahanara Banu, Dr. Rupert Thomas, Mr. Andrew Whitton and Ms. Katie Chamberlain for providing valuable discussion and advice to the author in making the microfluidic devices for bioelectrokinetic system applications until successful.

REFERENCES

1. ManzANGraberand H. MWidmerMiniaturized Total Chemical Analysis Systems: A Novel Concept for Chemical SensingSens. Actuators , 1990244248
2. ReyesD. Ret alMicro Total Analysis Systems. 1. Introduction, Theory, and Technology. Analytical Chemistry, 200226232636
3. LaoA. I. KDTrauand I. MHsingMiniaturized Flow Fractionation Device Assisted by a Pulsed Electric Field for Nanoparticle Separation.Analytical Chemistry200253645369
4. TerryS. CJ. HJermanand J. BAngellA Gas Chromatographic Air Analyzer Fabricated on A Silicon WaferIEEE Transactions on Electron Devices197918801886
5. ShirinyanA. SandMWauteletPhase Separation in NanoparticlesNanotechnology200417201731
6. GascoyneP. R. Cand J. VVykoukalDielectrophoresis-based Sample Handling in General-Purpose Programmable Diagnostic InstrumentsProceedings of the IEEE20042242
7. YunusN. A. Mand N. GGreenFabrication of Microfluidic Device Channel using A Photopolymer for Colloidal Particle SeparationMicrosystem Technologies201019

8. DarrigolOWorlds ofFlow: AHistory ofHydrodynamics From the Bernoullis to Prandtl.2005Oxford University Press, USA.
9. BatchelorG. KAn Introduction to Fluid Mechanics. 1967Cambridge University Press Cambridge.
10. NewtonIAMotteand JMachinThe Mathematical Principles of Natural Philosophy.Benjamin Motte.
11. StokesG. GOn the Effect of the Internal Friction of Fluids on the Motion of Pendulums.1851Pitt Press.
12. AchesonD. JElementary Fluid Dynamics. 1990Oxford University Press, USA.
13. MorganHand NGreenA. CElectrokineticsColloids and Nanoparticles. 2003Research Studies Press Ltd.
14. ReynoldsOAn Experimental Investigation of the Circumstances Which Determine Whether the Motion of Water Shall Be Direct or Sinuous, and of the Law of Resistance in Parallel Channels.Philosophical Transactions of the Royal Society of London, 1883935982
15. RottNNote on the History of The Reynolds NumberAnnual Review of Fluid Mechanics1990112
16. BrodyJ. Pet alBiotechnology at low Reynolds numbers.Biophysical Journal, 199634303441
17. DerveauxSet alSynergism between Particle-based Multiplexing and Microfluidics Technologies may bring Diagnostics Closer to the PatientAnalytical and Bioanalytical Chemistry200824532467
18. SuteraS. Pand RSkalakThe History of Poiseuille's LawAnnual Review of Fluid Mechanics1993120
19. ManzAet alPlanar Chips Technology for Miniaturization and Integration of Separation Techniques into Monitoring Systems: Capillary Electrophoresis on a Chip. Journal of Chromatography, 1992253258
20. HarrisonD. Jet alMicromachining a Miniaturized Capillary Electrophoresis-based Chemical Analysis System on a ChipScience1993895897
21. Von HeerenFet alMicellar Electrokinetic Chromatography Separations and Analyses of Biological Samples on a Cyclic Planar Microstructure.Analytical Chemistry199620442053
22. HuangYet alIntroducing Dielectrophoresis As A New Force Field for Field-Flow Fractionation.Biophysical Journal199711181129
23. PanYand RYangA Glass Microfluidic Chip Adhesive Bonding Method at Room Temperature-Journal of Micromechanics and Microengineering200626662672
24. HanA. Ret alMulti-layer Plastic/Glass Microfluidic Systems Containing Electrical and Mechanical Functionality.Lab on a Chip2003150157
25. SatyanarayanaSR. NKarnikand AMajumdarStamp-and-Stick Room-Temperature Bonding Technique for MicrodevicesJournal of Microelectromechanical Systems2005392399
26. YunusN. A. MHJaafarand JJasniThe Gradient of the Magnitude Electric Field Squared on Angled Microelectrode Array for Dielectrophoresis Applications (Special Issue: Asia-Pacific Symposium on Applied Electromagnetics and Mechanics (APSAEM10)). Journal of the Japan Society of Applied Electromagnetics and Mechanics (JSAEM), 2011S240S243
27. NieuwenhuisJ. Hand M. JVellekoopImproved Dielectrophoretic Particle Actuators for Microfluidics. in Proceedings of IEEE Sensors 2003
28. YunusN. A. Mand N. GGreenContinuous Separation of Submicron Particles Using Angled Electrode.Journal of Physics: Conference SeriesElectrostatics 2007012068
29. MetzSRHolzerand PRenaudPolyimide-based Microfluidic Devices. Lab on a Chip, 20012934
30. BurtR. JSemiconductor Chip Mounting System. 1986Google Patents

8 CHAPTER

Passive Mixing Capabilities of Micro- and Nanofibres When Used in Microfluidic Systems

Lauren Matlock-Colangelo [1], Nicholas W. Colangelo [2], Christoph Fenzl[3], Margaret W. Frey [4] and Antje J. Baeumner [1,3]

[1] *Department of Biological and Environmental Engineering, Cornell University, Ithaca, NY 14853, USA*

[2] *Department of Radiology, Rutgers New Jersey Medical School, Newark, NJ 07103, USA*

[3] *Institute for Analytical Chemistry, Chemo- and Biosensors, University of Regensburg, Regensburg 93040, Germany*

[4] *Department of Fibre Science and Apparel Design, Cornell University, Ithaca, NY 14853, USA*

ABSTRACT

Nanofibres are increasingly being used in the field of bioanalytics due to their large surface-area-to-volume ratios and easy-to-functionalize surfaces. To date, nanofibres have been studied as effective filters, concentrators, and immobilization matrices within microfluidic devices. In addition, they are frequently used as optical and electrochemical transduction materials. In this work, we demonstrate that electrospun nanofibre mats cause appreciable passive mixing and therefore provide dual functionality when incorporated within microfluidic systems. Specifically, electrospun nanofibre mats were integrated into Y-shaped poly(methyl methacrylate) microchannels and the degree of mixing was quantified using fluorescence microscopy and ImageJ analysis. The degree of mixing afforded in relationship to fibre diameter, mat height, and mat length was studied. We observed that the most mixing was caused by small diameter PVA nanofibres (450–550 nm in diameter), producing up to 71% mixing at the microchannel outlet, compared to up to 51% with polystyrene microfibres (0.8–2.7 μm in diameter) and 29% mixing in control channels containing no fibres. The mixing afforded by the PVA nanofibres is caused by significant inhomogeneity in pore size and distribution leading to percolation. As expected, within all the studies, fluid mixing increased with fibre mat height, which corresponds to the vertical space of the microchannel occupied by the fibre mats. Doubling the height of the fibre mat led to an average increase in mixing of 14% for the PVA nanofibres and 8% for the PS microfibres. Overall, mixing was independent of the length of the fibre mat used (3–10 mm), suggesting that most mixing occurs as fluid enters and exits the fibre mat. The mixing effects observed within the fibre mats were comparable to or better than many passive mixers reported in literature. Since the nanofibre mats can be further functionalized to couple analyte concentration, immobilization, and detection with enhanced fluid mixing, they are a promising nanomaterial providing dual-functionality within lab-on-a-chip devices.

Keywords: *nanofibres; fluid mixing; microfluidics; biosensors*

1. INTRODUCTION

Electrospinning is a well-understood and simple fibre fabrication process in which an electric field is applied to a polymer spinning dope to generate nano- and microscale fibres [1]. The nonwoven fibre mats produced by electrospinning are characterized by extremely large surface-area-to-volume ratios, high porosities, and small pore sizes. Additionally, fibres can be spun from a wide range of materials, including biocompatible polymers, and are easily functionalized through the incorporation of useful materials into the spinning dope or by post-spinning immobilization techniques [2].

Electrospun fibres have been used in a variety of applications, including the development of tissue engineering scaffolds [3,4,5,6], sensors [7,8,9,10,11], high-performance batteries [12,13,14], and vehicles for drug delivery [15,16]. Several review articles have been written about the many promising uses of electrospun nanofibers [17,18,19,20]. In particular, the incredibly high surface area afforded by electrospun fibre mats has been utilized to improve the performance of a wide range of analytical devices [2,21,22]. Fibre-based sensors feature an increased number of binding sites and faster mass transfer rates than conventional sensors, resulting in much lower limits of detection and faster analyte detection [22]. Recently, electrospun nanofibres have been incorporated into microfluidic sensing systems to allow for enhanced sample preparation and analyte detection [7,23,24,25,26,27]. In some applications, water-soluble nanofibers have been used to facilitate on-chip reagent storage in microfluidic biosensors [7,27]. Jin et al. used electrospinning to encapsulate a horseradish peroxidase tagged antibody into polyvinylpyrrolidone nanofibers [7]. Encapsulating the antibodies within the water-soluble fibres allowed for stable, long-term storage of the biorecognition element on-chip. This approach has also been used to allow on-chip enzyme storage [27]. Further, non-water-soluble electrospun nanofibers have been used as a substrate for microfluidic HIV immunoassays [26], as on-chip sample concentrators [23], and as a substrate for microfluidic *E.coli* detection [25]. In these applications, the non-water-soluble nanofibres were able to withstand the fluid flow within the channels and dramatically increased the functional surface area available within the devices. However, to date, the effects of electrospun fibre mats on the fluid flow within these microfluidic systems has not been studied.

Microfluidic channels generally feature low Reynold's number fluid flow, resulting in laminar flow patterns and limited fluid mixing [28,29,30]. Achieving rapid and reliable fluid mixing is essential for facilitating (bio)chemical reactions and allowing for adequate access of analytes to functional surfaces within microfluidic sensors [28]. While some applications allow for pre-mixing of reagents prior to injection into the microfluidic device, often it is necessary for mixing to occur on-chip [29]. Rapid chemical reactions that take place on a sub-millisecond time scale, such as protein folding, are often studied using microfluidic devices and require on-chip reagent mixing [31]. Additionally, microfluidic devices are well-suited for performing chemical reactions with hazardous or expensive chemicals because of the small internal volume of the devices and relative safety compared to bench-top reactions [29]. Finally, on-chip mixing is crucial to the success of microfluidic assays, both to facilitate access of analytes to all the functional sites within the sensor and to allow for multiple-step assays [32,33]. In general, mixing within microfluidic channels can be improved through increasing the contact surface and lowering the diffusion path between fluids [28,34].

Many micromixers have been proposed and developed to address this need for fluid mixing in miniaturized devices [35,36,37]. These mixers can either be integrated directly within the microfluidic channel or function as a separate component that interfaces with the microfluidic device. Mixers are classified either as passive or active mixers [28,29]. Passive mixers utilize diffusion and chaotic advection to achieve mixing [29,30]. Active mixers, on the other hand, apply an external energy field to enhance mixing within the channel [29]. In general, active mixers produce reliable, complete mixing but are more complicated and expensive than passive mixers due to the need to integrate the external force field [28]. Therefore, passive mixers are often much simpler than active mixers and can be easier to fabricate and utilize. Several common passive mixers include T-type, Y-type, split and recombine, multi-laminating and jet colliding mixers [29]. In each of these mixers, the flow pattern is manipulated by different channel geometries or by incorporating microstructures within the channel. Furthermore, in order for analyte detection to be performed within the microchannel, additional modifications must be made to immobilize the biorecognition element on the channel surfaces. Currently, most mixers are fabricated using photolithography or micromachining, which require access to a cleanroom and/or expensive machinery [38].

Electrospun fibres, on the other hand, are relatively easy to fabricate and do not require a cleanroom. A typical electrospinning apparatus consists of a high voltage power source, a syringe pump, a spinneret (typically a syringe) filled with a polymer spinning dope, and a grounded collector plate (Figure S1) [1,2]. During electrospinning, the viscous polymer solution is slowly pumped out of the spinneret, which is placed across from the grounded collector plate. When voltage is applied to the spinning solution, it forms a Taylor cone at the tip of the spinneret [2]. Once the electrostatic attraction between the charged polymer solution and the grounded collector plate overcomes the surface tension at the tip of the spinneret, fibres will accelerate from the tip of the Taylor cone and collect on the grounded collector plate [1]. Previously, we have demonstrated that electrospun poly(vinyl alcohol) (PVA) nanofibres can be used for sample purification and analyte concentration within microfluidic systems [23]. However, due to their complex nonwoven three-dimensional structure, it is hypothesized that electrospun fibre mats can also be used to enhance fluid mixing within microfluidic channels through manipulation of the flow patterns within the mats. Additionally, in contrast to micromixers that utilize channel geometry or two-dimensional microstructures, fibre-based mixers can be fabricated directly from spinning dopes functionalized with a wide range of additives such as conductive materials [39,40], enzymes [41], biotin [42,43], and aptamers [44]. These functionalized fibres could be used to perform analyte detection and fluid mixing without requiring further modifications of the microchannel. In this work, two different polymers, PVA and PS, were used to produce fibres with different diameters and morphologies to determine their effect on mixing within PMMA microfluidic channels. The work here was done using a Y-shape design, however, nanofiber mats can be introduced into any microchannel system as long as the height and width allow for successful bonding of the device while the nanofiber mats are within the channels.

2. MATERIALS AND METHODS

2.1. Electrospinning

The PS fibres were spun using a Spraybase vertical electrospinning system (Avectas, Dublin, Ireland), while the PVA nanofibres were spun using a homemade horizontal electrospinning system as previously described [23,24]. Positively charged PVA fibres were produced by adding hexadimethrine bromide (polybrene, Sigma-Aldrich, St. Louis, MO, USA) into the PVA spinning dope. The PVA used had a molecular weight of 78,000 g/mol and is 99.7% hydrolyzed (Polysciences Inc., Warrington, PA, USA) [24]. A 10% w/v PVA spinning dope was first created by dissolving PVA in deionized (DI) water at 95 °C for four hours. Then, polybrene was dissolved in DI water at room temperature for 10 min, and was added to the PVA solution to make a final spinning dope with a 90/10 w/w PVA/polybrene ratio. The nonionic surfactant Triton X-100 was added to the spinning dope to improve spinnability, and the resulting solution was mixed for two minutes using a vortex at its highest speed. The spinning dope was then loaded into a 5 mL BD plastic syringe with an 18 gauge needle, which was placed horizontally into a syringe pump. The syringe pump was set to 0.01 mL/min. A high voltage power source was connected to the spinning needle and set to 15 kV (Gamma High Voltage Research Inc., Ormond Beach, FL, USA). A piece of copper wrapped in aluminum foil was used as a grounded collector plate and was placed 15 cm from the syringe tip. After spinning, the aluminum foil was removed from the copper plate and was cut into strips using a razor blade. The nanofibre mats were then peeled off the foil strips using tweezers and were placed on pieces of poly(methyl methacrylate) (PMMA) that had undergone treatment in an UV-ozone (UVO) oven for 15 min (Jelight Company Inc., Irvine, CA, USA). The UVO oven contained a low pressure mercury vapor grid lamp, which had an output of 28,000 mW/cm^2 at 254 nm. During UVO treatment, oxygen was flowed through the oven at 0.5 L/min. The PVA fibres were initially spun as 20, 30, and 40 μm thick mats, however the nanofibre distribution varied along the heights of the mats, producing inconsistent mixing within the channels. Therefore, nanofibre mats of approximately 10 μm height were spun and stacked to create thicker nanofibre mats with more uniform nanofibre distributions. The mats were cut into 3 mm, 5 mm, and 10 mm long strips for the studies examining the effect of nanofibre mat length on mixing.

The PS microfibres were spun using a solvent solution composed of 50/50 v/v tetrahydrofuran/dimethylformamide. Polystyrene with a molecular weight of 280,000 g/mol (Sigma-Aldrich) was dissolved in the solvent solution at room temperature for 24 h on a stir plate on its medium setting. Different fibre morphologies were produced by using three different PS concentrations in the spinning solutions: 12.5%, 15%, and 17.5% w/v.

The PS spinning solution was loaded into a 5 mL glass syringe (BD) and was spun using a Spraybase vertical electrospinning system with an 18 gauge needle. The fibres were spun using a 20 cm vertical distance, 10.6 kV applied voltage, a 0.01 mL/min flow rate, and a circular metal collector plate. The fibres were then manually transferred to PMMA squares that had been treated in the UVO oven for 4 min at an oxygen flow rate of 0.5 L/min and lamp power of 28,000 mW/cm^2 (Jelight Company Inc.). The fibre mats were cut into 10 mm long strips using a scalpel and unwanted fibres on the PMMA surfaces were removed using a strip of adhesive tape. The mats were incorporated onto the PMMA chips in one-layer or two-layer configurations.

2.2. Analysis of Fibre Morphology

The PVA nanofibre mats were assessed using a SP2 confocal microscope (Leica, Wetzlar, Germany) to measure the thickness of the mats prior to incorporation into the microfluidic channels [23]. The polybrene/PVA nanofibres produced in our lab have previously been demonstrated to have an average diameter between 450 and 550 nm [24].

Micrographs of the three different types of PS microfibres were taken using a transmission electron microscope (TEM, type CM 12 from Philips, Hamburg, Germany) at 120 kV and a magnification ranging from 800-fold. In order to facilitate TEM analysis, the fibres were spun directly onto carbon-coated copper grids (400 mesh from Plano, Wetzlar, Germany) for 15 s. The diameter of the cylindrical portion of the different nanofibre types was measured in six points per fibre type and averaged. The 12.5% w/v PS fibres had a 0.8 ± 0.14 μm diameter, 15% w/v PS fibres had a 1.5 ± 0.2 μm diameter, and 17.5% w/v PS fibres had a 2.7 ± 0.5 μm diameter (Figure 1). Additionally, the 12.5% w/v PS fibres had a heavily beaded morphology (Figure 1), the 15% w/v fibres had sporadic beading, and the 17.5% w/v had large cylindrical diameters with no beads. The beads were not considered when measuring the microfibre diameters.

2.3. Microchannel Fabrication

Y-shaped microfluidic channels were stamped into PMMA using hot embossing with a copper template (Figure S2) [24]. The Y-shaped channel was used to allow injection of two different fluid solutions into the microchannel in order to simplify visualization of fluid mixing within the channel. The copper template was fabricated at the Cornell Nanoscale Facility (CNF) using KMPR 1050 photoresist (MicroChem Corp., Westborough, MA, USA) and copper electroplating to produce raised copper channels on a smooth copper plate [45]. The channels used in this study were 31 μm deep, 1 mm wide, and 29 mm long. The channels used with the PVA nanofibres were embossed using a Carver Laboratory Hot Press (Carver, Inc., Wabash, IN, USA) at 130 °C and 10,000 lbs (44,482 N) of force. The channels used with the PS microfibres were embossed using a Specac Hot Press (Specac Ltd, Orpington, Kent, UK) at 100 °C and 0.1 ton of force for 5 min. The inlet and outlet holes were drilled at each end of the channel with a 0.8 mm steel drill bit.

UVO-assisted thermal bonding was used to create the completed microfluidic devices [23]. First, the pieces of PMMA that had been embossed with the microfluidic channels were UVO treated using an oxygen flow rate of 0.5 L/min and lamp power of 28,000 mW/cm^2. The PMMA channels used with the PS microfibres was treated for 4 min, while the PMMA channels used with the PVA nanofibres was treated for 15 min.

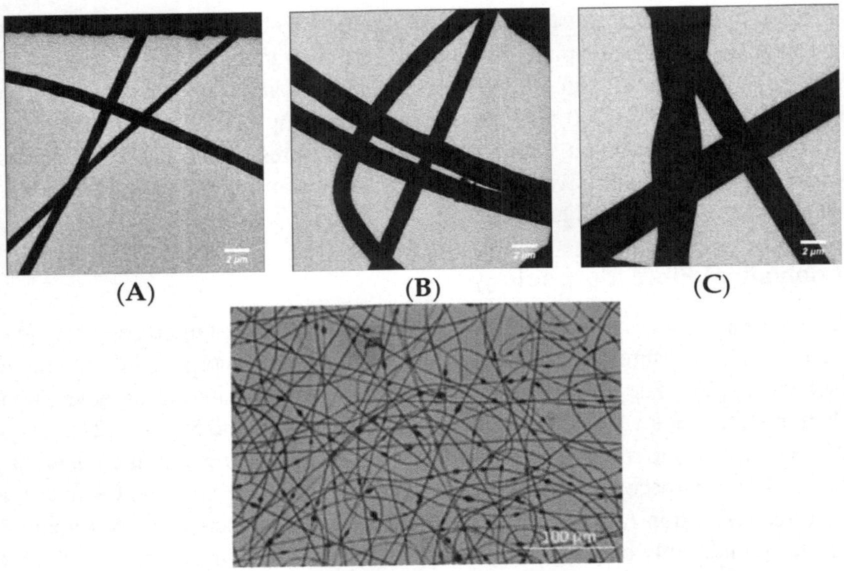

Figure 1. *(Top) TEM image of (A) 12.5% w/v (B) 15% w/v and (C) 17.5% w/v PS microfibres. Fibres were spun onto carbon-coated grids for 15 s. Micrographs taken using a type CM 12 Philips TEM at 120 kV; (Bottom) Morphology of 12.5% w/v PS microfibres. Fibres were spun onto a metal collector plate and transferred to a UVO-treated piece of PMMA. Image taken with a Nikon Digital Eclipse C1 confocal microscope in brightfield setting.*

Then, the pieces of PMMA with the fibre mats were UVO treated for 4 min. The two pieces of PMMA were sandwiched together so that the fibres faced the microchannels. The sandwich assembly was placed between two blank pieces of copper and pressed on the hot press. PS fibres were bonded into channels at 80 °C and 0.1 ton force, while PVA nanofibres were bonded into channels at 100 °C and 10,000 lbs (44,482 N) of force. The different UVO treatment times, temperatures and applied forces were necessary due to differences between the two hot presses used. Polyvinyl chloride tubing (Tygon, Saint-Gobain Performance Plastics, Paris, France) with a 0.508 mm external diameter was glued into the inlet and outlet holes with Quicktite instant adhesive gel (Loctite, Henkel Corporation, Westlake, OH, USA).

2.4. Fluid Mixing

Each Y-shaped microfluidic channel was filled with water in one inlet and a 0.03 M sulforhodamine B (SRB) in water solution in the other inlet (Figure 2).

Because channels filled with nanofibre mats are prone to form air bubbles at low flow rates, each channel was initially filled with fluid at 20 μL/min for 5 min to remove air bubbles from the nanofibre mats and ensure that they didn't impact the mixing observed. Then, the flow rate was dropped to 5 μL/min and a steady flow profile was allowed to develop for 5 min before a fluorescent microscopy image was taken of the channel (Leica). This was repeated for flow rates between 4 μL/min and 1 μL/min. Lower flow rates were not used as pulsatile flow was observed at rates below 1 μL/min.

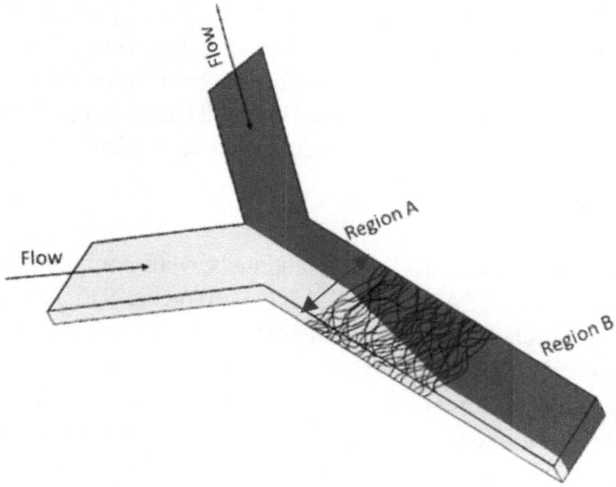

Figure 2. *Angled top view of the Y-channel mixer with embedded nanofiber mats. One inlet was used to fill the channel with DI water, the other filled the channel with SRB in water. Nanofibre mats with different lengths and thicknesses were placed in the center of the channel to encourage fluid mixing. The extent of fluid mixing in Region A of the channel (before fibre mat) was compared to the extent of fluid mixing in Region B (after fibre mat) using ImageJ to measure how the pixel intensity changed along a 1 mm wide vertical column that spanned the channel (blue arrow in picture).*

Several fluid mixing experiments were recorded using an Eclipse 90i confocal microscope (Nikon, Tokyo, Japan) to confirm that the observed mixing was consistent along the height (z direction) of the channel (Figure S3 in Supplementary Information). The fluid mixing experiments were conducted on the stage of the confocal microscope. During fluid flow, the confocal was used to perform a z-scan of the channel in order to examine the fluid flow profile along the z direction of the channel. Images of the channels were taken at 1 μm intervals in the z direction.

2.5. Data Analysis

ImageJ was used to measure the pixel intensity of each pixel in a column along the 1 mm width of the channel (Supplementary Information). In control channels, the pixel intensities along this column will vary from high values (in SRB portion of channel) to low values (in water portion of the channel) (Figure 2). Well-mixed channels will have uniform pixel intensities along the column, as the entire channel is filled with the same fluid mixture. The mixing index of each column of pixels was determined using the following formula [46,47]:

$$mixing\ index = \sqrt{\frac{1}{N}\sum\left(\frac{I_K - I_O}{I_O}\right)^2}$$

where I_K = intensity of a single pixel in a column; I_0= average intensity of all pixels in a column; N = number of pixels in a column.

The average mixing index at the inlet of each channel was determined by averaging the mixing index of each column in a 50 pixel wide section of the channel before the nanofibre mats where a steady flow profile existed (region A in Figure 2). The average mixing index of the channel after the nanofibre mat was determined by averaging the mixing index of each column in a 50 pixel wide section of the channel near the outlet of the channel where again a steady flow profile existed (region B in Figure 2). Each set of parameters was tested using between 3 and 5 different channels to determine reproducibility of the results.

2.6. Statistics

Average Mixing Index data for all the channels were compiled in Microsoft Excel (Microsoft, Redmond, WA, USA). A MATLAB (MathWorks, Natick, MA, USA) code was written to perform multiple linear regression analysis on the data and determine the statistical significance of mixing within channels containing the nanofibre mats (as compared to mats containing no nanofibres) (Supplementary Information Code S1, Code S2, Code S3, and Code S4). Multiple comparisons were made using the Holm test, which controls for the accumulation of error that occurs when multiple t tests are performed [48]. The Holm test sequentially compares the p values of multiple comparisons (from smallest p value to largest p value), with each comparison getting progressively less conservative to account for the number of comparisons that have already been done [49]. The roles of nanofibre mat length, nanofibre mat height (i.e., number of fibre layers), and flow rate on observed fluid mixing was analyzed for the PVA nanofibres. The influence of nanofibre mat morphology (diameter and number of fibre layers) and flow rate on observed fluid mixing was determined for the PS microfibres. Each condition was run in at least triplicate and the mixing values used by the MATLAB code were the average of 50 pixel-wide columns. Significance was set as $p < 0.05$.

3. RESULTS

3.1. Thick Nanofibre Mats

Initially, PVA nanofibre mats that were 20, 30 or 40 μm thick were incorporated into the microfluidic channels (31 μm deep) to determine if thicker nanofibre mats produced better mixing by increasing the volume of the channel occupied by nanofibres. Each nanofibre thickness was tested as a 3 mm, a 5 mm, and a 10 mm wide strip of fibres. While mixing was indeed observed in many of the channels tested, there was significant variability in the flow profiles produced from the same nanofibre mat morphology (Figure S4 in Supplementary Information). Confocal microscopy of the nanofibre mats demonstrated that the nanofibre distribution along the thickness of the mats was not uniform. The mats were most dense at the bottom of the mat and frequently had a sparse nanofibre distribution in the top of the mat (Figure S4 in Supplementary Information). However, while the fibre distribution between nanofibre mats with the same thickness also varied significantly along their depths, the first ~10 μm showed a consistent morphology both within a single nanofibre mat and between different fibre mats.

Several groups have previously demonstrated that electrospun nanofibres are not evenly collected on the grounded collector plate due to the shape of the electric field that exists between the charged spinneret and the collector plate [50,51,52]. During electrospinning, the thickness uniformity of the nanofibre mats has been shown to decrease with longer spinning times, with short spinning times yielding the most uniform nanofibre mats [50]. Therefore, it was determined that the variations seen in the flow profiles in the thick nanofibre mats were caused by these non-uniformities.

3.2. Layered PVA Nanofibre Mats

In order to address the need for nanofibre filters with uniform morphologies, Zhang et al. [50] and Podgórski et al. [52] created multi-layered nanofibre mats by stacking thin nanofibre mats. These multi-layered filters demonstrated both more uniform fibre distribution and improved filtration performance. Therefore, to improve the reproducibility of our mats, we spun ~10 μm thick nanofibre layers and stacked them to obtain thicker nanofibre mats with more uniform fibre distributions. This layered approach was used with varying fibre mat length (3, 5 and 10 mm) and thickness (one layer or two layer). Images taken of the different channels during fluid flow demonstrated that the mixing produced by all of the nanofibre mat morphologies was nearly identical (Figure 3).

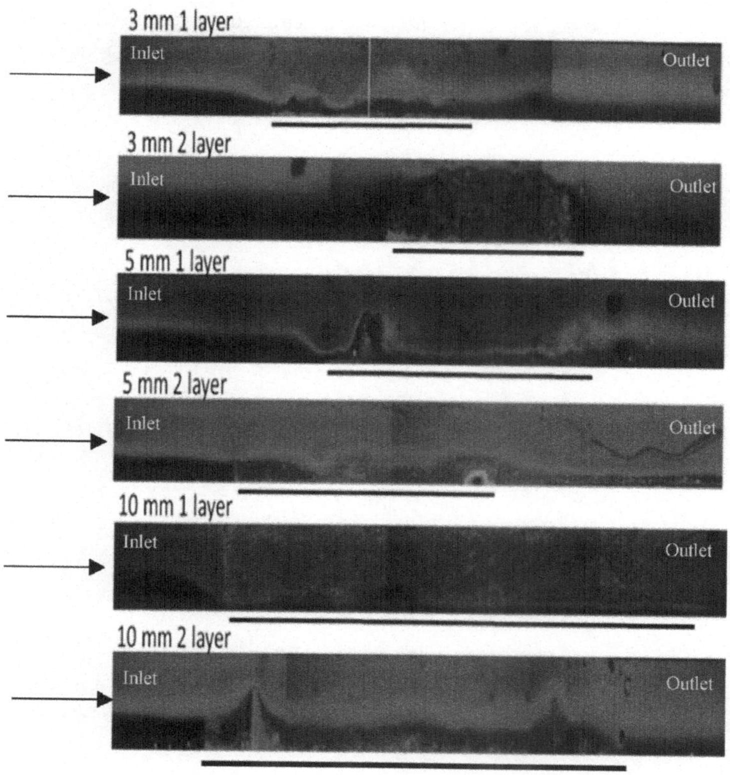

Figure 3. *Example images of fluid flow through nanofibre mats within microfluidic channels. Each set of nanofibre mat morphologies was investigated at a flow rate of 1 μL/min. Arrows indicate direction of fluid flow through the channels. Black lines indicate the location and length of the nanofibre mats within the channels. The inlet of each channel consisted of a red fluid stream (SRB) and a black fluid stream (water). After the nanofiber mats, the two fiber steams have mixed, producing a more homogeneous diluted SRB/water outlet fluid solution. Fluorescent microscope, 5× objective.*

For data analysis, ImageJ (U.S. National Institutes of Health, Bethesda, MD, USA) was used to measure the pixel intensity of each pixel in a vertical column that spanned the 1 mm width of the channel (Figure 2). In the unmixed regions before the nanofibre mats, the pixel intensities along a column had a binary profile, indicating that the channel was divided into two different flows that were not mixed (Figure 4). This is expected for laminar flow patterns. The half of the channel filled with water had a measured pixel intensity of ~10, the half of the channel filled with SRB had a measured pixel intensity of ~50. A peak was observed at the interface of the two solutions. As the SRB concentration chosen exhibited a slight quenching effect, its diluted form at the interface therefore results in a higher fluorescence signal. This two-phase system enters the nanofibre mats. Mixing occurring within the nanofibre mat will result in a decline of the signal difference observed between the two halves of the channel. For quantification, the pixel intensity distribution is also measured after the solution exits the nanofibre mat. As can be seen in Figure 4, the profile becomes more uniform, with the pixel intensity increasing rapidly and plateauing where the solution in the channel is mixed. Furthermore, it can be observed that a pure buffer solution is no longer present as the minimum pixel intensity within the channel increased significantly to ~25. The highest pixel intensity is reached more quickly in the region after the nanofibres and is more consistent, as the SRB dye has spread to a larger portion of the channel and has mixed into a more uniform profile.

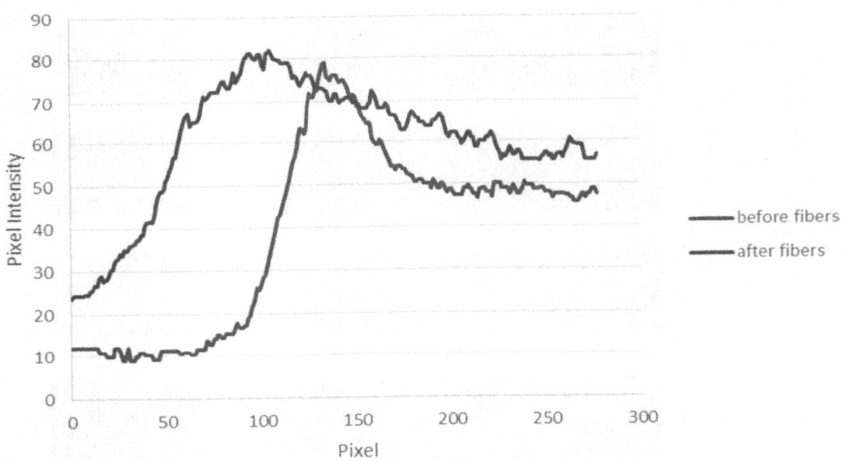

Figure 4. *Pixel intensities of a region before the nanofibre mat and after the nanofibre mat for a 5mm, two-layer mat with flow rate of 1 μL/min. Pixel intensity values shown represent the average value of 50 vertical columns across the 1 mm width of the channel.*

The average mixing index at the inlet and outlet of each type of channel was calculated using ImageJ (Table 1). The change in outlet mixing index between the control channels (no nanofibres) and the nanofibre channels was calculated to determine whether the channels containing nanofibre mats exhibited more fluid mixing than empty control channels (Table 1 and Table S1 in Supplementary Information). Based on their outlet mixing indexes, the nanofibre mats studied produced the most mixing in the two-layer mat configuration

(Table 1). As the one-layer mat will only occupy approximately one third of the total height of the microchannel, the fluid can partly pass above the nanofibre mat without having to flow directly through the mat. On the other hand, the two-layer mat will fill most of the channel height, forcing the fluid to pass through the nanofibre mats. Additionally, the outlet mixing index values for the two-layer nanofibre mats had lower standard deviations than the one-layer mats or the control channels, indicating that the mixing observed in the two-layer mat morphology is more reproducible.

Table 1. Average mixing index in channels containing different PVA nanofibre mats. Each mixing index was calculated for channels with a flow rate of 1 μL/min. Values shown are calculated as the average mixing index of at least three channels.

Morphology	Mixing Index (inlet)	Standard Deviation	Mixing Index (outlet)	Standard Deviation	Difference from Control Outlet	Standard Deviation
Control: No Fibres	0.8	0.17	0.71	0.12	-	-
3 mm 1 layer	0.56	0.20	0.41	0.13	0.30	0.17
3 mm 2 layer	0.53	0.07	0.36	0.09	0.35	0.15
5 mm 1 layer	0.64	0.08	0.45	0.11	0.26	0.16
5 mm 2 layer	0.45	0.08	0.29	0.09	0.42	0.15
10 mm 1 layer	0.74	0.22	0.52	0.07	0.19	0.14
10 mm 2 layer	0.57	0.21	0.32	0.06	0.39	0.13

It was also observed that for the two-layer mat morphologies, the average mixing index at the inlet of the channels was markedly lower than the inlet mixing index of the channels containing no nanofibres (Table 1). We assume that this is caused by back pressure produced by the mats. In turn, this pressure would slow the fluid velocity at the inlet of the channel, allowing for more diffusive mixing before the fluid even enters the nanofibre mat and enhancing the overall desired mixing of the two solutions.

The effect of fluid flow rate on observed mixing was also determined. As expected, the amount of mixing observed in the channels increased slightly as the flow rate decreased (Figure S5 and Table S2 in Supplementary Information).

Multiple linear regression was used to determine if the mixing observed in the channels containing PVA nanofibres was statistically different from the diffusive mixing observed in empty control channels (Table S1 in Supplementary Information). The analysis showed that the increase in fluid mixing observed in all the PVA nanofibre mats was statistically significant ($p < 0.05$), with the greatest increase in mixing observed with the two-layer mat morphologies as expected. The p value for mixing observed in the 3 mm 2 layer morphology is 2.35×10^{-10} with values below 0.05 generally considered to show statistically significant difference in data sets [53].

In order to better understand the influence of nanofibre mat height (number of layers) and length on the observed mixing, additional statistical analysis was performed. Based on the outlet mixing index values for the channels containing PVA nanofibre mats, the number of fibre mats and the flow rate appear to have an effect on the mixing produced in the microchannels, while the length of the nanofibre mat does not appear to play a role. Therefore, multiple linear regression was used to test these hypotheses with significance set as $p < 0.05$ (Table S2 in Supplementary Information). Overall, the hypothesis that the number of fibre mats affects the mixing index was highly significant ($p = 5.15 \times 10^{-8}$). On average, the two-layer mats reduced the mixing index by 0.14 relative to the one-layer mat. The flow rate was also found to play a statistically significant role in the observed mixing, with p = 0.0045. The mixing index increased by an average of 0.03 for each 1 μL/min increase in flow rate. However, the effect of nanofibre mat length was not significant (p = 0.24).

3.3. Layered PS Microfibre Mats

In order to study the effects of fibre diameter and shape on fluid mixing within the channels, PS microfibres of different diameters were incorporated into Y-shaped channels (Figure S6 in Supplementary Information). The PS fibres were chosen for this study since they afforded a much larger diameter (2–7 times lager) than the positively charged PVA nanofibres used in the previous parts of this work. Several labs have successfully used micropillars with diameters in the 10 μm range to encourage fluid mixing within microfluidic systems [54,55]. Therefore, investigation of mixing within the PS mats would determine if fibres with micrometer diameters could similarly be used as obstacles to enhance fluid mixing. Fibres with diameters larger than 2.7 μm were not investigated as they could not be successfully bonded into the microfluidic channels. Additionally, the PS fibres have a beads-on-a-string morphology when spun at lower polymer concentrations and a smooth, cylindrical morphology at high concentrations. Therefore, PS fibres were also used to determine which fibre shape (beaded or smooth) produces more mixing within the channels. In the end, three different PS fibre types were used in this study (Figure 1). Fibres spun from a 12.5% w/v PS solution had a beads-on-a-string morphology and a diameter of 0.8 ± 0.14 μm. Fibres spun from a 15% w/v PS solution have few beads and a smooth cylindrical morphology with a diameter of 1.5 ± 0.2 μm. Finally, fibres spun from a 17.5% w/v PS solution have no beads and a cylindrical morphology with a diameter of 2.7 ± 0.5 μm.

Based on the previous results, which indicated that mat length does not play a significant role in mixing, the microfibre mats were always 10 mm long and stacked in one or two layers within the channels. The 10 mm mat length was chosen as PS fibres are difficult to cut into smaller mats and the most reproducible mats could be made at this length as shorter mats were prone to tearing while being cut. SRB and water were pumped through the channels at flow rates between 1 μL/min and 5 μL/min (Figure S6 in Supplementary Information). Similar flow profiles at the inlet (Figure S7 in Supplementary Information) and outlet (Figure S8 in Supplementary Information) of the channels were observed for the PS systems as described above for the PVA nanofibre mats.

The average mixing index values at the inlets and outlets of the different channels were calculated using ImageJ (Table 2). The difference in outlet mixing index for the microfibre channels and the control channels was used as a metric of how the various PS fibre morphologies affected mixing within the channels.

Table 2. Average mixing index in channels containing different PS fibre mats. Calculated for channels with a flow rate of 1 μL/min. Each mixing index is calculated as the average mixing index of at least three channels.

Morphology	Mixing Index (Inlet)	Standard Deviation	Mixing Index (Outlet)	Standard Deviation	Difference from Control Outlet Mixing Index	Standard Deviation
Control: No Fibres	0.80	0.17	0.71	0.12	-	-
1 layer 12.5%	0.84	0.12	0.64	0.17	0.07	0.21
2 layer 12.5%	0.77	0.05	0.56	0.09	0.15	0.15
1 layer 15%	0.87	0.02	0.78	0.05	−0.07	0.13
2 layer 15%	0.67	0.05	0.49	0.03	0.22	0.12
1 layer 17.5%	0.71	0.28	0.56	0.19	0.15	0.22
2 layer 17.5%	0.72	0.23	0.57	0.18	0.14	0.22

The mixing in the PS fibres exhibited many of the same trends as the mixing in the PVA nanofibre mats, though the average mixing indexes were higher for the PS fibres (indicating less mixing). As with the PVA nanofibres, the two-layer PS fibre morphologies produced a larger change in outlet mixing index than the one-layer morphologies (Table 2). Additionally, the two-layer morphologies had smaller standard deviations in their outlet mixing index values, indicating that their mixing is more reproducible than the one-layer mats. Finally, the inlet mixing index values for the two-layer mats were also smaller than the inlet mixing index of the control channels, indicating that the increase in mixing observed in two-layer fibre mats is due both to an increase of diffusive mixing in the inlet region of the channel and mixing from fluid flow through the fibre mat itself. The PS fibres produced a smaller change in outlet mixing index than the PVA nanofibres, indicating that it is was more effective to use the nanoscale fibres than the larger microfibres to induce mixing in the microfluidic channels.

In order to determine if the mixing in the different PS mats was statistically significant (when compared to empty control channels), multiple linear regression was used. Multiple comparisons within the mat type group were made using the Holm test with the overall p-value set as $p < 0.05$. The different fibre weight percentages and layer combinations were represented as individual variables and compared for significance relative to their respective control of no fibre mat (Table S3 in Supplementary Information).

There were three PS microfibre morphologies that produced a statistically significant increase in mixing when compared to empty control channels: 1 layer 17.5% w/v PS, 2 layer 15% w/v PS, and 2 layer 17.5% w/v PS. Therefore, though overall mixing was highest with PVA nanofibre mats (which have the smallest diameters of all the fibres used in this study), within the PS microfibres mixing increased with increased diameter. This suggests that the mixing observed in the microfibres may be caused by a different mechanism than the mixing observed in the nanofibre mats. Multiple linear regression analysis was performed once more to determine the effect of fibre mat thickness (number of layers) and morphology (12.5%, 15%, or 17.5% w/v PS) on the mixing observed in the PS mats (Table S4 in Supplementary Information). Once again, the fibre mat thickness had a very significant effect on fluid mixing, though the magnitude of this effect was decreased for the PS fibres (−0.084 change to the outlet mixing index) when compared to using two layers of PVA fibres (−0.142 change to the outlet mixing index). Additionally, increasing the diameter of PS fibres also increased the mixing observed in the channels. Finally, flow rate once again affected the mixing observed, with decreasing mixing with increasing flow rate (Table S4).

4. DISCUSSION

Microfluidic analytical systems depend on efficient fluid mixing to ensure that the desired reactions can take place under optimal conditions. This work examines the mixing effect caused by nanofibre mats that have been integrated into a Y-shaped polymer microchannel (Figure 2). The question at hand is whether electrospun fibre mats used primarily for immobilization, separation, or transduction purposes also afford enough mixing capabilities so that the additional integration of passive or active mixers into a microanalytical system is unnecessary. Many passive micromixers reported in the literature utilize obstacles to split inlet streams into narrow flow streams, thus increasing the interfacial areas of the fluids and decreasing mixing time [56,57]. The obstacles used within these studies all have diameters on the order of 10 or 100 μm, making them substantially larger than the fibres used in this study [28,29,35,57]. Indeed, the individual fibres themselves are likely too small to serve as obstacles that redirect the flow pattern within the channels. However, it was postulated that the dense, porous structure of the fibre mats would lead to enhanced mixing of solutions flowing through the mats. In order to investigate which type of fibrous media was most capable of causing increased mixing within microfluidic channels, both electrospun nanofibres (PVA) and small electrospun microfibres (PS) were incorporated into the microchannels. Parameters investigated were fibre mat density, mat thickness (these two are closely related in the electrospinning process) [23], mat length, fibre diameter, fibre shape, and flow rate.

4.1. Comparison of PVA and PS Fibre Mixing

The amount of fluid mixing observed was highest in the PVA nanofibre mats, which had a much smaller diameter (450–550 nm) than the PS microfibres (0.8 μm, 1.5 μm, and 2.7 μm). However, within the PS microfibre mats, mixing increased with increasing fibre diameter, indicating that the mechanism of mixing within the PS and PVA fibres is different.

We assume that the largest PS microfibres can cause some flow manipulation through the volumetric presence of the microfibre, similar to using micropillar obstacles within the channel. However, the microfibres used within this study are still much smaller than the 10–100 μm diameter obstacles frequently used within micromixers, resulting in less mixing than reported in other obstacle-based micromixers (Table 3). In contrast, with the PVA nanofibre mat, it is the porous, inhomogeneous 3D shape of the mat itself that causes the increased dispersive mixing. It has been reported that micromixers that utilize an asymmetric arrangement of obstacles yield significantly more mixing than a symmetric obstacle arrangement [34,57]. An asymmetrical obstacle distribution within the channel results in different resistances to flow in the lateral direction of the channel, causing the fluids to find paths of least resistance through the obstacles [34]. The fluid flow is then repeatedly distorted and redirected as it flows through the obstacles, which in turn increases mixing [34]. The inhomogeneous pore size, pore density and fibre distribution of the PVA nanofibre mats similarly force the fluid to find paths of least resistance as it enters and travels through the fibre mat, producing the increase in mixing reported in this work. On the other hand, the PS microfibre mats used in this study have a more homogeneous pore size and distribution within the channels and thus produce less mixing within the channels than the PVA nanofibre mats. While the average pore size of the mats used in this study was not measured in order to prevent damage to the mats prior to incorporation within the microfluidic devices, average pore size within nanofibre mats typically correlates with the nanofiber diameter [58]. Our lab has previously measured the mean pore size of non-layered PVA mats using an 1100-AEHXL capillary flow porometer. The PVA mats had a measured mean pore size between 0.62 and 0.72 μm, which agrees with predictions for nanofibre mats with diameters of 300–550 nm [58]. Because fibre mat pore size and distribution appear to have a significant effect on mixing, future studies will look at optimizing pore size and distribution within the fibre-based micromixers. Several labs have reported methods for controlling the size, shape, and distribution of electrospun nanofibre pores, which could be used to examine the relationship between mat porosity and fluid mixing [59,60,61,62].

Table 3. Comparison of passive micromixers.

Mixer	Setting	Mixing Index	Reference
Split and Recombine	Re = 10	0.9	Ansari et al. [65]
	Re = 60	0.7	Ansari et al. [65]
Diamond Obstacles	Asymmetric Distribution	0.2	Bhagat et al. [57]
Zigzag		0.1	Jeon et al. [66]
Circular Baffles		0.3	Jeon et al. [66]
PVA Nanofibres	2 layer 5 mm	0.3	This work

The chemistry of the two polymers used likely also contributed to their different mixing behaviours. The hydrophilicity of PVA [63] may facilitate movement of water into the pores of the PVA fibre mats, thus producing the increased mixing observed. Further, the hydrophobic PS fibres [64] would likely have repelled the mixing fluids, making it more difficult for fluid to fill the pores of the PS fibre mats and mix together.

For all the fibres studied, mixing increased with mat height and with decreasing flow rate (Table S2 in Supplementary Information). Additionally, the fibre mat length did not have a significant impact on the final mixing observed, which leads to the conclusion that, for mixing in the PVA fibre mats, most of the observed mixing effects are from the fluid entering and exiting the fibre mats. Finally, the mixing index within the channels containing two layers of PS or PVA fibre mat was lower than the control mixing index at both the inlet and outlet of the channels, suggesting that the observed mixing was also caused by increased back pressure in the channels containing fibre mats.

4.2. Comparison to Conventional Micromixers

To date, passive fluid mixing in microfluidic channels is typically accomplished through patterning of microstructures into polymer channels or by creating complex channel geometries that alter the flow pattern [29]. While these passive mixers can be very effective, they often utilize multi-step lithography or require aligned assembly of multilayered channel geometries [57]. Additionally, simple channel geometries such as T-junction and Y-junction channels have also been used to allow for complete diffusive mixing by optimizing channel length [67,68]. While this can be very effective for thin microchannel designs, the length of channel required to allow for mixing in these simple channel designs is dependent on channel width and mixer aspect ratio [69,70], with wider channel designs requiring channel lengths on the order of tens of centimetres to allow for complete mixing [70]. Therefore, microanalytical systems that have to find strategies for effective mixing, immobilization, separation, and transduction can take advantage of the apparent dual functionality of nanofibres and don't have to use additional mixing structures when already using nanofibres as functional components in their systems. While expertise in electrospinning is required to spin and functionalize the fibres, a basic electrospinning system consists only of a syringe pump, a high voltage source, and a grounded collector plate and can be carried out in any lab setting. It is thus much simpler than most microfluidic fabrication methods. Additionally, fibres produced by electrospinning can be spun out of many different polymers and with many different functionalities [71,72,73]. Furthermore, the two-layer PVA micromixers described in this work produced an average outlet mixing index of between 0.29 and 0.36, corresponding to 71% and 64% fluid mixing, respectively.

These mixing index values are comparable to or better than several previously reported passive micromixers, such as circular baffle, diamond obstacle, and split and recombine mixers, though it is less than the 90% mixing observed using zigzag mixing channels (Table 3). However, unlike these other micromixers, electrospun nanofibre mats can easily be further functionalized to allow for coupling of fluid mixing directly with analyte detection or sample preparation as demonstrated in our research group [21,23,24]. The increased surface area provided by the nanofibre mats would also increase the sensitivity of detection when compared to conventional, two-dimensional sensors due to an increase in binding sites within the device.

While electrospun fibre mats are a promising alternative to conventional micromixers, the electrospinning process depends on many different parameters (such as solution viscosity, ambient temperature and humidity, feed rate, applied voltage, etc.) which can make precise control of fibre morphology difficult. Further, because electrospinning typically uses a spinneret to produce ultrathin fibres, it can have a lower yield than would be ideal for some industrial applications [74]. Therefore, future work will seek to optimize the fibre formation process used to produce the fibre-based micromixers to allow for more precise control over fibre morphology and deposition. The use of temperature and humidity controlled spinning systems could be examined to reduce beading and better control solvent evaporation. Further, alternative collector plates could be used to allow for better fibre alignment. In particular, rotating drum [75], aligned electrode [76], and air-flow impedance [77] collector plates can be used to more precisely control fibre alignment and distribution. Finally, alternative fibre formation processes could be utilized to address the reproducibility and scalability of the electrospinning process including meltblowing, bi-component spinning, forcespinning and flashspinning [78]. Mahalingam et al. have developed a pressurised gyration fibre formation process that could be used for mass production of fibres such as those used in this work [79].

5. CONCLUSIONS

Multiple morphologies of electrospun poly(vinyl alcohol) and polystyrene fibres were incorporated into poly(methyl methacrylate) microchannels to study their ability to enhance fluid mixing. The most mixing was observed in the PVA nanofibre mats, which had a diameter of 450–550 nm. The PVA nanofibre mats had an inhomogeneous morphology, which is likely the cause of the increased fluid mixing when compared to the more homogeneous PS fibre mats. On the other hand, the PS fibres likely produce mixing due to their volumetric presence within the channel, acting much like traditional microstructure-based mixers. Further, the electrospun fibre mats can easily be functionalized to allow for sensitive analyte detection as well as fluid mixing. The reproducible mixing and relatively easy fabrication of electrospun fibre mats makes them a highly interesting functional component for microanalytical systems in which reaction, separation, immobilization, and transduction capabilities can be combined with their inherent passive mixing performance. Amazingly, they even outperform many passive mixers described in literature. As previously demonstrated in our lab, nanofibres can function as a concentration matrix [23] and as an immobilization matrix [42]—the added benefit of mixing produced by the nanofibres makes them thus an even more interesting new material to study for use in bioanalytical sensing systems.

Acknowledgments

The authors are grateful for the partial support provided by the National Science Foundation (NSF) under grant No. CBET-0852900. Also, this work was performed in part at the Cornell NanoScale Facility, a member of the National Nanotechnology Infrastructure Network, which is supported by the National Science Foundation (Grant ECCS-0335765). Finally, the authors also acknowledge partial funding through the Multistate Federal Formula Grant "Development of a novel rapid-on-site biosensor for food safety" Project #2012-13-132.

Author Contributions

Lauren Matlock-Colangelo, Antje Baeumner, and Margaret Frey conceived and designed the experiments; Lauren Matlock-Colangelo performed the experiments; Nicholas Colangelo and Lauren Matlock-Colangelo analyzed the data; Lauren Matlock-Colangelo and Christoph Fenzl performed microscopy and analyzed images.

REFERENCES

1. Doshi, J.; Reneker, D.H. Electrospinning process and applications of electrospun fibers. In Proceedings of the Conference Record of the 1993 IEEE Industry Applications Society Annual Meeting, Toronto, ON, Canada, 2–8 October 1993; Volume 3, pp. 1698–1703.
2. Matlock-Colangelo, L.; Baeumner, A.J. Recent progress in the design of nanofiber-based biosensing devices. *Lab. Chip* 2012, *12*, 2612–2620.
3. Santoro, M.; Shah, S.R.; Walker, J.L.; Mikos, A.G. Poly(lactic acid) nanofibrous scaffolds for tissue engineering. *Adv. Drug Deliv. Rev.* 2016, in press.
4. Jing, X.; Mi, H.-Y.; Peng, J.; Peng, X.-F.; Turng, L.-S. Electrospun aligned poly(propylene carbonate) microfibers with chitosan nanofibers as tissue engineering scaffolds. *Carbohydr. Polym.* 2015, *117*, 941–949.
5. Kucinska-Lipka, J.; Gubanska, I.; Janik, H.; Sienkiewicz, M. Fabrication of polyurethane and polyurethane based composite fibres by the electrospinning technique for soft tissue engineering of cardiovascular system. *Mater. Sci. Eng. C* 2015, *46*, 166–176.
6. Ru, C.; Wang, F.; Pang, M.; Sun, L.; Chen, R.; Sun, Y. Suspended, Shrinkage-Free, Electrospun PLGA Nanofibrous Scaffold for Skin Tissue Engineering. *ACS Appl. Mater. Interfaces* 2015, *7*, 10872–10877.
7. Jin, S.; Dai, M.; Ye, B.; Nugen, S.R. Development of a capillary flow microfluidic Escherichia coli biosensor with on-chip reagent delivery using water-soluble nanofibers. *Microsyst. Technol.* 2013, *19*, 2011–2015.
8. Zhang, P.; Zhao, X.; Ji, Y.; Ouyang, Z.; Wen, X.; Li, J.; Su, Z.; Wei, G. Electrospinning graphene quantum dots into a nanofibrous membrane for dual-purpose fluorescent and electrochemical biosensors. *J. Mater. Chem. B* 2015, *3*, 2487–2496.
9. Bourourou, M.; Holzinger, M.; Bossard, F.; Hugenell, F.; Maaref, A.; Cosnier, S. Chemically reduced electrospun polyacrilonitrile-carbon nanotube nanofibers hydrogels as electrode material for bioelectrochemical applications. *Carbon* 2015, *87*, 233–238.
10. Adabi, M.; Saber, R.; Faridi-Majidi, R.; Faridbod, F. Performance of electrodes synthesized with polyacrylonitrile-based carbon nanofibers for application in electrochemical sensors and biosensors. *Mater. Sci. Eng. C* 2015, *48*, 673–678.

11. Zhang, Y.; Liu, S.; Li, Y.; Deng, D.; Si, X.; Ding, Y.; He, H.; Luo, L.; Wang, Z. Electrospun graphene decorated $MnCo_2O_4$ composite nanofibers for glucose biosensing. *Biosens. Bioelectron.* 2015, *66*, 308–315.

12. Chung, S.-H.; Han, P.; Singhal, R.; Kalra, V.; Manthiram, A. Electrochemically Stable Rechargeable Lithium-Sulfur Batteries with a Microporous Carbon Nanofiber Filter for Polysulfide. *Adv. Energy Mater.* 2015, *5*.

13. Aravindan, V.; Sundaramurthy, J.; Suresh Kumar, P.; Lee, Y.-S.; Ramakrishna, S.; Madhavi, S. Electrospun nanofibers: A prospective electro-active material for constructing high performance Li-ion batteries. *Chem. Commun.* 2015, *51*, 2225–2234.

14. Singhal, R.; Chung, S.-H.; Manthiram, A.; Kalra, V. A free-standing carbon nanofiber interlayer for high-performance lithium-sulfur batteries. *J. Mater. Chem. A* 2015, *3*, 4530–4538.

15. Hu, X.; Liu, S.; Zhou, G.; Huang, Y.; Xie, Z.; Jing, X. Electrospinning of polymeric nanofibers for drug delivery applications. *J. Controlled Release* 2014, *185*, 12–21.

16. Mendes, A.C.; Gorzelanny, C.; Halter, N.; Schneider, S.W.; Chronakis, I.S. Hybrid electrospun chitosan-phospholipids nanofibers for transdermal drug delivery. *Int. J. Pharm.* 2016, *510*, 48–56.

17. Pramanik, S.; Pingguan-Murphy, B.; Abu Osman, N.A. Progress of key strategies in development of electrospun scaffolds: Bone tissue. *Sci. Technol. Adv. Mater.* 2012, *13*.

18. Jiang, T.; Carbone, E.J.; Lo, K.W.-H.; Laurencin, C.T. Electrospinning of polymer nanofibers for tissue regeneration. *Prog. Polym. Sci.* 2015, *46*, 1–24.

19. Rezaei, B.; Ghani, M.; Shoushtari, A.M.; Rabiee, M. Electrochemical biosensors based on nanofibres for cardiac biomarker detection: A comprehensive review. *Biosens. Bioelectron.* 2016, *78*, 513–523.

20. Rezvani, Z.; Venugopal, J.R.; Urbanska, A.M.; Mills, D.K.; Ramakrishna, S.; Mozafari, M. A bird's eye view on the use of electrospun nanofibrous scaffolds for bone tissue engineering: Current state-of-the-art, emerging directions and future trends. *Nanomed. Nanotechnol. Biol. Med.* 2016, *12*, 2181–2200.

21. Li, D.; Frey, M.W.; Baeumner, A.J. Electrospun polylactic acid nanofiber membranes as substrates for biosensor assemblies. *J. Membr. Sci.* 2006, *279*, 354–363.

22. Luo, Y.; Nartker, S.; Miller, H.; Hochhalter, D.; Wiederoder, M.; Wiederoder, S.; Setterington, E.; Drzal, L.T.; Alocilja, E.C. Surface functionalization of electrospun nanofibers for detecting E. coli O157:H7 and BVDV cells in a direct-charge transfer biosensor. *Biosens. Bioelectron.* 2010, *26*, 1612–1617.

23. Matlock-Colangelo, L.; Cho, D.; Pitner, C.L.; Frey, M.W.; Baeumner, A.J. Functionalized electrospun nanofibers as bioseparators in microfluidic systems. *Lab Chip* 2012, *12*, 1696–1701.

24. Cho, D.; Matlock-Colangelo, L.; Xiang, C.; Asiello, P.J.; Baeumner, A.J.; Frey, M.W. Electrospun nanofibers for microfluidic analytical systems. *Polymer* 2011, *52*, 3413–3421.

25. Matlock-Colangelo, L.; Coon, B.; Pitner, C.L.; Frey, M.W.; Baeumner, A.J. Functionalized electrospun poly(vinyl alcohol) nanofibers for on-chip concentration of E. coli cells. *Anal. Bioanal. Chem.* 2016, *408*, 1327–1334.

26. Yang, D.; Niu, X.; Liu, Y.; Wang, Y.; Gu, X.; Song, L.; Zhao, R.; Ma, L.; Shao, Y.; Jiang, X. Electrospun Nanofibrous Membranes: A Novel Solid Substrate for Microfluidic Immunoassays for HIV. *Adv. Mater.* 2008, *20*, 4770–4775.

27. Dai, M.; Jin, S.; Nugen, S.R. Water-Soluble Electrospun Nanofibers as a Method for on-Chip Reagent Storage. *Biosensors* 2012, *2*, 388–395.

28. Nguyen, N.T.; Wu, Z. Micromixers—A review. *J. Micromec. Microeng.* 2005, *15*, R1–R16.
29. Hessel, V.; Löwe, H.; Schönfeld, F. Micromixers—A review on passive and active mixing principles. *Chem. Eng. Sci.* 2005, *60*, 2479–2501.
30. Li, J. *Computational Analysis of Nanofluid Flow in Microchannels with Applications to Micro-heat Sinks and Bio-MEMS*; ProQuest: Ann Arbor, MI, USA, 2008.
31. Bökenkamp, D.; Desai, A.; Yang, X.; Tai, Y.-C.; Marzluff, E.M.; Mayo, S.L. Microfabricated Silicon Mixers for Submillisecond Quench-Flow Analysis. *Anal. Chem.* 1998, *70*, 232–236.
32. Ducrée, J.; Haeberle, S.; Lutz, S.; Pausch, S.; von Stetten, F.; Zengerle, R. The centrifugal microfluidic bio-disk platform. *J. Micromech. Microeng.* 2007, *17*, S103–S115.
33. Vijayendran, R.A.; Motsegood, K.M.; Beebe, D.J.; Leckband, D.E. Evaluation of a Three-Dimensional Micromixer in a Surface-Based Biosensor. *Langmuir* 2003, *19*, 1824–1828.
34. Wang, H.; Iovenitti, P.; Harvey, E.; Masood, S. Optimizing layout of obstacles for enhanced mixing in microchannels. *Smart Mater. Struct.* 2002, *11*, 662–667.
35. Yang, Z.; Matsumoto, S.; Goto, H.; Matsumoto, M.; Maeda, R. Ultrasonic micromixer for microfluidic systems. *Sens. Actuators Phys.* 2001, *93*, 266–272.
36. Wong, S.H.; Ward, M.C.L.; Wharton, C.W. Micro T-mixer as a rapid mixing micromixer. *Sens. Actuators B Chem.* 2004, *100*, 359–379.
37. Kim, D.S.; Lee, S.W.; Kwon, T.H.; Lee, S.S. A barrier embedded chaotic micromixer. *J. Micromech. Microeng.* 2004, *14*, 798.
38. Marle, L.; Greenway, G.M. Microfluidic devices for environmental monitoring. *TrAC Trends Anal. Chem.* 2005, *24*, 795–802.
39. Chronakis, I.S.; Grapenson, S.; Jakob, A. Conductive polypyrrole nanofibers via electrospinning: Electrical and morphological properties. *Polymer* 2006, *47*, 1597–1603.
40. Jang, S.-Y.; Seshadri, V.; Khil, M.-S.; Kumar, A.; Marquez, M.; Mather, P.T.; Sotzing, G.A. Welded Electrochromic Conductive Polymer Nanofibers by Electrostatic Spinning. *Adv. Mater.* 2005, *17*, 2177–2180.
41. Patel, A.C.; Li, S.; Yuan, J.-M.; Wei, Y. In situ Encapsulation of Horseradish Peroxidase in Electrospun Porous Silica Fibers for Potential Biosensor Applications. *Nano Lett.* 2006, *6*, 1042–1046.
42. Li, D.; Frey, M.W.; Vynias, D.; Baeumner, A.J. Availability of biotin incorporated in electrospun PLA fibers for streptavidin binding. *Polymer* 2007, *48*, 6340–6347.
43. Wang, D.; Sun, G.; Xiang, B.; Chiou, B.-S. Controllable biotinylated poly(ethylene-co-glycidyl methacrylate) (PE-co-GMA) nanofibers to bind streptavidin–horseradish peroxidase (HRP) for potential biosensor applications. *Eur. Polym. J.* 2008, *44*, 2032–2039.
44. Kim, J.H.; Hwang, E.T.; Kang, K.; Tatavarty, R.; Gu, M.B. Aptamers-on-nanofiber as a novel hybrid capturing moiety. *J. Mater. Chem.* 2011, *21*, 19203–19206.
45. Nugen, S.R.; Asiello, P.J.; Baeumner, A.J. Design and fabrication of a microfluidic device for near-single cell mRNA isolation using a copper hot embossing master. *Microsyst. Technol.* 2008, *15*, 477–483.
46. Lu, L.-H.; Ryu, K.S.; Liu, C. A magnetic microstirrer and array for microfluidic mixing. *J. Microelectromec. Syst.* 2002, *11*, 462–469.
47. Chang, S.T.; Beaumont, E.; Petsev, D.N.; Velev, O.D. Remotely powered distributed microfluidic pumps and mixers based on miniature diodes. *Lab Chip* 2007, *8*, 117–124.
48. Holm, S. A Simple Sequentially Rejective Multiple Test Procedure. *Scand. J. Stat.* 1979, *6*, 65–70.

49. Glantz, S.; Slinker, B. *Primer of Applied Regression & Analysis of Variance*; McGraw-Hill Education: New York, NY, USA, 2000.

50. Zhang, Q.; Welch, J.; Park, H.; Wu, C.-Y.; Sigmund, W.; Marijnissen, J.C.M. Improvement in nanofiber filtration by multiple thin layers of nanofiber mats. *J. Aerosol. Sci.* 2010, *41*, 230–236.

51. Przekop, R.; Gradon, L. Deposition and Filtration of Nanoparticles in the Composites of Nano- and Microsized Fibers. *Aerosol. Sci. Technol.* 2008, *42*, 483–493.

52. Podgórski, A.; Bałazy, A.; Gradoń, L. Application of nanofibers to improve the filtration efficiency of the most penetrating aerosol particles in fibrous filters. *Chem. Eng. Sci.* 2006, *61*, 6804–6815.

53. Bland, J.M.; Altman, D.G. Multiple significance tests: The Bonferroni method. *BMJ* 1995, *310*, 170.

54. Lin, Y.; Gerfen, G.J.; Rousseau, D.L.; Yeh, S.-R. Ultrafast Microfluidic Mixer and Freeze-Quenching Device. *Anal. Chem.* 2003, *75*, 5381–5386.

55. Chen, L.; Wang, G.; Lim, C.; Seong, G.H.; Choo, J.; Lee, E.K.; Kang, S.H.; Song, J.M. Evaluation of passive mixing behaviors in a pillar obstruction poly(dimethylsiloxane) microfluidic mixer using fluorescence microscopy. *Microfluid. Nanofluid.* 2008, *7*, 267–273.

56. Johnson, T.J.; Ross, D.; Locascio, L.E. Rapid Microfluidic Mixing. *Anal. Chem.* 2002, *74*, 45–51.

57. Bhagat, A.A.S.; Peterson, E.T.K.; Papautsky, I. A passive planar micromixer with obstructions for mixing at low Reynolds numbers. *J. Micromech. Microeng.* 2007, *17*, 1017–1024.

58. Li, D.; Frey, M.W.; Joo, Y.L. Characterization of nanofibrous membranes with capillary flow porometry. *J. Membr. Sci.* 2006, *286*, 104–114.

59. Nedjari, S.; Schlatter, G.; Hébraud, A. Thick electrospun honeycomb scaffolds with controlled pore size. *Mater. Lett.* 2015, *142*, 180–183.

60. Park, C.H.; Bae, H.; Kwak, S.J.; Jang, M.S.; Lee, J.-H.; Lee, J. Interconnection of electrospun nanofibers via a post co-solvent treatment and its open pore size effect on pressure-retarded osmosis performance. *Macromol. Res.* 2016, *24*, 314–322.

61. Krifa, M.; Yuan, W. Morphology and pore size distribution of electrospun and centrifugal forcespun nylon 6 nanofiber membranes. *Text. Res. J.* 2016, *86*, 1294–1306.

62. Kerr-Phillips, T.E.; Woehling, V.; Agniel, R.; Nguyen, G.T.M.; Vidal, F.; Kilmartin, P.; Plesse, C.; Travas-Sejdic, J. Electrospun rubber fibre mats with electrochemically controllable pore sizes. *J. Mater. Chem. B* 2015, *3*, 4249–4258.

63. Park, J.-C.; Ito, T.; Kim, K.-O.; Kim, K.-W.; Kim, B.-S.; Khil, M.-S.; Kim, H.-Y.; Kim, I.-S. Electrospun poly(vinyl alcohol) nanofibers: Effects of degree of hydrolysis and enhanced water stability. *Polym. J.* 2010, *42*, 273–276.

64. Lu, X.; Zhou, J.; Zhao, Y.; Qiu, Y.; Li, J. Room Temperature Ionic Liquid Based Polystyrene Nanofibers with Superhydrophobicity and Conductivity Produced by Electrospinning. *Chem. Mater.* 2008, *20*, 3420–3424.

65. Ansari, M.A.; Kim, K.-Y.; Anwar, K.; Kim, S.M. A novel passive micromixer based on unbalanced splits and collisions of fluid streams. *J. Micromech. Microeng.* 2010, *20*, 055007.

66. Jeon, W.; Shin, C.B. Design and simulation of passive mixing in microfluidic systems with geometric variations. *Chem. Eng. J.* 2009, *152*, 575–582.

67. Cama, J.; Bajaj, H.; Pagliara, S.; Maier, T.; Braun, Y.; Winterhalter, M.; Keyser, U.F. Quantification of Fluoroquinolone Uptake through the Outer Membrane Channel OmpF of *Escherichia coli*. *J. Am. Chem. Soc.* 2015, *137*, 13836–13843.

68. Cama, J.; Chimerel, C.; Pagliara, S.; Javer, A.; Keyser, U.F. A label-free microfluidic assay to quantitatively study antibiotic diffusion through lipid membranes. *Lab Chip* 2014, *14*, 2303–2308.

69. Gobby, D.; Angeli, P.; Gavriilidis, A. Mixing characteristics of T-type microfluidic mixers. *J. Micromech. Microeng.* 2001, *11*, 126–132.

70. Kirby, B.J. *Micro- and Nanoscale Fluid Mechanics: Transport in Microfluidic Devices*; Cambridge University Press: New York, NY, USA, 2010.

71. Min, B.-M.; Lee, S.W.; Lim, J.N.; You, Y.; Lee, T.S.; Kang, P.H.; Park, W.H. Chitin and chitosan nanofibers: Electrospinning of chitin and deacetylation of chitin nanofibers. *Polymer* 2004, *45*, 7137–7142.

72. Zhang, M.; Wang, Z.; Wang, Z.; Feng, S.; Xu, H.; Zhao, Q.; Wang, S.; Fang, J.; Qiao, M.; Kong, D. Immobilization of anti-CD31 antibody on electrospun poly(ε-caprolactone) scaffolds through hydrophobins for specific adhesion of endothelial cells. *Colloids Surf. B Biointerfaces* 2011, *85*, 32–39.

73. Xie, J.; Hsieh, Y.-L. Ultra-high surface fibrous membranes from electrospinning of natural proteins: Casein and lipase enzyme. *J. Mater. Sci.* 2003, *38*, 2125–2133.

74. Luo, C.J.; Stoyanov, S.D.; Stride, E.; Pelan, E.; Edirisinghe, M. Electrospinning versus fibre production methods: From specifics to technological convergence. *Chem. Soc. Rev.* 2012, *41*, 4708–4735.

75. Katta, P.; Alessandro, M.; Ramsier, R.D.; Chase, G.G. Continuous Electrospinning of Aligned Polymer Nanofibers onto a Wire Drum Collector. *Nano Lett.* 2004, *4*, 2215–2218.

76. Li, D.; Wang, Y.; Xia, Y. Electrospinning Nanofibers as Uniaxially Aligned Arrays and Layer-by-Layer Stacked Films. *Adv. Mater.* 2004, *16*, 361–366.

77. McClure, M.J.; Wolfe, P.S.; Simpson, D.G.; Sell, S.A.; Bowlin, G.L. The use of air-flow impedance to control fiber deposition patterns during electrospinning. *Biomaterials* 2012, *33*, 771–779.

78. Nayak, R.; Padhye, R.; Kyratzis, I.L.; Truong, Y.B.; Arnold, L. Recent advances in nanofibre fabrication techniques. *Text. Res. J.* 2012, *82*, 129–147.

79. Mahalingam, S.; Edirisinghe, M. Forming of Polymer Nanofibers by a Pressurised Gyration Process. *Macromol. Rapid Commun.* 2013, *34*, 1134–1139.

9 CHAPTER

Frequency-Switchable Microfluidic CSRR-Loaded QMSIW Band-Pass Filter Using a Liquid Metal Alloy

Seunghyun Eom, Muhammad Usman Memon and Sungjoon Lim
School of Electrical and Electronics Engineering, College of Engineering, Chung-Ang University, 84 Heukseok-ro, Dongjak-gu, Seoul 156-756, Korea

ABSTRACT

In this paper, we have proposed a frequency-switchable complementary split-ring resonator (CSRR)-loaded quarter-mode substrate-integrated-waveguide (QMSIW) band-pass filter. For frequency switching, a microfluidic channel and liquid metal are used. The liquid metal used is eutectic gallium-indium (EGaIn), consisting of 24.5% indium and 75.5% gallium. The microfluidic channels are built using the elastomer polydimethylsiloxane (PDMS) and three-dimensional-printed microfluidic channel frames. The CSRR-loaded QMSIW band-pass filter is designed to have two states. Before the injection of the liquid metal, the measured center frequency and fractional bandwidths are 2.205 GHz and 6.80%, respectively. After injection, the center frequency shifts from 2.205 GHz to 2.56 GHz. Although the coupling coefficient is practically unchanged, the fractional bandwidth changes from 6.8% to 9.38%, as the CSRR shape changes and the external quality factor decreases. After the removal of the liquid metal, the measured values are similar to the values recorded before the liquid metal was injected. The repeatability of the frequency-switchable mechanism is, therefore, verified.

Keywords: *band-pass filter; QMSIW; 3D printing; microfluidic channel; CSRR; EGaIn; frequency-switchable*

1. INTRODUCTION

Improvements in frequency-switchable devices, such as PIN diodes [1,2,3,4], varactor diodes [5,6,7,8], radio frequency microelectromechanical systems (RF MEMS)

[9,10,11,12], and field-effect-transistor (FET) switches [13,14] have enabled the growth of frequency-switchable technology, owing to its reliability and versatility. However, these switchable mechanisms need additional DC bias networks. Isolating the biasing and control circuits can be a challenging task.

Microfluidic technologies are emerging as one of the alternative switching mechanisms because these methods do not require additional DC bias networks [15]. A conduction path can be established by filling the microfluidic channels with a conductive material [16,17,18]. In the case of frequency-switchable applications, to fill microchannels with conducting materials, a liquid metal, such as eutectic gallium-indium (EGaIn, indium (In) 24.5% and gallium (Ga) 75.5%) is extensively used, since it is easy to inject and non-toxic in nature [19,20,21]. Injecting the liquid metal inside microfluidic channels constructed over an elastomeric substrate (e.g., polydimethylsiloxane (PDMS)) has proven to be an easier technique for the fabrication of microwave antennas and circuits and making them reconfigurable. For example, dipole antennas have been designed by using liquid metal [22]. In addition, ground planes were fabricated by using liquid metal-filled meshed microfluidic paths [23]. Unlike conventional microwave structures which are made of solid metal elements, these fluidic RF structures incorporate the mechanical properties of the sheathing material (e.g., PDMS); therefore, they are flexible and mechanically durable. Injecting liquid metal into microfluidic channels offers a simple method for shaping the metal into useful structures such as switches, to achieve reconfigurability in filters, antennas, and other similar devices. Since the liquid metal is a low-viscosity liquid, these structures adapt the mechanical properties of the PDMS microchannels. Consequently, the structures return to their original shape after being deformed. Mercury, a well-known liquid metal, cannot be used in microfluidics because it has the tendency to lead up to minimize its surface energy and, hence, draws out unexpectedly from the microchannels. In contrast to mercury, EGaIn forms a thin oxide skin which provides a mechanically stable formation of the liquid after it is inserted into the microchannels [24].

The substrate-integrated-waveguide (SIW) cavity resonator is a well-known technology that provides advantages of compact size, easy fabrication, and low leakage loss [25,26]. In particular, we introduced a quarter mode substrate-integrated-waveguide (QMSIW) cavity resonator, with approximately 75% reduction in size compared to a full SIW resonator, while maintaining the same resonant frequency [27,28,29]. Previously, the microfluidic eight-mode SIW (EMSIW) resonator was proposed, where the microfluidic channel was loaded on the open sides of the EMSIW for reconfigurable antenna applications [30]. Although an EMSIW resonator is more compact than a QMSIW resonator, it is not a good candidate for a filter application owing to the high radiation loss.

In this study, we propose a frequency-switchable CSRR-loaded QMSIW band-pass filter. A complementary split-ring resonator (CSRR) is a squared-slot structure with the split on the metallic pattern. When the CSRR is loaded on the top plane of the QMSIW structure, we can achieve a further size reduction without requiring additional space. For the frequency-switchable feature, we utilized a microfluidic channel using a polydimethylsiloxane (PDMS) substrate with EGaIn. The microfluidic channel was fabricated using a three-dimensional (3D) printer which provided simpler and faster fabrication compared to conventional lithography [31,32].

2. CSRR-LOADED QMSIW BAND-PASS FILTER DESIGN

For designing the proposed band-pass filter, the finite-element-method (FEM)-based ANSYS high-frequency structure simulator (HFSS) was used. It has also been used in a previous study on band-pass filter design [33].

The equation used to determine the resonant frequency of the rectangular SIW cavity resonator is [34]:

$$f_{mn0} = \frac{1}{2\pi\sqrt{\mu\varepsilon}}\sqrt{\left(\frac{m\pi}{W_{SIW}}\right)^2 + \left(\frac{n\pi}{L_{SIW}}\right)^2} \tag{1}$$

where ε and μ represent the permittivity and permeability of the dielectric material inside the cavity, respectively. L_{SIW} and W_{SIW} are the length and the width of the SIW, respectively, while m and n are the mode numbers. The lowest resonant frequency corresponds to the TE_{100} mode.

Figure 1 shows the simulated magnitude of the electric field distributions of the full SIW, half-mode SIW (HMSIW), and QMSIW resonators. When the SIW cavity resonator is cut on the AA' plane along the magnetic wall, an HMSIW resonator can be realized [35,36]. Similarly, the QMSIW resonator can be realized by cutting on the OB plane of the HMSIW [37,38].

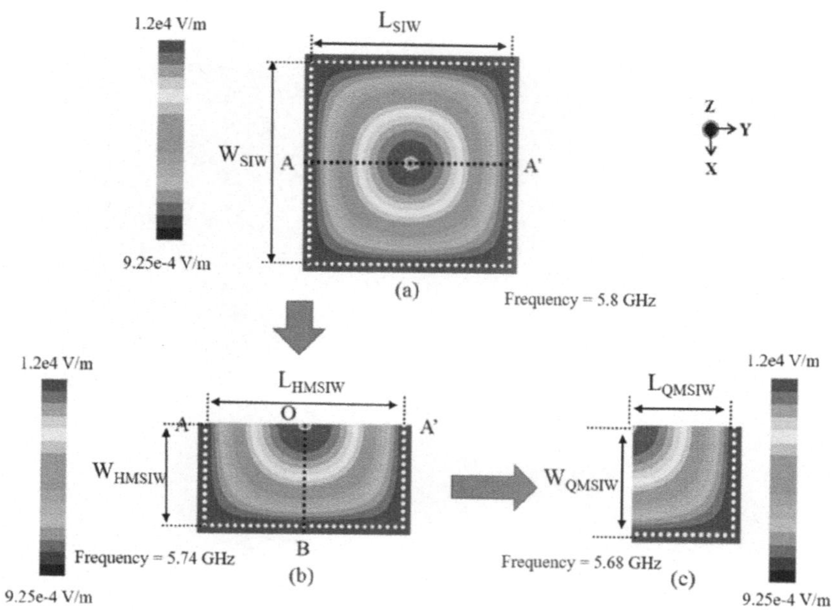

Figure 1. *The simulated magnitude of E-field distributions for (a) full-mode SIW; (b) half-mode SIW; and (c) quarter-mode SIW at their dominant resonant frequencies.*

Ideally, the QMSIW is half that of the HMSIW size. However, fringe fields are generated from the open sides along W_{QMSIW} and L_{QMSIW}. ΔL_{QMSIW} and ΔW_{QMSIW} are the effects of length and width extensions owing to fringe fields, respectively. Therefore, the resonant frequency of the QMSIW cavity at the TE_{100} dominant mode is given by:

$$f_{100,QMSIW} = \frac{1}{2\pi\sqrt{\mu\varepsilon}}\sqrt{\left(\frac{\pi}{W_{QMSIW}+\Delta W}\right)^2+\left(\frac{\pi}{L_{QMSIW}+\Delta L}\right)^2} \tag{2}$$

The magnitude of the E-field distribution of the QMSIW is kept similar to that of the full-mode SIW, as shown in Figure 1. Thus, the QMSIW cavity resonator is 1/4 the size of the SIW, while maintaining almost the same resonant frequency as that of the SIW. In this study, the QMSIW resonator is employed to design a compact filter.

Figure 2a shows the geometry and dimensions of a single QMSIW cavity, while Figure 2b shows the same QMSIW cavity with a CSRR installed over it. It can be seen that the structure is planar, the gray part is the dielectric material, which is a Rogers Duroid 5880 substrate having a thickness of 0.51 mm. The orange part is copper and the white part is the integrated gold vias connecting the top and bottom parts of the structure. As illustrated in Figure 2c, the resonant frequency of the CSRR-loaded QMSIW is reduced from 5.68 to 2.28 GHz. This represents a 59.86% reduction. As observed, the electric field is confined around the CSRR rather than at the edges of the open sides of the QMSIW. The CSRR interacts with the QMSIW structure to create a new propagation mode [29]. The physical size of the CSRR-loaded QMSIW is 14 mm × 24 mm. Its electrical length is $0.1188\lambda_g \times 0.1283\lambda_g$, which is calculated at 2.28 GHz.

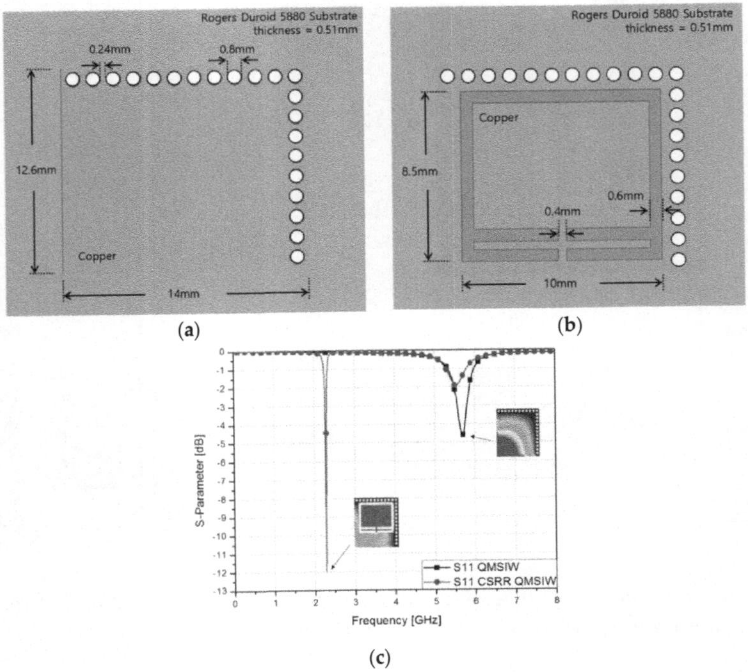

Figure 2. *Geometry of (a) QMSIW and (b) CSRR-loaded QMSIW; (c) simulated S-parameters of CSRR-loaded QMSIW and QMSIW with the E-field distribution layout.*

The external quality factor (Q_e) can be controlled by the offset distance t from the feeding line, as shown in Figure 3a. The coupling coefficient K_{12} can be controlled by the offset distance s, as shown in Figure 3b. The coupling coefficient can be calculated using the following coupling coefficient method [39]:

$$K_{1,2} = \frac{f_{p2}^2 - f_{p1}^2}{f_{p2}^2 + f_{p1}^2} \tag{3}$$

where f_{p1} and f_{p2} are the two split resonant frequencies of a pair of coupled resonators. The band-pass filter design has been studied in many previously reported studies [28,29,36]. The CSRR-loaded QMSIW is designed using the n-order Chebyshev low-pass prototype value. The coupling coefficient and external quality factor can be determined by using the following equation:

$$K_{i,i+1}\Big|_{i=1\,to\,n-1} = \frac{FBW}{\sqrt{g_i g_{i+1}}} \quad Q_e = \frac{g_0 g_1}{FBW} \tag{4}$$

where g and *FBW* are the Chebyshev low-pass filter prototype value and fractional bandwidth, respectively.

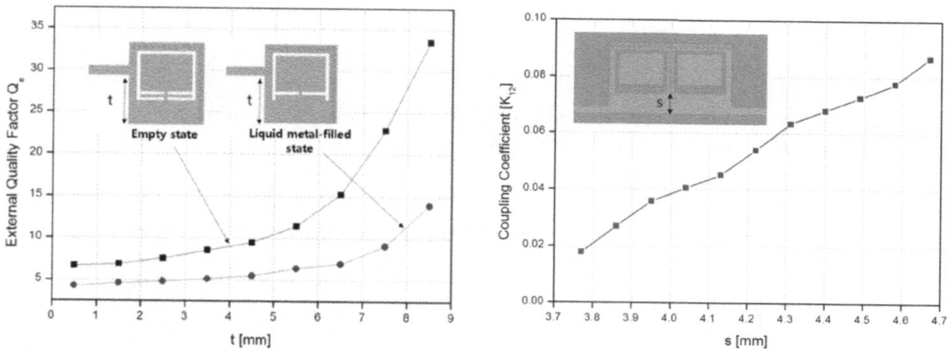

Figure 3. *(a) External quality factor variation with distance t for the CSRR-loaded QMSIW resonator. (b) The coupling coefficient of coupled cavities with distance s.*

Figure 4 shows the layout of the proposed CSRR-loaded QMSIW band-pass filter, which is composed of two different substrates. The relative permittivity, thickness, and dielectric loss of the Duroid-5880 substrate are 2.2, 0.51 mm, and 0.0009, respectively. The PDMS substrate is attached on top of the Duroid-5880 substrate to form the microfluidic channels. The relative permittivity, thickness, and dielectric loss of PDMS are 2.8, 1 mm, and 0.02, respectively.

A second-order band-pass filter was implemented using CSRR-loaded QMSIW cavities with a center frequency of 2.205 GHz, fractional bandwidth of 7.71%, and two transmission zeros at 2.19 GHz and 2.25 GHz. An insertion loss of less than 1.65 dB with an in-band and a return loss greater than 19 dB were achieved. Figure 5 shows the simulated

S-parameter results when the microfluidic channels are empty and filled with liquid metal. After injecting the liquid metal, the center frequency shifted from 2.205 GHz to 2.545 GHz. An insertion loss of less than 1.3 dB with an in-band and a return loss greater than 24 dB were achieved. Although the coupling coefficient was practically the same, the fractional bandwidth changed from 7.17% to 10.61% because of the change in the CSRR shape and a decrease in the external quality factor, as shown in Figure 3a. To demonstrate the reliability of the idea, a third-order band-pass filter was also designed using the same mechanism. The design of the third-order bandpass filter and its frequency response are shown in Figure 6.

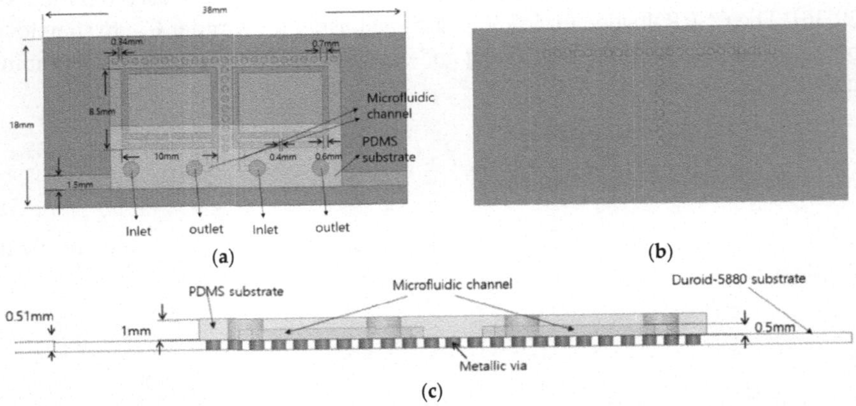

Figure 4. *(a) Top view of the CSRR-loaded QMSIW filter with a PDMS substrate; (b) bottom view of the proposed filter; and (c) side view of the proposed filter.*

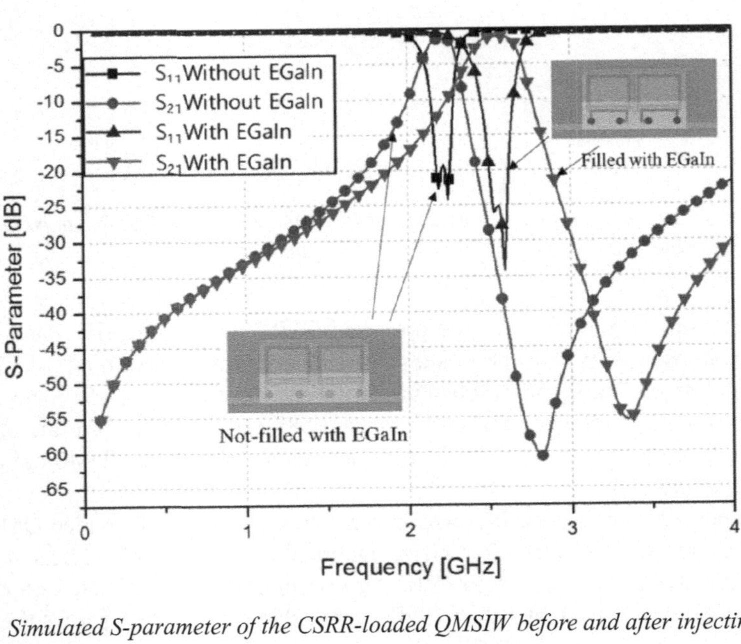

Figure 5. *Simulated S-parameter of the CSRR-loaded QMSIW before and after injecting EGaIn.*

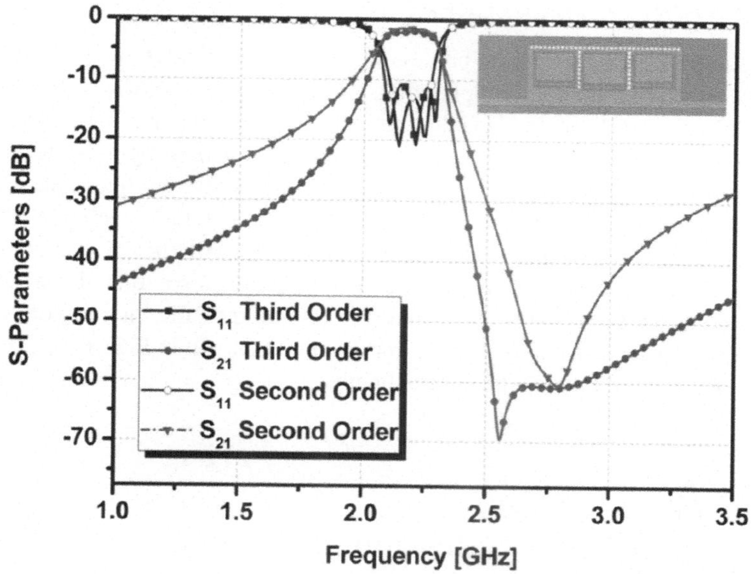

Figure 6. *Designs of second-order and third-order CSRR-loaded QMSIW band-pass filters and their simulated S-parameter results.*

In Figure 6, the S-parameter results of second-order and third-order CSRR-loaded QMSIW band-pass filters are compared. Thus, the proposed filter design can be extended to higher orders by using the demonstrated approach. The equivalent circuits for the proposed second-order and demonstrated third-order bandpass filter designs are shown in Figure 7.

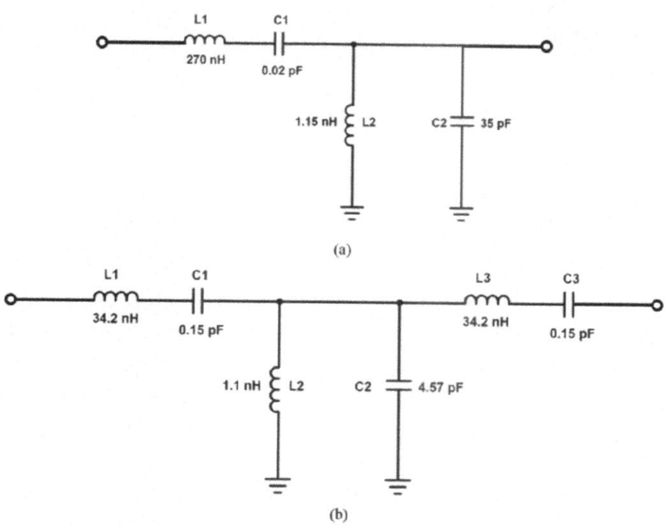

Figure 7. *Equivalent circuits of (a) proposed second-order band-pass filter design and (b) third-order bandpass filter design shown in Figure 6.*

3. FABRICATION AND MEASUREMENT

Figure 8 shows the fabrication process of the PDMS substrate with the microfluidic channel. As shown in Figure 8a, a mold of the microfluidic channel was initially fabricated by using a 3D printer (Ultimaker 2, Geldermalsen, The Netherlands). After considering the printing resolution, a mold with a thickness of 0.5 mm was built by using the 3D printer, which became the thickness of the microfluidic channel where the liquid metal would be sustained. Further, as shown in Figure 8b, a PDMS silicone elastomer base (Sylgard184 A) and a curing agent (Sylgard184 B) were mixed in a ratio of 10:1. A vacuum pump was then used to remove unnecessary air bubbles from the mixed solution. To make a PDMS substrate, the uncured solution was poured into the mold of the microfluidic channel as shown in Figure 8c. It was cured at 80 °C for 30 min as shown in Figure 8d,e. Finally, the PDMS with the microfluidic channel was obtained.

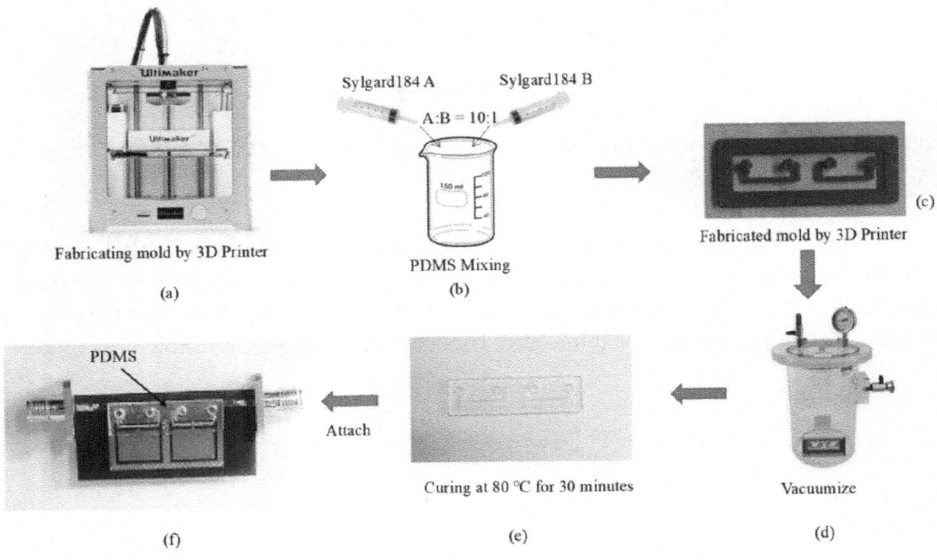

Figure 8. *Fabrication process of the microfluidic channel: (a) Fabricating the mold using a 3D printer; (b) mixing of the PDMS solution; (c) pouring of the PDMS solution into the mold/frame; (d) vacuuming; (e) curing; and (f) Final sample (PDMS attached to the Duroid substrate).*

The QMSIW CSRR pattern was realized by using a conventional PCB manufacturing process on a 0.51-mm-thick Duroid-5880 substrate. To attach the Duroid-5880 and PDMS substrates, the uncured PDMS solution was used as an adhesive. Figure 8f shows the final sample, where PDMS is attached at the position indicated (high E-field position of the CSRR).

Figure 9a shows the fabricated QMSIW CSRR filter prototype. The return loss and insertion loss of the filter were recorded using an Anritsu MS2038C vector network analyzer. Figure 9b shows the simulated and measured S-Parameters of the proposed filter along with their comparison. The measured center frequency and fractional bandwidths were 2.205 GHz and 6.80%, respectively. Two transmission zeros were measured at 2.18

GHz and 2.25 GHz. An insertion loss of less than 1.8 dB with an in-band and a return loss greater than 14 dB were achieved.

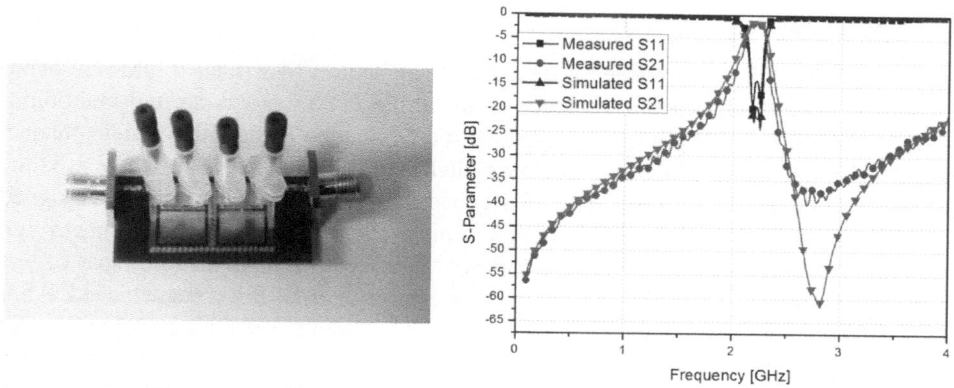

(a) (b)

Figure 9. *(a) Fabricated CSRR-loaded QMSIW filter prototype and (b) simulated and measured S-parameter of the proposed CSRR-loaded QMSIW filter.*

Figure 10a shows the measured S-parameter results before and after the microfluidic channels were filled with liquid metal. The liquid metal was injected using a micropipette. After the injection, the center frequency shifted from 2.205 GHz to 2.56 GHz. An insertion loss of less than 1.5 dB with an in-band and a return loss greater than 15 dB were achieved. The fractional bandwidth changed from 6.8% to 9.38%.

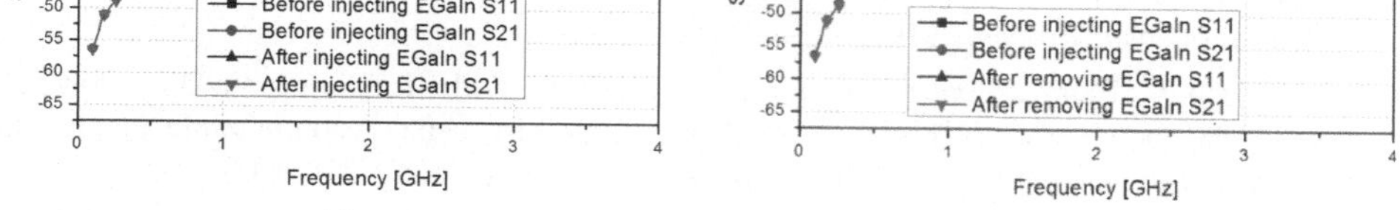

(a) (b)

Figure 10. *Measured S-parameter of the fabricated CSRR-loaded QMSIW (a) before and after injecting the microfluidic channels with EGaIn; and (b) before injecting EGaIn and after the removal of EGaIn.*

After the removal of the liquid metal, a small amount of liquid metal remained on the PDMS surface. However, it did not come in contact with the CSRR directly. Therefore, as shown in Figure 10b, the measured S-parameter results before and after injecting the liquid metal are practically the same.

4. CONCLUSIONS

In this research, we proposed a frequency-switchable CSRR-loaded QMSIW band-pass filter. Microfluidic channels were built using the PDMS elastomer and microfluidic channel frames. A 3D printer was used for the fabrication of the microfluidic channel frames. The size of the proposed band-pass filter is $0.2376\lambda_g \times 0.1283\lambda_g$, where λ_g is the guided wavelength at 2.28 GHz. The CSRR-loaded QMSIW band-pass filter is designed to have two states. Before injecting the liquid metal, the measured center frequency and fractional bandwidths were 2.205 GHz and 6.80%, respectively. An insertion loss of less than 1.8 dB with an in-band and a return loss greater than 14 dB were achieved. After injecting the liquid metal, the center frequency shifted from 2.205 GHz to 2.56 GHz. An insertion loss of less than 1.5 dB with an in-band and a return loss greater than 15 dB were achieved. Although the coupling coefficient was practically the same, the fractional bandwidth changed from 6.8% to 9.38% as the CSRR shape changed and the external quality factor decreased. After the removal of the liquid metal, the measured results were practically the same as those before injection. Consequently, the repeatability of the frequency-switchable mechanism was verified.

Acknowledgments

This research was supported by the MSIP (Ministry of Science, ICT and Future Planning), Korea, under the ITRC (Information Technology Research Center) support program (IITP-2016-H8501-16-1007) supervised by the IITP (Institute for Information and Communications Technology Promotion) and the National Research Foundation of Korea (NRF) grant funded by the Korea government (MSIP) (No. 2014R1A2A1A11050010).

REFERENCES

1. Lim, J.H.; Back, G.T.; Ko, Y.I.; Song, C.W.; Yun, T.Y. A reconfigurable PIFA using a switchable PIN-diode and a fine-tuning varactor for USPCS/WCDMA/m-WiMAX/WLAN. *IEEE Trans. Antennas Propag.* 2010, *58*, 2404–2411.
2. Nikolaou, S.; Bairavasubramanian, R.; Lugo, C.; Carrasquillo, I.; Thompson, D.C.; Ponchak, G.E.; Papapolymerou, J.; Tentzeris, M.M. Pattern and frequency reconfigurable annular slot antenna using pin diodes. *IEEE Trans. Antennas Propag.* 2006, *54*, 439–448.
3. Brito-Brito, Z.; Llamas-Garro, I.; Pradell-Cara, L.; Corona-Chavez, A. Microstrip switchable bandstop filter using PIN diodes with precise frequency and bandwidth control. In Proceedings of the 38th European Microwave Conference (EuMC 2008), Amsterdam, The Netherlands, 27–31 October 2008; pp. 1707–1710.
4. Liang, C.; Li, L.; Zhai, H. Variational stability form for the capacitance of an arbitrarily shaped conducting plate. *Chin. J. Electron.* 2004, *13*, 714–718.
5. Aghdam, S.A. A Novel UWB monopole antenna with tunable notched behavior using varactor diode. *IEEE Antennas Wirel. Propag. Lett.* 2014, *13*, 1243–1246.

6. Khidre, A.; Yang, F.; Elsherbeni, A.Z. A Patch Antenna with a Varactor-Loaded Slot for Reconfigurable Dual-Band Operation. *IEEE Trans. Antennas Propag.* 2015, *63*, 755–760.
7. Lin, F.; Rais-Zadeh, M. A tunable 0.6 GHz–1.7 GHz bandpass filter with a constant bandwidth using switchable varactor-tuned resonators. In Proceedings of the 2015 IEEE MTT-S International Microwave Symposium (IMS 2015), Phoenix, AZ, USA, 17–22 May 2015; pp. 2–5.
8. Liu, B.; Wei, F.; Zhang, H.; Shi, X.; Lin, H. A Tunable Bandpass Filter with Switchable Bandwidth. *J. Electromagn. Waves Appl.* 2011, *25*, 223–232.
9. Erdil, E.; Topalli, K.; Unlu, M.; Civi, O.A.; Akin, T. Frequency tunable microstrip patch antenna using RF MEMS technology. *IEEE Trans. Antennas Propag.* 2007, *55*, 1193–1196.
10. Chan, K.Y.E.; Ramer, R.; Guo, Y.J. RF MEMS millimeter-wave switchable bandpass filter. In Proceedings of the 2013 IEEE International Wireless Symposium (IWS), Beijing, China, 14–18 April 2013; pp. 1–4.
11. Park, S.; Lee, K.; Rebeiz, G.M. Low-loss 5.15–5.70 GHz RF MEMS switchable filter for wireless lan application. *IEEE Trans. Microw. Theory Tech.* 2006, *54*, 3931–3939.
12. Yuk, K.Y.; Fouladi, S.; Ramer, R.; Mansour, R.R. RF MEMS switchable interdigital bandpass filter. *IEEE Microw. Wirel. Compon. Lett.* 2012, *22*, 44–46.
13. Aboufoul, T.; Alomainy, A.; Parini, C. Reconfiguring UWB monopole antenna for cognitive radio applications using GaAs FET switches. *IEEE Antennas Wirel. Propag. Lett.* 2012, *11*, 392–394.
14. Kodera, T.; Caloz, C. Multi-function reconfigurable microwave component based on a switchable FET circuit. In Proceedings of the 2013 IEEE MTT-S International Microwave Symposium Digest (IMS), Seattle, WA, USA, 2–7 June 2013; Volume 2, pp. 9–11.
15. Wang, Y.; Yoon, K.; Lee, J. A Frequency Tunable Double Band-Stop Resonator (BSR) with Voltage Control by Varactor Diodes. *J. Electromagn. Eng. Sci.* 2016, *16*, 159–163.
16. Aïssa, B.; Nedil, M.; Habib, M.A.; Haddad, E.; Jamroz, W.; Therriault, D.; Coulibaly, Y.; Rosei, F. Fluidic patch antenna based on liquid metal alloy/single-wall carbon-nanotubes operating at the S-band frequency. *Appl. Phys. Lett.* 2013, *103*.
17. Cheng, S.; Wu, Z. Microfluidic electronics. *Lab Chip* 2012, *12*, 2782–2791.
18. Hayes, G.J.; So, J.-H.; Qusba, A.; Dickey, M.D.; Lazzi, G. Flexible Liquid Metal Alloy (EGaln) Microstrip Patch Antenna. *TAP IEEE Trans. Antennas Propag.* 2012, *60*, 2151–2156.
19. Kelley, M.; Koo, C.; McQuilken, H.; Lawrence, B.; Li, S.; Han, A.; Huff, G. Frequency reconfigurable patch antenna using liquid metal as switching mechanism. *Electron. Lett.* 2013, *49*, 1370–1371.
20. Ling, K.; Kim, H.K.; Yoo, M.; Lim, S. Frequency-switchable metamaterial absorber injecting eutectic gallium-indium (EGaln) liquid metal alloy. *Sensors* 2015, *15*, 28154–28165.
21. Zhu, W.M.; Zhang, W.; Huang, R.F.; Ting, S.K.; Lo, G.Q.; Kwong, D.L.; Liu, A.Q. Metamaterial tunable filter with liquid metal. In Proceedings of the 26th International Conference on Micro Electro Mechanical Systems (MEMS), Taipei, Taiwan, 20–24 January 2013; pp. 725–728.
22. So, J.-H.; Thelen, J.; Qusba, A.; Hayes, G.J.; Lazzi, G.; Dickey, M.D. Reversibly Deformable and Mechanically Tunable Fluidic Antennas. *Adv. Funct. Mater.* 2008, 3632–3637.
23. Cheng, S.; Rydberg, A.; Hjort, K.; Wu, Z. Liquid metal stretchable unbalanced loop antenna. *Appl. Phys. Lett.* 2009, *94*, 144103.
24. Dickey, M.D.; Chiechi, R.C.; Larsen, R.J.; Weiss, E.A.; Weitz, D.A.; Whitesides, G.M. Eutectic gallium indium (EGaln): A liquid metal alloy for the formation of stable structures in microchannels at room temperature. *Adv. Funct. Mater.* 2008, *18*, 1097–1104.

25. Bozzi, M.; Georgiadis, A.; Wu, K. Review of substrate-integrated waveguide circuits and antennas. *IET Microw. Antennas Propag.* 2011, *5*, 909–920.

26. Sabri, S.S.; Ahmad, B.H.; Othman, A.R.B. A review of Substrate Integrated Waveguide (SIW) bandpass filter based on different method and design. In Proceedings of the 2012 IEEE Asia-Pacific Conference on Applied Electromagnetics (APACE), Melaka, Malaysia, 11–13 December 2012; pp. 210–215.

27. Guo, Z.; Chin, K.-S.; Che, W.; Chang, C.-C. Cross-coupled bandpass filters using QMSIW cavities and S-shaped slot coupling structures. *J. Electromagn. Waves Appl.* 2013, *27*, 160–167.

28. Senior, D.E.; Rahimi, A.; Jao, P.; Yoon, Y.K. Flexible Liquid Crystal Polymer based complementary split ring resonator loaded quarter mode substrate integrated waveguide filters for compact and wearable broadband RF applications. In Proceedings of the 64th Electronic Components and Technology Conference (ECTC 2014), Orlando, FL, USA, 27–30 May 2014; pp. 789–795.

29. Zhang, Z.; Yang, N.; Wu, K. 5-GHz bandpass filter demonstration using quarter-mode substrate integrated waveguide cavity for wireless systems. In Proceedings of the 4th International Conference on Radio and Wireless Symposium (RWS 2009), San Diego, CA, USA, 18–22 January 2009; pp. 95–98.

30. Seo, Y.; Memon, M.U.; Lim, S. Microfluidic Eighth-Mode Substrate-Integrated-Waveguide Antenna for Compact Ethanol Chemical Sensor Application. *IEEE Trans. Antennas Propag.* 2016, *64*, 3218–3222.

31. Khondoker, M.A.H.; Sameoto, D. Fabrication methods and applications of microstructured gallium based liquid metal alloys. *Smart Mater. Struct.* 2016, *25*, 93001.

32. Li, X.; Abe, T.; Esashi, M. Deep reactive ion etching of Pyrex glass using SF6 plasma. *Sens. Actuators A Phys.* 2001, *87*, 139–145.

33. Kwon, K.; Kim, H.K.; Yun, S.W. Design of Dual-Band Bandpass Filters for Cognitive Radio Application of TVWS Band. *J. Electromagn. Eng. Sci.* 2016, *16*, 19–23.

34. Memon, M.U.; Lim, S. Review of reconfigurable substrate-integrated-waveguide antennas. *J. Electromagn. Waves Appl.* 2014, *28*, 1815–1833.

35. Cheng, Y.J.; Hong, W.; Wu, K. Millimeter-wave half mode substrate integrated waveguide frequency scanning antenna with quadri-polarization. *IEEE Trans. Antennas Propag.* 2010, *58*, 1848–1855.

36. Wang, Y.; Hong, W.; Dong, Y.; Liu, B.; Tang, H.J.; Chen, J.; Yin, X.; Wu, K. Half mode substrate integrated waveguide (HMSIW) bandpass filter. *IEEE Microw. Wirel. Compon. Lett.* 2007, *17*, 265–267.

37. Zhang, S.; Bian, T.-J.; Zhai, Y.; Liu, W.; Yang, G.; Liu, F.-L. Quarter substrate integrated waveguide resonator applied to fractal-shaped BPFs. *Microw. J.* 2012, *55*, 200–208.

38. Memon, M.U.; Lim, S. Frequency-Tunable Compact Antenna Using Quarter-Mode Substrate Integrated Waveguide. *IEEE Antennas Wirel. Propag. Lett.* 2015, *14*, 1606–1609.

39. Hong, J.S.; Lancaster, M.J. *Microstrip Filters for RF/Microwave Applications*; John Wiley & Sons: Hoboken, NJ, USA, 2004; p. 167.

10 CHAPTER

3D-Printed Dip Slides Miniaturize Bacterial Identification and Antibiotic Susceptibility Tests Allowing Direct Mastitis Sample Analysis

Tai The Diep *, Samuel Bizley and Alexander Daniel Edwards

**Department of Pharmacy, Reading School of Pharmacy, University of Reading, Reading RG6 6UR, Berkshire, UK*

ABSTRACT

The early detection of antimicrobial resistance remains an essential step in the selection and optimization of antibiotic treatments. Phenotypic antibiotic susceptibility testing including the measurement of minimum inhibitory concentration (MIC) remains critical for surveillance and diagnostic testing. Limitations to current testing methods include bulky labware and laborious methods. Furthermore, the requirement of a single strain of bacteria to be isolated from samples prior to antibiotic susceptibility testing delays results. The mixture of bacteria present in a sample may also have an altered resistance profile to the individual strains, and so measuring the susceptibility of the mixtures of organisms found in some samples may be desirable. To enable simultaneous MIC and bacterial species detection in a simple and rapid miniaturized format, a 3D-printed frame was designed for a multi-sample millifluidic dip-slide device that combines panels of identification culture media with a range of antibiotics (Ampicillin, Amoxicillin, Amikacin, Ceftazidime, Cefotaxime, Ofloxacin, Oxytetracycline, Streptomycin, Gentamycin and Imipenem) diluted in Muëller–Hinton Agar. Our proof-of-concept evaluation confirmed that the direct detection of more than one bacterium parallel to measuring MIC in samples is possible, which is validated using reference strains E. coli ATCC 25922, Klebsiella pneumoniae ATCC 13883, Pseudomonas aeruginosa ATCC 10145, and Staphylococcus aureus ATCC 12600 and with mastitis milk samples collected from Reading University Farm. When mixtures were tested, a MIC value was obtained that reflected the most resistant organism present (i.e., highest MIC), suggesting it may be possible to estimate a minimum effective antibiotic concentration for mixtures directly from samples containing multiple pathogens. We conclude that this simple miniaturized approach to the rapid simultaneous identification and antibiotic susceptibility testing may be suitable for directly testing agricultural samples, which is achieved through shrinking conventional tests into a simple "dip-and-incubate" device that can be 3D printed anywhere.

Keywords: *microfluidic microbiology; antimicrobial resistance; dip slide; minimum inhibitory concentration; 3D printing; miniaturized microbiology*

1. INTRODUCTION

Rising antimicrobial resistance remains one of the biggest global threats to public health, animal health, and the environment. For example, within an agricultural setting, bovine mastitis is a major challenge to the dairy industry both economically through reduced milk production and affecting animal well-being. Antibiotic use to treat mastitis is associated with increasing antibiotic resistance. In addition, there is a great diversity of both pathogens and commensal microbial communities present in bovine mastitis [1,2]. Recent studies have indicated that microbes interacting in species and communities have contributed to individual bacterial growth and their subsequent evolution of antibiotic resistance [3,4], although the relevance of their resistance regarding selection in the microbiome is still unclear [5]. This highlights the importance of exploring the association between host–pathogen, pathogen–pathogen and pathogen–commensals, including the positive or negative impact of antibiotics on the natural microbiota [6,7]. One way to improve this understanding in the face of microbial complexity is to scale up our ability to measure antibiotic sensitivity.

Microbiology testing remains central, contributing to diagnosis and the effective treatment of infections, and surveillance to guide empirical antibiotic selection is important as well. Measuring minimum inhibitory concentration (MIC) using antibiotic susceptibility testing assays remains vital, and it is included in many standardized and clinical reference methods, with more accuracy compared to disk diffusion [8]. The MIC is still an important value for making clinical decisions. Combined with a pharmacokinetic-pharmacodynamic profile, MIC values guide the selection of the effective dose for an antibiotic [9,10]. Analysis of MIC remains critical for setting breakpoint threshold concentrations that define resistance vs. susceptibility, underpinning the surveillance of resistance levels among bacteria important to human and animal health [11]. Monitoring MIC shows new bacterial resistance trends as they emerge, and they can be used to modify clinical doses administered to achieve a required therapeutic concentration in patients or animals [12]. MIC measurements form a vital quality control and standardization tool between different laboratories across the world. Every year, MIC values are reported to reference agencies including CLSI, EUCAST, and BSAC and are used to define standards for clinical and veterinary laboratories and to assess global microbiological susceptibility trends [5,13].

MIC is measured by a range of antibiotic susceptibility assays in vitro, including broth microdilution, agar dilution, and antimicrobial gradient methods such as E-test (bioMerieux, Marcy–L'Etoile), MIC Test Strip (Lio-filchem Inc., Waltham, MA, USA),

M.I.C. Evaluator (Oxoid, Basingstoke, UK) and Ezy MIC Strip (HiMedia Laboratories Pvt. Ltd., Mumbai, India). Many automated systems adapt these methods for higher throughput. These are classically conducted as routine methods within a microbiology laboratory, but they are associated with high costs, a need for skilled workforce with careful quality control systems and substantial labor time [14]. The interpretation of MIC results can be influenced by several factors including the concentration of bacterial inoculum, the way the concentration of antibiotics is prepared, and the quality of media. The current approach detects MIC for a single bacterial isolate and as a result typically takes 2–3 days to complete and analyze as the pure strains need to be isolated after overnight culture before performing identification and then finally antibiotic susceptibility to determine MIC. In well-resourced situations, rapid identification (e.g., MALDI-TOF) combined with rapid susceptibility testing systems can reduce this to 1–2 days. The impact of this for people without rapid identification or susceptibility testers is that it delays the physician or veterinary surgeon's decision making, resulting in a possible delay in patient treatment. Alternatively, antibiotics are used empirically without testing. In addition, with current methods, there are a number of other factors that reduce/restrict the efficiency of the testing process, including the following: antibiotic panels of testing from commercial suppliers are not customizable, significant time is required for agar media preparation, and supply chain challenges remain regarding the import of specific antibiotics in some countries. A separate weakness of current testing protocols is a focus on single isolates that may not reflect the progress of any infection caused by a combination of multiple bacterial strains, either multiple pathogens or a single pathogen growing amongst commensal organisms that may influence antibiotic responsiveness [15].

To address the high level of antibiotic resistance, a variety of methods have been developed and commercialized such as clinically oriented automated systems including VITEK 2® [16] and MALDI-TOF VITEK MS® systems [17]. Several factors must be considered when developing suitable methods including time saving, cost saving, reliability and robustness of the method, alongside convenience for the end-user. The majority of molecular biology-based methods offer the rapid detection of specific targets including identification and a panel of common resistance genes, alongside the expansion of next-generation sequencing that has improved our understanding of microbiome and the resistome. We still rely heavily on phenotypic assays, and these continue to focus on single strains, yet the importance of understanding the response of mixed microbial populations to antibiotics is driving the development of direct sample testing methods.

Recently, the use of digital cameras for timelapse imaging has contributed to improved analytical microbiology methods such as the early detection of antimicrobial resistance (AMR). Based on the falling cost and rapidly improving performance of digital cameras driven by consumer products (smartphones), the process of bacterial growth and response to antibiotics can be recorded without requiring large colonies clearly visible to the eye after overnight growth. Many miniaturization efforts focus on the microscopic analysis of growth exploiting digital cameras for time-resolved microscopy [18,19]. Others have directly imaged bacterial particles in liquid using digital cameras imaging fluorescence [20] or scatter [21]. Conventional culture methods have been improved through digital imaging, giving us a far more detailed understanding of colony growth on solid media [22]. These innovations have reached clinical microbiology guidelines, with EUCAST protocols allowing disc-diffusion antibiotic susceptibility to be recorded in as little as 4 h of culture [23].

At the same time that high-performance cheap digital cameras permit smaller colonies to be detected earlier, additive manufacturing methods such as 3D printing have become inexpensive and accessible, with fused-filament deposition methods cost-effective for producing routine microbiology labware [24]. Applying 3D printing to rapid custom microbiology labware production, we designed a new frame to create dip slides that combine two main functional goals: (1) directly identify and (2) determine the antibiotic MIC of microbes from samples. By shrinking solid culture format by >100-fold, a set of 6 mm well diameters in each dip slide frame uses far less media but still allows single colonies of bacterial species to grow and be identified. The detection of smaller colonies on these miniaturized versions of Petri dishes is aided by digital imaging using inexpensive cameras (e.g., smartphone, compact consumer camera), thereby avoiding expensive laboratory scanning instruments. The 3D-printed frame holding multiple antibiotic concentrations in agar is easy to prepare and has a fully customizable panel, making it easier to use compared to the current agar dilution methods. The aims of this study are to firstly optimize the 3D-printed frame for making dip slides for the detection of reference bacterial strains and to analyze the antibiotic susceptibility profile, to secondly observe the kinetics of bacterial growth on the different agar surfaces, and finally to evaluate the analysis of bacteria in bovine mastitis samples as a proof-of-concept application.

2. MATERIALS AND METHODS

2.1. Experimental Approach

The design of the 3D printed frame used in this study includes two parts to identify bacterial species with two 35 mm × 7 mm rectangles; and to determine MIC, we used two rows of 6 mm diameter round wells. All designed files were published as open source models for customization or 3D printing by anyone [24]. The two rows of 10× round wells include up to 10 antibiotic concentrations, which are typically prepared as doubling dilutions. The size of the round-shaped well was >150 times smaller than current conventional Petri dishes (Figure 1), allowing each sample to be tested on a large panel of conditions.

2.2. Bacterial Strains, Mastitis Samples, Media, and Measurement of MIC

Bovine mastitis samples were collected from the Centre for Dairy Research (CEDAR) at the University of Reading. All samples were transported to the laboratory and processed/analyzed on the collection day. Milk samples were diluted at 3 serial dilutions until $1{:}10^{-3}$, and 100 μL of these dilutions was applied to streak out onto the CHROMagarTM Mastitis agar plates prepared following manufacturers guidance (CHROMagar™, Paris, France) to identify the bacteria. Antibiotic susceptibility tests were performed by agar dilution on Muëller–Hinton Agar (MHA) (Sigma). All reference strains and isolates, including *K. pneumoniae* ATCC 13883, *E. coli* ATCC 25922, *P. aeruginosa* ATCC 10145, *S. aureus* ATCC 12600, and isolates from mastitis samples such as *Staphylococcus aureus*, *E. coli*, *Pseudomonas* spp., *Klebsiella* spp., *Streptococcus uberis*, and *Streptococcus agalactiae* were inoculated overnight to reach at 10^8 CFU/mL in Muëller–Hinton Broth (MHB) (Sigma for microbiology, Dorset, UK). Conventional MIC measurement was performed with CLSI guideline (CLSI M100-ED31:2021) and also by the E-test methods. Examples of different specific culture media including MacConkey agar (MC), Baird Parker Agar (BP), Muëller Hinton Agar (MHA) (Sigma Aldrich, Dorset, UK), and CHROMagar™ Mastitis (CHROMagar™, Paris, France) were used to identify bacteria.

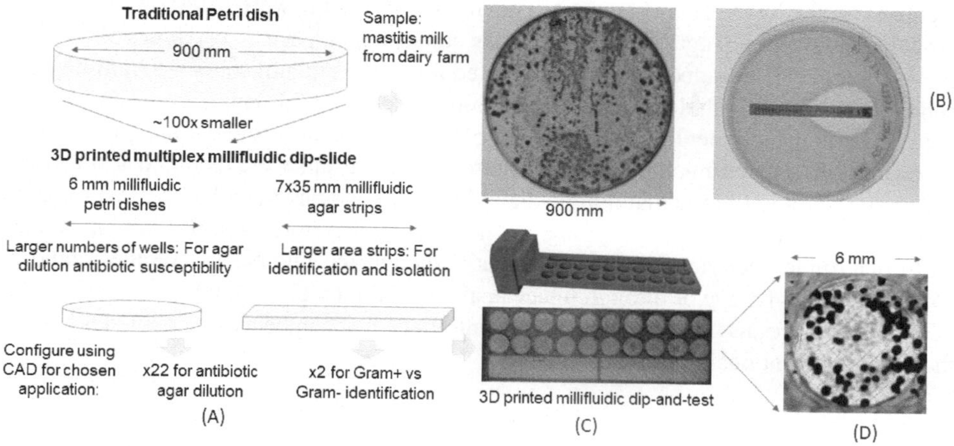

Figure 1. *Shrinking Petri dishes. Miniaturized 3D-printed arrays for solid microbiological media. (A) Illustration of how a conventional Petri dish can be shrunk into a frame designed with different areas depending on required function. Smaller circular wells are used for antibiotic agar dilution, whereas longer strips are better suited to identification and colony isolation. (B) Examples of identification of bacteria and antibiotic susceptibility testing using a conventional Petri dish. (C) CAD illustration and photograph of 3D-printed frame for customized multiplex dip slide. (D) Zoomed-in image showing small bacterial colonies grown on individual wells of the 3D printed dip-slide frame.*

2.3. Bacterial Identification and MIC Measurement Using 3D-Printed Dip Slides

2.3.1. Preparation of 3D Printed Dip-Slide Frame

The Standard Tessellation Language (STL) design of the frame dip slide is available online [24], the frame dips slide were 3D printed, sterilized with 70% alcohol and dried before use. CHROMagar™ Mastitis (CHROMagar™, Paris, France) was prepared according to the manufacturer's instructions and added to long rectangles to identify bacterial species. Muëller–Hinton (MH) Agar with triphenyl tetrazolium chloride (TTC) dye added to make colonies appear more strongly in digital photographs was added with serial dilutions of antibiotics to determine MIC. The antibiotic powders of Ampicillin, Amoxicillin, Amikacin, Ceftazidime, Cefotaxime, Ofloxacin, Oxytetracycline, Streptomycin, Gentamycin and Imipenem were purchased from Sigma Aldrich (Dorset, UK). The range of 9 antibiotic concentrations prepared in MH agar for testing was from 0.125 to 32 μg/mL for each antibiotic in a doubling dilution series.

2.3.2. Detection of Bacterial Species and Determining MIC Using 3D-Printed Dip Slides

After collecting, milk samples were diluted into two serial 10-fold dilutions and considered as testing samples. The dip slides were directly dipped into this diluted milk, kept submerged in the samples for 1 min, and removed, allowing excess sample to drain off the dip slide. Finally, the dip slide was placed in a sealed box to prevent drying out and incubated at 37 °C for 12 h (Figure 2). The interpreted results were based on the bacterial growth on two types of agar media. Firstly, the colony appearance on two types of commercial chromogenic and selective agars was noted for species identification—Gram-negative and Gram-positive, with different colored colonies identified following the manufacturer's guidance. Alongside identification, a panel of antibiotic concentrations in agar was used for agar dilution determination of MIC (Figure 2). MIC results were recorded as the lowest concentration of antibiotic that inhibits the growth of bacteria. Note that the exact volume of sample deposited following dipping was not known; therefore, colonies were not counted to determine bacterial cell density. Thus, although the number of colonies could be counted visually by eye or through image analysis of digital photographs, the smaller area of solid media means that the range of concentrations that can be easily counted is smaller than for a conventional Petri dish.

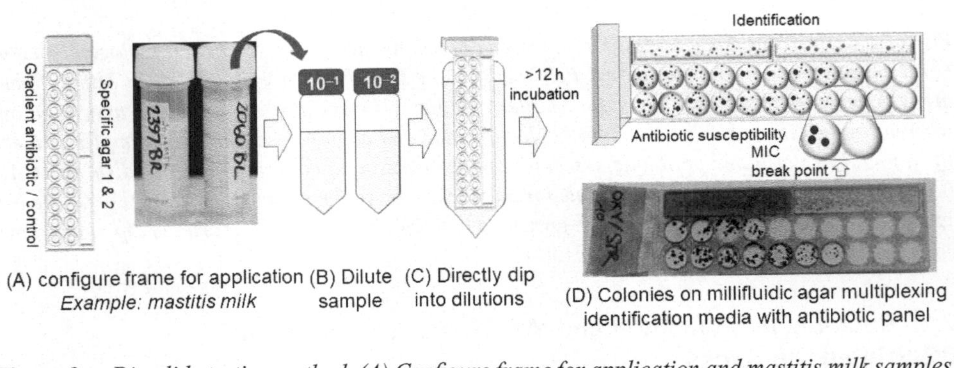

Figure 2. *Dip-slide testing method. (A) Configure frame for application and mastitis milk samples. (B) Dilute sample into 2 fold, (C) Directly dip frame into dilutions, (D) Observe the results on frame.*

Muëller–Hinton Agar was used for the determination of MIC according to standard agar dilution methodology, and CHROMagarTM Mastitis (CHROMagar™, Paris, France) was used for identifying species. The colonies with specific colors helped to identify specific bacterial species, and at the point with and without bacterial colonies, growth was used to determine MIC breakpoint (Figure 2).

2.4. Time-Lapse Imaging to Check Kinetics of Bacterial Growth on Millifluidic Solid Microbiological Media

With the same preparation as that of the dip slide, MH, MacConkey agar, and CHROMagarTM Mastitis (CHROMagar™, Paris, France) were applied to observe the kinetics of bacterial growth for *E. coli* ATCC 25922. The Raspberry camera 2.0 and Python code was used to take the photo, and ImageJ was used to build the time-lapse of bacterial growth. The distance from camera to 3D frame was 10.5 cm, and one cover with 3.5 cm height was used to prevent the agar from drying during time-lapse imaging. In other experiments, a digital Canon EOS 1300D with a Canon EF-S 60 mm f/2.8 Macro USM Lens was used to capture images representing the bacterial morphology on agar on the dip slides.

2.5. Comparison of MIC Breakpoints with E-Test Trip and Broth Microdilution in a Single Strain and a Mixture of Bacterial Strains

We compared our frame dip slide to broth microdilution methods. To do so, the density of inoculum was standardized at 10^8 CFU/mL equivalence; i.e., McFarland 0.5 was used for testing with the frame dip slide and broth microdilution method. The panel of antibiotic and dye was used the same for frame and broth microdilution methods; the final volume for each test was 200 μL in 96-well plates. Both the frame dip slide and microwell were kept at 37 °C overnight for endpoint analysis (12–16 h). The MIC was read at the points where antibiotics stopped the growth of bacteria. The whole protocol to evaluate the novel dip-slide test was mapped in Figure 3.

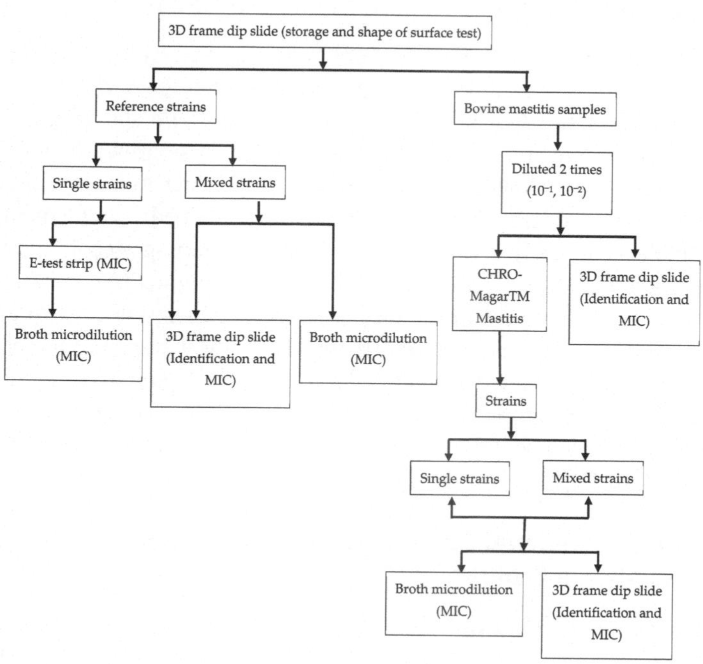

Figure 3. *Flowchart showing isolation, identification and sensitivity testing for reference and mastitis strains.*

3. RESULTS

3.1. Use of 3D-Printed Multiplex Dip Slide for Bacterial Identification and Antibiotic Susceptibility Testing

The 3D-printed frame allowed us to combine two types of solid media on a single dip slide. Longer wells provided enough space to identify colonies of bacterial species on specific chromogenic agar for Gram negatives or Gram positives simultaneously. Alongside this on the same device, agar dilution antibiotic susceptibility testing was possible. Even with only 6 mm diameter wells, single separate colonies were visible on these tiny solid media wells, with a similar appearance to a conventional Petri dish culture (Figure 1). Customization of the CAD design allows the user to decide the type of agar media required for testing to match the test purpose in the laboratory. Colonies visible on the miniaturized 6 mm wells could be counted by eye or from digital images, but this device was not designed for colony counting. Whilst we found testing very repeatable, the dip-slide format does not deposit a precise known volume onto the agar. Here, we focused on mastitis milk sample testing and adapted frames for two different identification agars provided in the CHROMagar™ Mastitis product (CHROMagar™, Paris, France). Using this frame, <1 mL of each type of agar is needed instead of 20 mL of agar for a standard 100 mm Petri dish. The MIC agar dilution wells require just 140 μL of media per slide, which is >70 times less than standard test tube agar dilution protocols and >150 times less than Petri dish, alongside a major reduction in incubator space.

3.2. Characteristics of Multiplex Multifluidic Dip Slides

By dipping directly frame into the sample (e.g., diluted milk) and incubation at 37 °C (Figure 2), the results were able to be reported in just 8 h for *E. coli* and around 10–12 h for other bacteria such as *Staphylococcus* spp. and *Klebsiella* spp. (Figure 4). Bacterial colonies grown on the dip slide appeared comparable to those grown on a Petri dish for a wide range of agar types. However, it was significantly easier to record smaller bacterial colonies with a digital camera when using media that contained a supplement of TTC dye rather than Muëller–Hinton agar without dye, where the yellow-gray colony color was only weakly visible on the background of the agar color itself (Figure 4A).

To monitor speed of colony appearance, a Raspberry Pi equipped with the v2 camera module was used to capture images every 10 min. This time-lapse analysis showed that the results of MIC can be interpreted around 7–8 h for *E. coli* (Figure 4B) and more than 10 h for other bacteria such as *Staphylococcus aureus*, *Pseudomonas aeruginosa*. This means that the results of MIC can be interpreted during 7–8 h compared to routine methods needed overnight (Figure 4B).

3.3. Detection of Individual Bacteria Isolates and Their Antibiotic Resistance Profile

Three-dimensional (3D)-printed dip slides with sets of miniaturized 6 mm diameter dip-slide wells plus rectangular identification strips were compared to conventional broth microdilution for the capacity to detect individual strains and measure their antibiotic susceptibility. Antibiotics such as Ampicillin, Amoxicillin, Amikacin,

Ceftazidime, Cefotaxime, Ofloxacin, Oxytetracycline, Streptomycin, Gentamycin and Imipenem were used for this purpose, with Figure 5 showing images with four representative antibiotics. There was an agreement of MIC results between our dip slide to broth dilution methods.

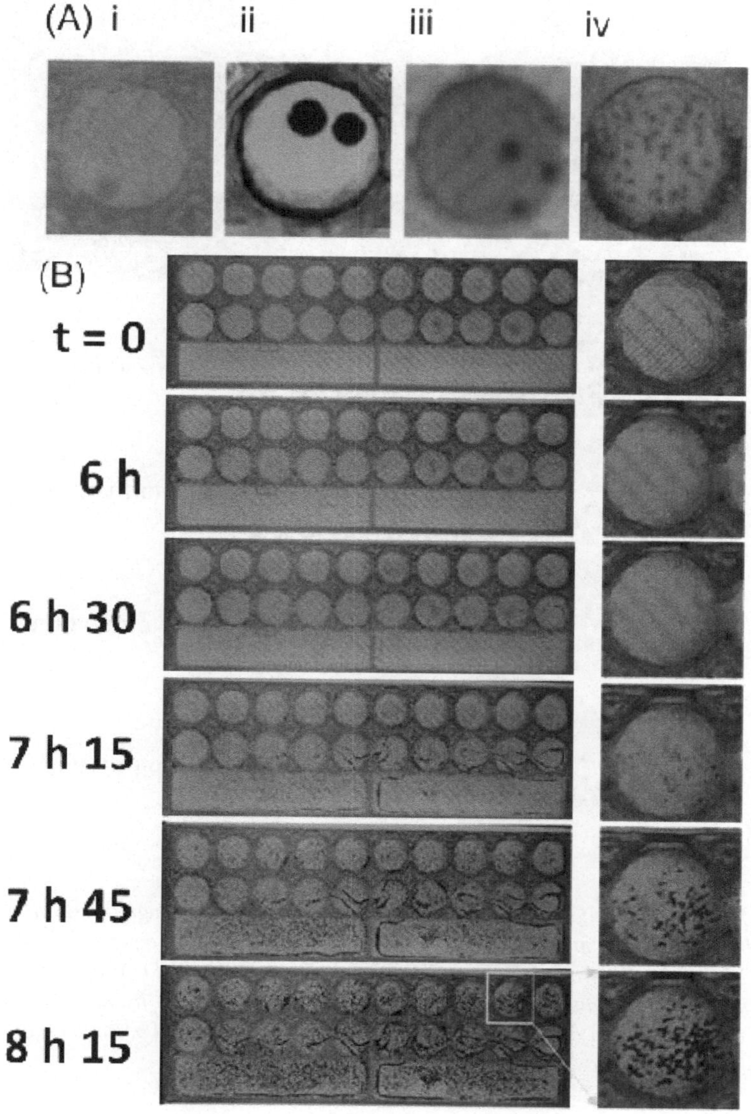

Figure 4. *Different media appearance and kinetics of growth on millifluidic solid microbiological media. (A) The appearance of colonies on 6 mm 3D-printed discs containing the following agar types: Muëller–Hinton Agar (MHA) (i), MHA with triphenyl tetrazolium chloride (TTC) dye added (ii), MacConkey (iii), and CHROMagar™ Mastitis Gram– (iv). (B) Selected timepoints from time-lapse analysis of colony growth, showing the emergence of colonies around 7–8 h on MHA with TTC.*

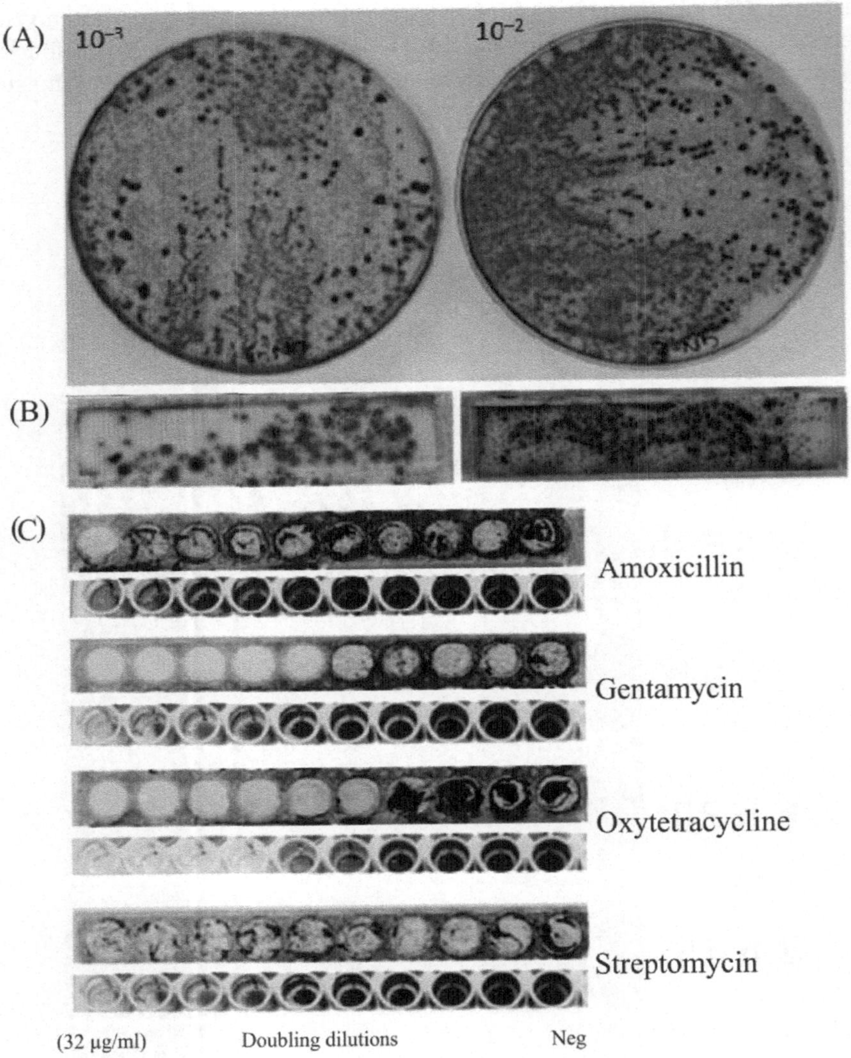

Figure 5. *Comparison of identification and determination of MIC through frame dip slide vs. Petri dish combined with broth microdilution on mastitis milk sample. (A). Bacteria (E. coli—pink colonies, Klebsiella—blue colonies) grown on CHROMagar™ Mastitis (CHROMagar™, Paris, France) at two dilutions on Petri dish, (B) Bacteria grown at two dilutions on frame, (C) Comparison of the results of MIC between frame (upper image for each antibiotic) and broth microdilution (lower image).*

Next, bacterial isolation was compared between the agar dish and miniaturized dip slides with specific agar media—Mackonkey agar (MC) and CHROMagarTM Mastitis (CHROMagar™, Paris, France). Notably, the data indicated 100% accordance of individual strain using different agar media for testing the bacterial growth. The results showed *E. coli*—pink colonies and *Klebsiella* spp.—blue colonies grown on CHROMagarTM

Mastitis (CHROMagar™, Paris, France) were the same results on the frame and agar dish (Figure 5).

As expected, although the overall pattern of antibiotic sensitivity was the same, there were some minor disagreements in MIC results between the dip slide and conventional broth dilution for some strain/antibiotic combinations. For example, Gentamycin differed by one doubling dilution of MIC (Figure 5). However, this is common for MIC determination and can result from different interpretations of intermediate growth, where it can be hard to decide which dilution constitutes inhibition vs. growth. For example, when capturing the photo, some small colonies can be seen on agar by eye, but these are not obvious on the digital photo, whereas the microplate shows intense coloration from microbial growth at this concentration, possibly resulting in the difference seen in the MIC of one doubling dilution. Despite the small surface agar (6 mm), bacterial colonies have still been separately seen, allowing the inhibition by antibiotics to be seen. This is important as after performing MIC, many resistant strains will be kept for further testing (e.g., for surveillance or reporting) research in the future such as genotype and prediction of resistant

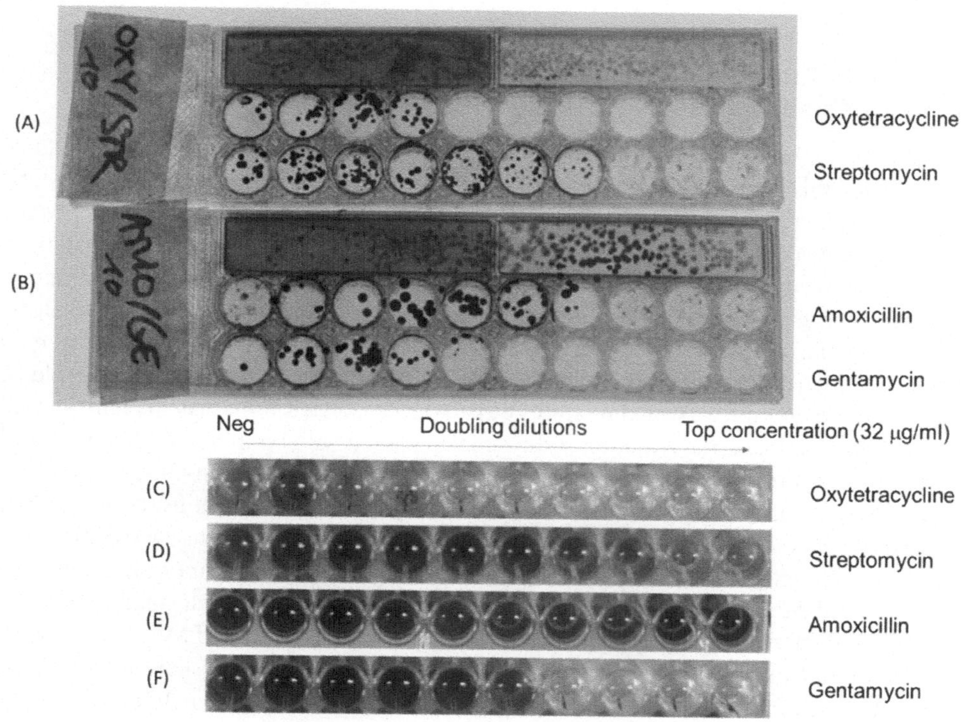

Figure 6. *Comparison of frame dip slide and broth dilution methods to determine* ***E. coli*** *isolated from the bovine sample and its MIC. (A) The results of dip-slide tests of E. coli on MC/CHROMagar™ Mastitis Gram negative and its MIC for Oxytetracycline and Streptomycin, (B) The results of E. coli on MC/CHROMagar™ Mastitis Gram negative and its MIC for Amoxicillin and Gentamycin, (C) The results of E. coli on broth microdilution of Oxytetracycline, (D) The results of E. coli on broth microdilution of Streptomycin, (E) The results of E. coli on broth microdilution of Amoxicillin, (F) The results of E. coli on broth microdilution of Gentamycin.*

trends of bacterial collection. With such purpose, an individual colony was collected and kept in the biobank. There was no difference in picking individual colonies between tube test, Petri dish or dip-slide frame (Figure 5 and Figure 6). For example, when comparing the MIC breakpoint and the *E. coli* growth on specific agar—MC and CHROMagarTM Mastitis (CHROMagar™, Paris, France), the data showed that the *E. coli* grew on MC and CHROMagarTM Mastitis (CHROMagar™, Paris, France) with typical pink colony, and its MIC to Oxytetracycline was 1 μg/mL. At this concentration, there was no bacterial growth on this concentration. However, for Streptomycin and Amoxicillin, the MIC breakpoint was over 32 μg/mL because *E. coli* kept growing at this concentration (Figure 6). The same results have been found on the broth dilution methods with the same strains and antibiotics. With the current MIC methods, strains needed to be isolated and identified, and the purity needs to be checked before performing MIC. Alongside that, an extra test tube—just bacteria without antibiotic—was needed to conduct parallel with MIC. The main purpose for doing this is to make sure the right clone of bacteria was relevant to its MIC. Our frame, however, does not need to perform this step because resistant strains still grew on the agar media, and it was easy to collect the right clone for storage in terms of further research.

3.4. Direct Testing of Mixtures of Bacteria and Their Antibiotic Resistance Profile

In reality, bacteria live in their niches or community, not existing as pure individual strains, and this can affect both response to antibiotics and the measurement of antibiotic MIC. Many samples contain mixtures of microbes; it is rare to find a sample with only one single species or strain, although clinical samples often contain one dominant pathogen that can be readily identified, which forms the basis for established conventional microbiology testing. A second reason for testing mixtures is to reduce the time-to-result by directly testing samples without initial overnight culture and colony isolation. Therefore, it remains important to explore if antibiotic susceptibility of mixtures of pathogens could be measured. To do this, we tested mixtures of pathogens isolated from mastitis milk as well as reference strains, using a multiplex dip slide compared with broth microdilution as well as testing mixtures of reference strains as control groups. With our dip-slide frame, there were always two parts: identification and MIC breakpoints. On the identification panel, the aim was to identify different bacteria present as single colonies present on specific types of identification agar. However, the MIC was not identical if a single strain was compared to two mixed strains. To prove this concept, we performed MIC for each single strain (Table 1) and then mixed them at equal cell concentrations and performed MIC again for the mixture. In the experiment of mixture strains, the MIC results always showed the highest concentration of antibiotic resistance. For example, the MIC of amoxicillin of *S. aureus* isolated from milk samples was 16 μg/mL and was 32 μg/mL for *Pseudomonas aeruginosa* isolated from milk as well (Figure 7). When combing two strains into a mixture, the results showed an MIC of 32 μg/mL for both strains together (Table 2). These findings implicated that the final MIC value was the highest concentration of pathogens in the mixture. However, to conclude resistance or susceptibility for individual species or isolates, we need to follow guidelines such as CLSI or EUCAST that specify single colonies must be picked prior to testing.

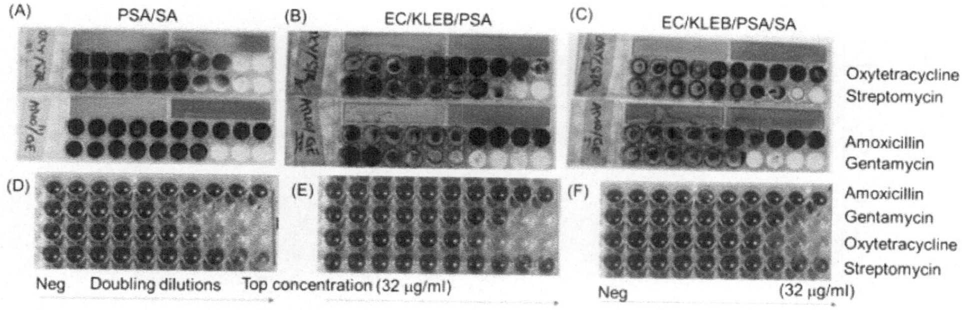

Figure 7. *Antibiotic susceptibility by agar dilution for mixtures of microbes, simulating multiple organisms present in some samples. Mixtures of the following strains were prepared and split for testing in parallel using 3D-printed multiplex millifluidic dip slides (A–C) vs. conventional microplate broth microdilution (D–F). Naming of microbial species: Pseudomonas aeruginosa (PSA), Staphylococcus aureus (SA), E. coli (EC), Klebsiella spp. (KLEB).*

Interestingly, when observing the interaction of bacterial mixtures to antibiotic resistance, the results showed that there was no significant difference between using the frame dip slide and broth dilution, even performing on reference strains or isolates from bovine mastitis samples. We also tested the frame with *Streptoccocus* species and found an agreement of bacterial growth on two methods. The small compact design makes this ideal for use for obligate anaerobes within an anaerobic cabinet and incubator. Furthermore, the dip slide can be applied for both single strain and mixture strains (Table 1 and Table 2). Table 1 summarizes the MIC results for single reference and pure isolated strains; the abbreviation of "MRF" for Mastitis Reading Farm was used to name individual isolates from milk samples from cows suffering from mastitis. In Table 2, the MIC results observed for dip slides vs. conventional methods are presented for various mixtures of different microbial strains and for directly tested mastitis milk samples.

4. DISCUSSION

We characterized a simple method to make and use a 3D-printed frame design that combines multiple miniaturized agar wells into a convenient dip slide. We asked if pathogens can be identified alongside MIC determination in one direct "dip-and-test" device. The proof-of-concept results presented here show it is feasible for bacterial species to be identified on miniaturized wells of chromogenic media alongside agar dilution for antibiotic susceptibility testing, with good agreement to conventional Petri dish methods. We also show that these dip slides can simultaneously measure mixed samples containing up to four different bacteria representing more complex clinical situations where multiple organisms are present in a single sample. The overall observed MIC was similar to the inhibitory concentration for the strain within the mixture with the highest inhibitory concentration. We also demonstrate the feasibility of using these "millifluidic" multiplex dip slides for the direct testing of milk samples from cows with mastitis. From such results, we suggest that the physician or the farmer can use these rapid direct test results to select an appropriate antibiotic, reducing the risk of treatment failure if the animal is infected with at least one

Table 1. Comparison frame dip slide vs. broth microdilution to individual strains.

Testing Samples	Methods	Antibiotic MIC (mg/mL)									
Reference Strains		**AMP**	**AMO**	**AMI**	**CEF**	**CEFO**	**OFL**	**OXT**	**STR**	**GEN**	**IMI**
Escherichia coli ATCC 25922 (EC)	Frame	4	>32	0.5	16	>32	2			1	0.125
	MIC broth	4	>32	0.5	16	>32	2			1	0.125
	E—strip test	4								1	0.125
Klebsiella pneuminae ATCC 13883 (KB)	Frame	>32	>32	1	16	>32	2			1	0.5
	MIC broth	>32	>32	1	16	>32	2			1	0.5
	E—strip test	>32								1	0.5
Pseudomonas aeruginosa ATCC 10145 (PSA)	Frame	>32	>32	2	16	>32	2			0.5	1
	MIC broth	>32	>32	2	16	>32	2			0.5	1
	E—strip test									0.5	1
Staphylococcus aureus ATCC 12600 (SA)	Frame	<0.125	>32	1	16	>32	2				
	MIC broth	<0.125	>32	1	16	>32	2				
	E—strip test	0.032									
Direct from milk mastitis samples											
19.02 MRF (EC)	Frame		>32					1	>32	4	
	MIC broth		>32					1	>32	4	
19.03 MRF (*Klebsiella* sp.—K)	Frame		>32					2	16	4	
	MIC broth		>32					2	16	4	
19.07 MRF (SA)	Frame		>32					>32	>32	4	
	MIC broth		>32					>32	>32	4	
19.08 MRF (SA)	Frame					16		8			
	MIC broth					16		8			
19.09 MRF (SA)	Frame		>32					16	>32	8	
	MIC broth		>32					16	>32	8	
19.09 MRF (*Klebsiella* sp.)	Frame	>32	>32	>32	16	4	2				
	MIC broth	>32	>32	>32	16	4	2				
19.10 MRF (EC)	Frame		>32					1	>32	2	
	MIC broth		>32					1	>32	2	
19.10 MRF (*Streptococcus uberis*)	Frame		>32			16					
	MIC broth		>32			16					
19.11 MRF (*Streptococcus agalactiae*)	Frame		8			16					
	MIC broth		8			16					

Table 1. *(Cont.)*

19.14 MRF (PSA)	Frame	>32	>32	1		
	MIC broth	>32	>32	1		
19.15 MRF (*S. aureus*—SA)	Frame	16	>32	2		
	MIC broth	16	>32	2		
19.16 MRF (EC)	Frame	16	8	2	>32	>32
	MIC broth	16	8	2	>32	>32
19.16 MRF (SA)	Frame	>32		1	8	16
	MIC broth	>32		1	8	16
19.17 MRF (*Streptococcus agalactiae*)	Frame	8	16			
	MIC broth	8	16			
19.17 MRF (SA)	Frame	>32		8	16	4
	MIC broth	>32		8	16	4
20. 20 MRF (*S. aureus*—SA)	Frame	8	32			
	MIC broth	8	32			
20. 20 MRF (PSA)	Frame	>32		>32	>32	8
	MIC broth	>32		>32	>32	8

Table 2. Measure the impact of mixtures pathogens to antibiotic resistance on reference strains and mastitis strains.

Testing Samples	Methods	Antibiotic MIC (mg/mL)									
		AMP	AMO	AMI	CEF	CEFO	OFL	OXT	STR	GEN	IMI
Reference Strains											
EC/SA/KB (Reference strains)	Frame	>32	>32	2	16	>32	2				
	MIC broth	>32	>32	2	16	>32	2				
EC/PSA/KB (Reference strains)	Frame	>32	>32	2	16	>32	2	>32	>32	>32	
	MIC broth	>32	>32	2	16	>32	2	>32	>32	>32	
EC/SA/PSA/KB (Reference strains)	Frame	>32	>32	2	16	>32	2	>32	>32	>32	
	MIC broth	>32	>32	2	16	>32	2	>32	>32	>32	
Strains Isolated from mastitis samples											
19.15 MRF (SA)/20.20 MRF (PSA)	Frame		>32					>32	>32	8	
	MIC broth		>32					>32	>32	8	
19.17 MRF (EC)/19.09 MRF (KB)/20.20 MRF (PSA)	Frame		>32					>32	>32	>32	
	MIC broth		>32					>32	>32	>32	
19.17 MRF (EC)/19.15 MRF (SA)/20.20 MRF (PSA)/19.09 MRF (K)	Frame		>32					>32	>32	>32	
	MIC broth		>32					>32	>32	>32	
19.17 MRF (EC)/19.09 MRF (K)	Frame		>32					16	>32	>32	
	MIC broth		>32					16	>32	>32	

Table 2. *(Cont.)*

19.17 MRF (*Streptococcus agalactiae*) + 20.20 MRF (*S. aureus*—SA)	Frame		8			>32				
	MIC broth		8			>32				
19.10 MRF (*Streptococcus uberis*) + 19.15 MRF (*S. aureus*—SA)	Frame		>32			>32				
	MIC broth		>32			>32				
Direct from milk mastitis samples										
19.07 MRF	Frame		>32					>32	>32	4
	MIC broth		>32					>32	>32	4
19.09 MRF	Frame	>32	>32	>32	16	4	2	16	>32	8
	MIC broth	>32	>32	>32	16	4	2	16	>32	8

resistant pathogen and avoiding the need for empirical use of broad-spectrum antibiotics without testing.

There is a trade-off in sampling simplicity achieved by miniaturization vs. absolute quantitation. A defined volume of sample can be plated on a large area with Petri dish culture, allowing precise determination of colony counts over a relatively high dynamic range (from a few colonies to several hundred on a 100 mm Petri dish). However, more space and media are required, plus skilled operation. In contrast, with the miniaturized culture area on the dip slide, it was less easy to quantify exact microbial load; however, identification and MIC determination were still possible with a much simpler procedure.

Likewise, direct testing represents a trade-off of speed to result vs. precisely measuring each organism present within the sample. Current methods to detect MIC breakpoints typically take two to three days mainly because the sample requires plating overnight, following which isolates are tested to identify organism, and finally antibiotic susceptibility tests are performed. By direct testing using a 3D-printed frame dip slide, we can reduce the turnaround time to overnight or even as little as 10 h culture. A major time-saving measure was avoiding isolating individual bacterial strains before performing the antibiotic susceptibility test; instead, the 'bulk' resistance of directly diluted samples is estimated. Whilst for some samples containing mixtures of organisms and/or pathogens, this could lead to false resistance, it is unlikely that false susceptibility will be observed, as when mixtures were tested, the MIC was always equivalent to the organism with the highest MIC, as would be expected. This reduces the risk of false susceptibility results that could lead to more significant clinical error, in which an ineffective antibiotic is selected. By combining two routine assays—the identification and determination of MIC—into one test with the frame dip slide, it should often be possible to interpret MIC through identification, as breakpoint resistance values can vary between Gram-negative vs. Gram-positive species. Digital imaging allows recording smaller colonies at earlier timepoints than conventional overnight incubation, again if the risk of missing slower-growing organisms is taken into account.

Facing relentless increases in antimicrobial resistance, many studies have investigated innovative alternative methods and technologies for antibiotic susceptibility testing and/or resistance detection including both genotype and phenotype [25]. However, most genotype-based methods are expensive (especially sequencing) and not easy to deploy widely in all laboratories. Detection methods such as LAMP and PCR can be fast but rely on the detection of specific resistance targets that can vary over time and location. For these reasons, phenotypic assays remain crucial techniques and contribute an important central role in the determination of antibiotic susceptibility for surveillance and treatment, especially the quantitation of MIC that can be used to identify breakpoint values for resistance vs. susceptibility [26]. As a result, significant advances in phenotypic assays have been continually developed such as microplate-based surface methods [27], nanoliter array [28], and multiplex fluidic chip [29]. Our 3D-printed dip slides add to this evolution, with an intermediate level of miniaturization between microfluidics and conventional large culture dishes.

Recently, significant focus has turned to the resistome and the relationship between the complex microbiome and its interaction with pathogen behavior, and whether the overall resistance of mixtures of bacteria found in the site of infection might affect the effectiveness of antibiotic treatment [13,30]. In this study, we explore the possibility of an alternative approach to the phenotype-based method that determines the MIC value directly from mixed bacterial population in a sample [3,9,31]. Whilst this may not provide the same standardized measure of antibiotic susceptibility of each individual strain present in the sample, it will be important to continue researching the link between "mixed MIC" found from direct susceptibility testing

Our assay is simpler than many microfluidic devices, and the open source 3D-printed frame design is flexible and customizable. The system is compact, allowing mobile operation especially if combined with other portable microbiology tools such as mobile incubators [32]. Three-dimensional (3D) printing has become cheaper; we used 3D printers costing £200–£750 (Creality Ender 3 and Prusa i3 MK3) with the materials cost per frame dip slide significantly under £2 to determine simultaneously Gram-negative and Gram-positive bacterial identification plus two different antibiotics. However, in spite of being simple and inexpensive, they do need to be freshly prepared due to the limited stability of antibiotic in agar medium, and challenges remain such as scaled-up mass-production or methods to make the addition of agar to the dip slides less laborious. Future innovations should focus on exploring ways to stabilize antibiotics within such devices, permitting longer-term storage, perhaps by a combination of the dip-slide format with antibiotic discs.

5. CONCLUSIONS

With this simple miniaturization of essential phenotypic microbiology testing, the initial identification and MIC profiles can be simultaneously detected in direct samples. Three-dimensional (3D)-printed open source designs make this "millifluidic microbiology" accessible across the world. Further insight into the antibiotic susceptibility of mixed organisms in some important samples may also be gained by permitting a larger-scale direct phenotypic testing of complex samples. We showed that the 3D-printed dip slide was able to be used for identification and MIC determination for a range of bacteria. The dip frame is easy to use and can be easily transport to the field, with no extensive training or introduction to the user required. Combing the dip frame with our previously developed mobile incubator [32] will increase the portability of these tests for broader application, and it may reduce the turnaround time for samples taken a long way from microbiology testing labs. Finally, direct testing with 3D-printed dip slides may contribute to an alternative and rapid way to measure mixed microbes including the microbiome alongside pathogens and measure the antibiotic resistance profiles associated with complex mixtures.

Author Contributions

Conceptualization—T.T.D. and A.D.E. Methodology and design—T.T.D. and A.D.E. Investigation—T.T.D. Writing—original draft—T.T.D. Writing—review and editing—S.B. and A.D.E. All authors have read and agreed to the published version of the manuscript.

Acknowledgments

We would like to thank Barney Jones for providing spare mastitis milk samples from the University of Reading farm. We also sincerely thank Partha Pratim Ray who gave many valuable comments for this work.

REFERENCES

1. Hoque, M.N.; Istiaq, A.; Clement, R.A.; Gibson, K.M.; Saha, O.; Islam, O.K.; Abir, R.A.; Sultana, M.; Siddiki, A.Z.; Crandall, K.A.; et al. Insights Into the Resistome of Bovine Clinical Mastitis Microbiome, a Key Factor in Disease Complication. *Front. Microbiol.* 2020, *11*, 860.
2. Falentin, H.; Rault, L.; Nicolas, A.; Bouchard, D.S.; Lassalas, J.; Lamberton, P.; Aubry, J.-M.; Marnet, P.-G.; Le Loir, Y.; Even, S. Bovine Teat Microbiome Analysis Revealed Reduced Alpha Diversity and Significant Changes in Taxonomic Profiles in Quarters with a History of Mastitis. *Front. Microbiol.* 2016, *7*, 480.
3. Baumgartner, M.; Bayer, F.; Pfrunder-Cardozo, K.R.; Buckling, A.; Hall, A.R. Resident microbial communities inhibit growth and antibiotic-resistance evolution of Escherichia coli in human gut microbiome samples. *PLoS Biol.* 2020, *18*, e3000465.
4. Perri, R.; Kolvenbach, B.A.; Corvini, P.F.X. Subsistence and complexity of antimicrobial resistance on a community-wide level. *Environ. Microbiol.* 2020, *22*, 2463–2468.
5. Murray, A.K.; Zhang, L.; Yin, X.; Zhang, T.; Buckling, A.; Snape, J.; Gaze, W.H. Novel Insights into Selection for Antibiotic Resistance in Complex Microbial Communities. *mBio* 2018, *9*, e00969-18.
6. Ribeiro, C.F.A.; Silveira, G.; Candido, E.S.; Cardoso, M.H.; Espinola Carvalho, C.M.; Franco, O.L. Effects of Antibiotic Treatment on Gut Microbiota and How to Overcome Its Negative Impacts on Human Health. *ACS Infect. Dis.* 2020, *6*, 2544–2559.
7. Zhang, L.; Forst, C.V.; Gordon, A.; Gussin, G.; Geber, A.B.; Fernandez, P.J.; Ding, T.; Lashua, L.; Wang, M.; Balmaseda, A.; et al. Characterization of antibiotic resistance and host-microbiome interactions in the human upper respiratory tract during influenza infection. *Microbiome* 2020, *8*, 39.
8. Turnidge, J.; Paterson, D.L. Setting and revising antibacterial susceptibility breakpoints. *Clin. Microbiol. Rev.* 2007, *20*, 391–408.
9. Mouton, J.W.; Muller, A.E.; Canton, R.; Giske, C.G.; Kahlmeter, G.; Turnidge, J. MIC-based dose adjustment: Facts and fables. *J. Antimicrob. Chemother.* 2018, *73*, 564–568.
10. Nielsen, E.I.; Cars, O.; Friberg, L.E. Pharmacokinetic/pharmacodynamic (PK/PD) indices of antibiotics predicted by a semimechanistic PKPD model: A step toward model-based dose optimization. *Antimicrob. Agents Chemother.* 2011, *55*, 4619–4630.
11. Li, J.; Xie, S.; Ahmed, S.; Wang, F.; Gu, Y.; Zhang, C.; Chai, X.; Wu, Y.; Cai, J.; Cheng, G. Antimicrobial Activity and Resistance: Influencing Factors. *Front. Pharmacol.* 2017, *8*, 364.
12. de Boer, M.; Heuer, C.; Hussein, H.; McDougall, S. Minimum inhibitory concentrations of selected antimicrobials against Escherichia coli and Trueperella pyogenes of bovine uterine origin. *J. Dairy Sci.* 2015, *98*, 4427–4438.
13. Klumper, U.; Recker, M.; Zhang, L.; Yin, X.; Zhang, T.; Buckling, A.; Gaze, W.H. Selection for antimicrobial resistance is reduced when embedded in a natural microbial community. *ISME J.* 2019, *13*, 2927–2937.
14. Benkova, M.; Soukup, O.; Marek, J. Antimicrobial susceptibility testing: Currently used methods and devices and the near future in clinical practice. *J. Appl. Microbiol.* 2020, *129*, 806–822.
15. Jenkins, S.G.; Schuetz, A.N. Current concepts in laboratory testing to guide antimicrobial therapy. *Mayo Clin. Proc.* 2012, *87*, 290–308.
16. Lemos, T.C.; Cogo, L.L.; Maestri, A.C.; Hadad, M.; Nogueira, K.D.S. Is it possible to perform bacterial identification and antimicrobial susceptibility testing with a positive blood culture bottle for quick diagnosis of bloodstream infections? *Rev. Soc. Bras. Med. Trop.* 2018, *51*, 215–218.
17. Pereira, D.C.; Goldani, L.Z. Integrating Bacterial Identification and Susceptibility Testing: A Simple and Rapid Approach to Reduce the Turnaround Time in the Management of Blood Cultures. *BioMed Res. Int.* 2019, *2019*, 8041746.

18. Choi, J.; Jung, Y.G.; Kim, J.; Kim, S.; Jung, Y.; Na, H.; Kwon, S. Rapid antibiotic susceptibility testing by tracking single cell growth in a microfluidic agarose channel system. *Lab Chip* 2013, *13*, 280–287.

19. Choi, J.; Jeong, H.Y.; Lee, G.Y.; Han, S.; Han, S.; Jin, B.; Lim, T.; Kim, S.; Kim, D.Y.; Kim, H.C.; et al. Direct, rapid antimicrobial susceptibility test from positive blood cultures based on microscopic imaging analysis. *Sci. Rep.* 2017, *7*, 1148.

20. Toosky, M.N.; Grunwald, J.T.; Pala, D.; Shen, B.; Zhao, W.; D'Agostini, C.; Coghe, F.; Angioni, G.; Motolese, G.; Abram, T.; et al. A rapid, point-of-care antibiotic susceptibility test for urinary tract infections. *J. Med. Microbiol.* 2020, *69*, 52.

21. Mo, M.; Yang, Y.; Zhang, F.; Jing, W.; Iriya, R.; Popovich, J.; Wang, S.; Grys, T.; Haydel, S.E.; Tao, N. Rapid Antimicrobial Susceptibility Testing of Patient Urine Samples Using Large Volume Free-Solution Light Scattering Microscopy. *Anal. Chem.* 2019, *91*, 10164–10171.

22. Bär, J.; Boumasmoud, M.; Kouyos, R.D.; Zinkernagel, A.S.; Vulin, C. Efficient microbial colony growth dynamics quantification with ColTapp, an automated image analysis application. *Sci. Rep.* 2020, *10*, 16084.

23. Åkerlund, A.; Jonasson, E.; Matuschek, E.; Serrander, L.; Sundqvist, M.; Kahlmeter, G. EUCAST rapid antimicrobial susceptibility testing (RAST) in blood cultures: Validation in 55 European laboratories. *J. Antimicrob. Chemother.* 2020, *75*, 3230–3238.

24. Diep, T.T.; Ray, P.P.; Edwards, A.D. Methods for rapid prototyping novel labware: Using CAD and desktop 3D printing in the microbiology laboratory. *Lett. Appl. Microbiol.* 2021, *74*, 247–257.

25. Needs, S.H.; Donmez, S.I.; Bull, S.P.; McQuaid, C.; Osborn, H.M.I.; Edwards, A.D. Challenges in Microfluidic and Point-of-Care Phenotypic Antimicrobial Resistance Tests. *Front. Mech. Eng.* 2020, *6*, 73.

26. Michael, A.; Kelman, T.; Pitesky, M. Overview of Quantitative Methodologies to Understand Antimicrobial Resistance via Minimum Inhibitory Concentration. *Animals* 2020, *10*, 1405.

27. Flentie, K.; Spears, B.R.; Chen, F.; Purmort, N.B.; DaPonte, K.; Viveiros, E.; Phelan, N.; Krebill, C.; Flyer, A.N.; Hooper, D.C.; et al. Microplate-based surface area assay for rapid phenotypic antibiotic susceptibility testing. *Sci. Rep.* 2019, *9*, 237.

28. Avesar, J.; Rosenfeld, D.; Truman-Rosentsvit, M.; Ben-Arye, T.; Geffen, Y.; Bercovici, M.; Levenberg, S. Rapid phenotypic antimicrobial susceptibility testing using nanoliter arrays. *Proc. Natl. Acad. Sci. USA* 2017, *114*, E5787–E5795.

29. Wistrand-Yuen, P.; Malmberg, C.; Fatsis-Kavalopoulos, N.; Lubke, M.; Tangden, T.; Kreuger, J. A Multiplex Fluidic Chip for Rapid Phenotypic Antibiotic Susceptibility Testing. *mBio* 2020, *11*, e03109-19.

30. Lee, K.; Kim, D.W.; Lee, D.H.; Kim, Y.S.; Bu, J.H.; Cha, J.H.; Thawng, C.N.; Hwang, E.-M.; Seong, H.J.; Sul, W.J.; et al. Mobile resistome of human gut and pathogen drives anthropogenic bloom of antibiotic resistance. *Microbiome* 2020, *8*, 2.

31. Letten, A.D.; Baumgartner, M.; Pfrunder-Cardozo, K.R.; Levine, J.M.; Hall, A.R. Human-associated microbiota suppress invading bacteria even under disruption by antibiotics. *ISME J.* 2021, *15*, 2809–2812.

32. Diep, T.T.; Bizley, S.; Ray, P.P.; Edwards, A.D. MicroMI: A portable microbiological mobile incubator that uses inexpensive lithium power banks for field microbiology. *HardwareX* 2021, *10*, e00242.

11 CHAPTER

3D-Printed Microneedles for Point-of-Care Biosensing Applications

Misagh Rezapour Sarabi [1,†], **Sattar Akbari Nakhjavani** [1,2,†] **and Savas Tasoglu** [1,2,3,4,5,6]

[1] *Mechanical Engineering Department, School of Engineering, Koç University, Istanbul 34450, Turkey*

[2] *Koç University Translational Medicine Research Center (KUTTAM), Koç University, Istanbul 34450, Turkey*

[3] *Physical Intelligence Department, Max Planck Institute for Intelligent Systems, 70569 Stuttgart, Germany*

[4] *Koç University Arçelik Research Center for Creative Industries (KUAR), Koç University, Istanbul 34450, Turkey*

[5] *Boğaziçi Institute of Biomedical Engineering, Boğaziçi University, I stanbul 34684, Turkey*

[6] *Koç University İş Bank Artificial Intelligence Lab (KUIS AI Lab), Koç University, Sariyer, Istanbul 34450, Turkey*

ABSTRACT

Microneedles (MNs) are an emerging technology for user-friendly and minimally invasive injection, offering less pain and lower tissue damage in comparison to conventional needles. With their ability to extract body fluids, MNs are among the convenient candidates for developing biosensing setups, where target molecules/biomarkers are detected by the biosensor using the sample collected with the MNs. Herein, we discuss the 3D printing of microneedle arrays (MNAs) toward enabling point-of-care (POC) biosensing applications.

Keywords: *microneedles; biosensing; 3D printing; point-of-care*

1. 3D PRINTING FOR FABRICATION OF MICRONEEDLE ARRAYS (MNAS)

The organization of needles in micro- and nano- scales on an array or patch, introduced the emerging term of microneedle arrays (MNAs), are prominent minimally invasive injection setups for medical and biomedical applications ranging from delivery of drugs and agents to body to extracting body fluids and bio signal acquisition, to name a few [1]. Minimizing insertion pain in comparison with conventional needle technologies, MNAs not only offer a solution for needle phobia, the anxiety linked to the injection or utilizing needles, but also pave the way for the fabrication of more convenient devices that decrease tissue damage [2]. On the other hand, while conventional needles are usually made of stainless steel, MNAs can be fabricated with a variety of alternatives such as polymers and inorganic materials in various geometries, with different heights, diameters, and needle shapes [3]. Also, based on the end application of MNAs, they can be prepared in solid, coated, hollow, or biodegradable/dissolvable forms.

Over the past decade, the fabrication method of MNAs has been advancing from molding, micro machining, and photolithography to the emerging method of 3D printing, where the structure is produced in a layer-by-layer manner using a computer-aided design (CAD) input [4]. 3D printing can be carried out using various technologies as its working mechanism, including, but not limited, to stereolithography (SLA), digital light processing (DLP), two- or multi- photon polymerization (TPP or MPP), fused deposition modeling (FDM), selective laser sintering (SLS), and selective laser melting (SLM). While conventional fabrication methods require cost- and time- consuming procedures of high labor, cleanroom facilities, and/or unusable material waste, 3D printing is an additive manufacturing method that can be used directly for fabrication of MNAs, offering ease of access, being capable of rapid prototyping, and enabling cell-encapsulated structures with integration with bioinks. For example, fabrication with lithography protocol in a cleanroom not only needs expensive facilities that are usually not highly accessible but also requires several steps such as wafer washing and preparation, applying a photoresist coat, soft baking, exposing, post-exposure baking, and development. On the other hand, lithography is not a convenient approach for rapid prototyping; i.e., all the steps should be retaken if only one feature of the product is changed. 3D printing, with comparatively high accessibility and a higher level of reproducibility, offers a solution to these challenges [5]. Furthermore, since there is no requisite for expensive consumables such as wafers and photoresists, 3D printing reinforces its place for a cost- and time-efficient fabrication approach for MNA-based biosensors. 3D printing methods have been used for fabrication of immense pieces such as automotive parts and turbines, as well as minute delicate objects such as vessel stents and MNAs. For the case of such delicate items, high-precision 3D printing is required to be able to print the tiny features in the desired form. Precision is usually enabled in 3D printers by smaller layer heights, optimized printing orientation, smaller spot diameter, optimized laser power, and more accurate nozzles. 3D-printed MNAs hence are emerging for fabrication of advanced healthcare setups [6], enabling personalized treatment options and point-of-care (POC) diagnosis, i.e., testing at or near the clinical point of patient care Figure 1.

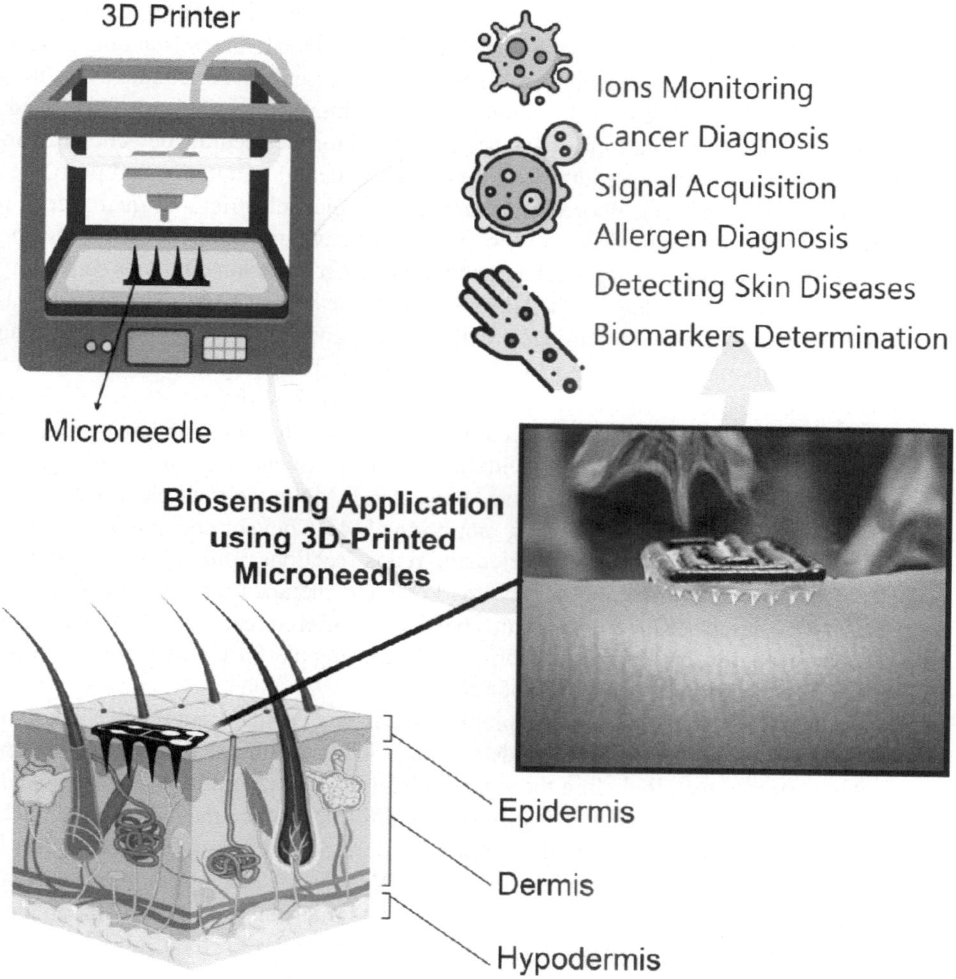

Figure 1. *Integration of microneedle arrays (MNAs) with point-of-care (POC) biosensor technologies will provide new insights in biosensing applications. Taking advantage of emerging manufacturing method of 3D printing, MNA-integrated biosensors can be fabricated with customizable properties rapidly, resulting in biomedical applications such as monitoring different particles (for instance, biomarkers, and ions) and diagnosis of diseases (for instance, cancers, and allergies). The figure in the box was adapted from the journal cover photo of ref. [7] with permission from John Wiley & Sons (2019).*

2. MNA-BASED POINT-OF-CARE (POC) BIOSENSING

Advancements in nanotechnology and biosensing sciences have facilitated utilizing POC biosensing platforms in various fields, including medicine, industry, and defense [8,9]. A biosensor is composed of two fundamental parts: a bio-recognition element and a signal transducer [10]. Applying specific biomolecules in the presence of noble nanostructures

(NSs) and proper signal transducers has caused POC biosensing platforms to demonstrate enhanced performance in terms of limit of detection (LOD). Due to the outstanding properties that NSs offer, such as increasing active surface area and enhancing conductivity, they have successfully been utilized for development of POC biosensing platforms [11]. In addition, different biosensing platforms have been designed for the accurate detection of various diseases, cancer biomarkers, drug residues, environmental pollutants, and pesticides [12]. Among electrochemical, fluorescent, luminescent, piezoelectric, and quartz crystal microbalance transducers, which have been applied for development of POC biosensing platforms, the electrochemical approach has attracted higher attention due to its sensitivity and simplicity. Taking advantage of the largest organ of the human body, skin, the integration of MNAs with biosensor technologies will provide new insights into POC biosensing applications. Additionally, modifying MNAs with specific bio-recognition elements such as aptamer can dramatically elevate the performance of the MNAs-based sensing platform. For example, hydrogel MNs were modified with a fluorescent-labeled aptamer probe for detection of glucose in a model animal, with minimum invasiveness, great specificity, and sensitivity [13]. Furthermore, another study presented a MN-supported electrochemical, aptamer-based sensor for continuous drug monitoring [14]. Conventional methods of MN sensing could only sense molecules experiencing redox reactions, limiting their capability to detect non-redox-active metabolites, biomarkers, and therapeutics. The proposed aptamer-based MN-based sensor overcame this barrier and offered real-time electrochemical detection. To reveal this capability, the fabricated biosensing platform was examined every 4.5 s, while flowing buffer at a rate of 0.66 mL/min.

Generally, MNA-based biosensing platforms utilize two main approaches for biomarker detection, i.e., direct and indirect detection. In brief, during the direct detection approach, whole procedures, including the sample obtaining and recognition of corresponding biomarkers and readout, occur in an all-in-one platform, while in the indirect approach, MNs are used for sampling and the detection procedure occurs in another instrument. It is worth mentioning that fabrication of the direct MNA-based biosensors needs various technologies such as microfluidics, nanotechnology and microfabrication which could be considered as a disadvantage regarding their application. However, their advantages exceed such limitations.

One of the main purposes for developing MNA-based biosensors is providing a pain-free, safe, and convenient sampling procedure, followed by the detection of desired biomarkers with the minimum amount of invasion. In this regard, one MNA-based biosensor was designed using an ultra-sensitive fluorescent optic fiber for transdermal glucose monitoring in which the glucose binding protein was immobilized on Nickel-Nitriloacetic acid agarose beads [15]. The obtained results revealed that the developed biosensing system would be a promising device for noninvasive glucose monitoring. Furthermore, continuous glucose monitoring (CGM) plays a promising role in the management of diabetic patients by providing real-time data of body glucose concentrations. Considering the exceptional correlation of blood glucose and ISF glucose concentrations, a MNA was microfabricated using DLP 3D printing technology for CGM applications [16]. Followed by microfabrication, electrochemical sensing electrodes including work, reference, and counter were produced by the electroplating of MNAs (Figure 2A). The biorecognition element, the glucose oxidase (GOD) enzyme, was immobilized on the desired surface of the working electrodes. Upon successful assembly of the 3D-printed biosensing platform, the setup was employed

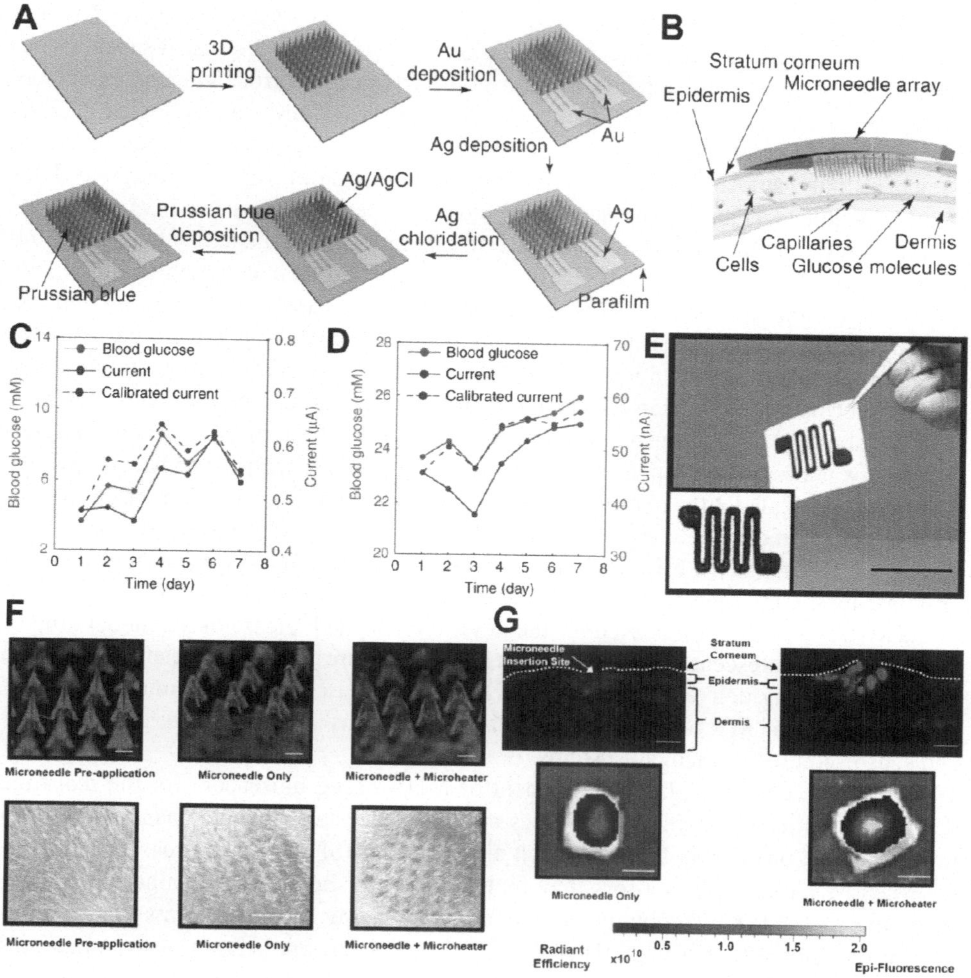

Figure 2. *Examples of MNA-based biosensors. (A) Fabrication steps of the MN biosensing device for glucose monitoring. (B) Schematic representation of the device insertion into the skin. (C) The device was used for in vivo monitoring of subcutaneous glucose in a normal mouse and (D) in a diabetic mouse. In both diagrams, the mouse was monitored for seven days. The blue lines indicate the MN biosensing device, and the red lines indicate a commercial blood glucose meter. (E) A 3D-printed microheater sensor-integrated MN patch was developed for pain management application. Optical image of the microheater printed on a paper (Scale bar: 35 mm). (F) The setup was used for in vitro controlled release of the model drug of near-infrared fluorescence dye Cy5 in rat skin (Scale bars, top row: 500 μm, bottom row: 5 mm). (G) Fluorescence images representative of the rat skin tissue slides, indicating wider and deeper diffusion of Cy5 (upper row) and more initial drug depot in skin (lower row), in microheater integrated group in comparison with the MN-only group (Scale bars, top row: 100 μm, bottom row: 5 mm). Subfigures (A–D) were adapted with permission of Springer Nature (2021) from ref. [16], and subfigures (E–G) were adapted with permission of John Wiley & Sons (2019) from ref. [7].*

to measure the subcutaneous glucose of the dermis layer of a mouse skin (Figure 2B). The comparison of the obtained results of this platform for *in vivo* monitoring of subcutaneous glucose in a normal mouse and in a diabetic mouse with a commercial glucometer demonstrated the accuracy of the device, facilitating future *in vivo* CGM studies (Figure 2C,D).

Also, a MN-based biosensor was designed for the continuous monitoring of ketone bodies, utilizing an electrochemical technique for detection of β-hydroxybutyrate (HB), as an indicator of ketones. The developed biosensor demonstrated sensitive selectivity for the detection of target molecules in the artificial interstitial fluid (ISF). It also presented the ability for the detection of HB besides glucose and lactate, which could be considered as a simultaneous MNA-based biosensor with the capability of sensing multiple biomarkers of diabetic ketosis [17]. Additionally, transdermal drug delivery has attracted the attention of physicians due to its promising advantages compared to oral administration, such as minimum side effects, gastrointestinal bleeding, and problems regarding liver metabolism. As a promising application, a 3D-printed microheater sensor-integrated MNA was designed for pain management in the patients with lower back pains [7]. A drug-encapsulated MN platform was integrated with a 3D-printed microheater for precise control of drug delivery in the injured area. A printing ink solution of multiwalled carbon nanotubes (MWCNTs) and polydimethylsiloxane (PDMS) was employed to microfabricate the microheater Figure 2E, and then it was capitalized on the back surface of the MNAs. The setup was used for the in vitro controlled release of the model drug of near-infrared fluorescence dye Cy5 in mouse skin (Figure 2F,G). The results of in vitro investigation drug delivery application of the developed platform facilitate a new way for application of MNA-integrated biosensing platforms in transdermal drug delivery and sensor-controlled medical systems.

During recent years, the development of MNA-based biosensors for the real-time detection/monitoring of various diseases such as cancer and diabetes has widely been investigated. Considering biofluids such as blood and ISF as rich sources of biomarkers, such MNA-based biosensors may be employed for both sample gathering and the read-out process [18]. To obtain accurate and convincing results in terms of determining various biomarkers using MNA-based biosensors, the application of nanomaterials such as gold and carbon nanotubes in their design and development is inevitable. The nano-coating of MNA not only improves the detection performance of biosensing platforms but also facilitates additional functionalization of MNA with bio-transducers such as antibodies and aptamers. It should be noted that the design of MNA patch in terms of size, material, geometry, and mechanical properties should also be considered in developing the MNA-based biosensors. Experimentally reported MNs with applications in sampling or extraction of body fluids such as ISF, blood, and glucose detection delineated the heights of MN to be in the range of 400 to 800 μm [2]. Also, commercial MN vaccine delivery platforms mostly have a height of 600 μm [19]. In addition, polymers with different degradation rates, swelling characteristics, and behaviors in response to biological and physical stimuli can be utilized to manufacture polymeric MNs with different mechanical properties and performance [20]. Conducting polymers can also enable the fabrication of more functional MNA-based biosensors by offering conductivity and modification capability.

A highly porous gold MNA-based biosensor was fabricated using electrochemical self-templating approach [21]. The MNA was microfabricated on a polycarbonate scaffold.

After electrochemically cleaning the MNA, the high-porous gold was electrodeposited on the surface of working electrodes. After stabilizing, washing, and activating the electrodes, the as-prepared highly porous MNA was modified with the acting enzyme glucose dehydrogenase (GDH). The characterization results revealed that fabricated MNA had a larger active surface area compared to bare MNA (100 times larger). As a concept of application, the designed system was applied for glucose monitoring in the artificial ISF in a skin model consisting of chitosan/agarose hydrogels. The performance of the designed MNA-based biosensor in terms of sensitivity, stability, and selectivity was evaluated satisfactory.

Skin ISF is a biofluid containing information-rich biomarkers which can be utilized for several purposes such as diagnosis and prognosis of diseases, analysis of tumors, and drug inspection. Hence, 3D printing MNAs for ISF-collection and -analysis biosensors can enable a great opportunity for developing POC systems. An attempt in this regard took advantage of photopolymerization 3D printing technology for the fabrication of MNAs with open groove channels, which were aimed to be used for extraction of body liquids [22]. The groove structure was designed for enabling fluid extraction with the help of capillary forces. The 3D-printed MNAs were tested on porcine skin and several parts of chicken tissues for their skin piercing capability. Afterwards, for demonstrating MNAs' ability for fluid extraction and biomarker analysis, an in vitro hydrogel model was used. This model mimicked skin and contained 1% agarose hydrogel premixed with several values of glucose ranging from 0.1 to 2 wt%. Commercial glucose strips were attached to the backside of the MNAs, and the colorimetric response of the strips were observed for the amount of the glucose that MNAs were capable of extraction within a period of 60 s. The results demonstrated convenient detection sensitivity and exhibited a setup for rapid detection upon extraction.

Thanks to the MNAs, various studies have initiated to evaluate and investigate the ISF-originated biomarkers such as nucleic acids, proteins, and small molecules. Compared to the other biofluids such as blood, saliva, or urine, these biomarkers could be easily obtained. Organophosphorus nerve agents (OPNAs) are one of deadliest chemical ever produced for warfare application. Skin is one of the sites that could extensively be exposed to such chemicals. In this regard, an electrochemical MNA-based biosensor was designed for rapid, sensitive, and specific detection of OPNAs from ISF [23]. Having used the transdermal drug delivery applications of MNAs, a hollow MNA with a pyramidal shape was fabricated using acrylate-based polymer as main material. Following the preparation of the MNAs, they were treated and covered with carbon paste to enhance their performance toward detecting OPNAs. It worth mentioning that 2 out of every 3 MNs were modified with carbon paste, while the third one in each array was modified with Ag/AgCl ink (to serve as the refence electrode). After the successful assembly of MNAs and their connectors, the hollow MNA-based biosensor was applied for detection of target analyte in a detection range of 20 to 180 μM. Furthermore, the applicability of the fabricated MNA-based biosensor was investigated by its application into a mouse skin that was previously exposed to OPNAs. The developed platform presented a great performance for the *in vivo* detection of the target analyte (with an effective recovery). The developed platform was suggested as a potential approach for the further development of on-body measurement.

For the purpose of sequence-specific sampling, isolation, and the detection of nucleic acid biomarkers from skin ISF, MNs coated with an alginate–peptide nucleic acid hybrid material were developed [24]. The device was demonstrated for 15 min, pressing on the skin for sufficient sampling. MNAs were functionalized with bespoke peptide nucleic acid (PNA) probes, which were covalently bound to an alginate hydrogel matrix via a photo-cleavable linker (PCL). The alginate-azide polymer was prepared as by ethyl (dimethyl-amino propyl) carbodiimide/N-hydroxysuccinimide (EDC/NHS) mediated peptide coupling between low-viscosity alginate and 11-azido-3,6,9-trioxyundecan-1-amine. All three hydrogel samples (unmodified alginate, alginate azide, and alginate-PNA) were characterized. The nuclear magnetic resonance (NMR) spectrum of alginate-azide was analyzed to characterize the amount of azide functionalization. The results demonstrated that isolating miRNA biomarkers efficiency is higher for alginate-PNA, alginate azide, and unmodified alginate, respectively.

Surface-enhanced Raman spectroscopy (SERS) is a robust method that has been applied for sensitive detection of various molecules based on Raman scattering of target analyte on a plasmonic surface. Considering the unique molecular specificity of Raman spectroscopy, so called biomolecule fingerprinting, SERS based biosensors could be employed for development of label-free POC MNA-based biosensors. In this regard, a biocompatible MNA-based sensor was fabricated for accurate detection of methylene blue (MB) as model analyte in mouse skin [25]. The histologic results of mouse skin presented insignificant immune response regarding insertion demonstrating the minimally invasiveness of MNA. The performance of the fabricated system for the detection of MB was calculated for 1 nM via a single insertion/withdraw, suggesting its potential application for developing label-free intradermal MNA-based biosensors. Advances in producing SERS substrates have attracted biomedical engineers to develop enhanced MNA-based biosensors for detection of various analytes such as pH, glucose, and alcohol using ISF as a pool of biomarkers. The Norland optical adhesive 65 polymer (NOA 65) was used for the fabrication of MNA using a commercial PDMS mold [26]. Due to the presence of mercaptoester groups and the transparency of the material on the other hand, gold nanorods could be functionalize on the surface of MNAs and hence, the collecting of SERS signals could be facilitated. The results showed that the fabricated MNA-based sensor could effectively quantify pH over a range of 5 to 9. Interestingly, and as proof of application, the fabricated system successfully detected pH levels in the artificial agar gel skin phantom and human skin in situ with a good robustness and proper stability even after one-month incubation in phosphate-buffered saline.

3. OUTLOOK AND CHALLENGES

Biocompatibility is one of the main prerequisites for the devices that are in close contact with human body, which means that such devices should not cause any adverse effect or attract responsive behavior from the immune system. This highlights that MNA-based biosensors that are proposed for translation and commercialization purposes considering their benefits should satisfy biocompatibility requirements as well. While using biocompatible materials in the fabrication of MNA-based biosensors is an obvious solution to achieve this goal, immunomodulatory coatings can be used as an alternative approach when there is a limitation for choosing biocompatible materials. Aimed to decrease triggering of the immune system of the human body and promote integration with the surrounding organs, immunomodulatory coatings can be applied on the needles in the same way that a drug is coated. On the other hand, in the case that biodegradable materials are used in the fabrication procedure, residues of the degradation should be harmless to human body as well. In this regard, polycaprolactone (PCL) and polylactic acid (PLA) are among the examples that have been studied to be not only biocompatible but also not leaving behind any irritating degradation products [6]. Another main concern regarding the translation to clinical practices is that most of the available reports of MNA-based biosensing procedures have been performed within phantoms that are mimicking human tissues or on animals. Utilizing organoids can be a helpful approach here to increase the pace of translation since organoids offer more realistic comprehension of the human body in comparison with the conventional methods.

Analysis of chemical data in biosensor systems usually requires several chemometrics methods dealing with mathematical and statistical models followed by optimization procedures, which can be time-consuming in systems that are aimed for smart rapid detection and/or sensing various parameters at once. One of the approaches to enable more advanced POC robust biosensor setups in the future can be enabled by the incorporation of various detection/sensing frameworks along with several analysis layouts. In this regard, the integration of artificial intelligence (AI) techniques, for instance, machine learning (ML) and deep learning (DL), can be a useful approach for big data analysis and automation of sensing. On the other hand, AI techniques can contribute to fabricating steps as well; for example, features of 3D-printed MNAs were reported to be optimized and predicted using ML and DL [27]. On the other hand, the information from different biomarkers, aptamers, and sensing materials can be utilized to create libraries for ML training models, which can pave the way for predictive systems, resulting in development of more advanced healthcare setups [28]. Furthermore, 3D printing is an emerging scalable fabrication method and hence can prepare the infrastructure for manufacturing and commercializing setups integrating MNAs with biosensors.

Advances in the development of non-invasive MNA-based biosensing platforms for monitoring of various biomarkers is an interesting area in the field of biosensors. Considering the great potential of electrochemical and SERS techniques, the development and commercialization of MNA-based POC biosensors for the multianalyte detection of biomarkers would be an asset in the management and monitoring of different diseases. Interestingly, such biosensing devices would be proper candidates for developing wearable patches for the online monitoring of the desired biomarker with the minimum amount of invasiveness in the patients. For instance, the design and fabrication of MNA-based POC biosensing platforms (wearable patches) would be suggested for real-time monitoring of glucose in diabetic patients or sensing of lactic acid in athletes. Although many advances regarding the development and application of MNA-based biosensors for the detection of various biomarkers have been achieved, still, some challenges in terms of real-time monitoring of biomarkers remain unsolved. In addition, in MNAs with nonconductive materials, effectively modifying the 3D-printed MNA-based biosensors with nanomaterials would be a concern. In this regard, MNs fabricated with metals can be considered as the initial solution for conductivity, however, fabrication of metals usually requires more time- and cost-consuming procedures. On the other hand, 3D printers are usually based on inks made with material family of polymers. Progress in the science of polymers is advancing to the development of novel conducting polymers such as polyaniline (PANI), polypyrrole (PPy), polythiophene (PTh), and polyacetylene [29]. Hence, conducting polymers can introduce a new approach for fabrication of more functional MNA-based biosensors.

The other limitation that may affect the translation of MNA-based biosensing platforms from benches to beds would be their high cost of assembly processes, which need advanced laboratories. It is believed that further investigations should be performed to address the shortcomings associated with the application of 3D-printed MNA-based biosensing platforms. However, by the advancement of technology, particularly the development of innovative printing approaches and the application of affordable materials for microfabrication of MNAs, these biosensing platforms will smoothly find their place in the clinical trials and will be commercially available in the near future. Focusing on the number of clinical investigations of MNA-based biosensing platforms reveals that there is a demanding interest toward their development as new diagnostic techniques for detection of various biomarkers.

4. CONCLUSIONS

The advances in technologies will revolutionize the field of POC biosensing platforms in terms of convenience, safety, non-invasiveness, biocompatibility, cost effectiveness, and ease of fabrication. In other words, by entering into the smart devices era, a new group of MNA-based biosensing platforms will emerge. Such health biosensors are expected to provide a real-time analysis reflecting the condition of the applied environment. It is believed that by integrating smartphone technologies into the MNA-based POC biosensors, a new generation of these devices will act as a game-changer in the field of detection. Initially, the monitoring and evaluation of glucose and its derivatives was the main purpose of developing MNA-based POC biosensors. Next, by understanding their advantages, their application in detection of other types of biomarkers has been initiated. An evaluation on the growing number of published papers about MNA-based POC biosensors demonstrates the increasing interest toward designing these devices, which are at the early stages of their road and will find their main role in the near future.

Author Contributions

Conceptualization, M.R.S., S.A.N. and S.T.; writing—original draft preparation, M.R.S. and S.A.N.; writing—review and editing, S.T.; visualization, M.R.S.; supervision, S.T.; project administration, S.T.; funding acquisition, S.T. All authors have read and agreed to the published version of the manuscript.

REFERENCES

1. Zhang, W.; Zhang, W.; Li, C.; Zhang, J.; Qin, L.; Lai, Y. Recent Advances of Microneedles and Their Application in Disease Treatment. *Int. J. Mol. Sci.* 2022, *23*, 2401.

2. Sarabi, M.R.; Ahmadpour, A.; Yetisen, A.K.; Tasoglu, S. Finger-Actuated Microneedle Array for Sampling Body Fluids. *Appl. Sci.* 2021, *11*, 5329.

3. Dabbagh, S.R.; Sarabi, M.R.; Rahbarghazi, R.; Sokullu, E.; Yetisen, A.K.; Tasoglu, S. 3D-printed microneedles in biomedical applications. *iScience* 2021, *24*, 102012.

4. Rahmani Dabbagh, S.; Rezapour Sarabi, M.; Birtek, M.T.; Mustafaoglu, N.; Zhang, Y.S.; Tasoglu, S. 3D bioprinted organ-on-chips. *Aggregate* 2022, e197.

5. Yigci, D.; Rezapour Sarabi, M.; Ustun, M.; Atceken, N.; Sokullu, E.; Onder, T.B.; Tasoglu, S. 3D Bioprinted Glioma Models. *Prog. Biomed. Eng.* 2022.

6. Sarabi, M.R.; Bediz, B.; Falo, L.D.; Korkmaz, E.; Tasoglu, S. 3D printing of microneedle arrays: Challenges towards clinical translation. *J. 3D Print. Med.* 2021, *5*, 65–72.

7. Yin, M.; Xiao, L.; Liu, Q.; Kwon, S.-Y.; Zhang, Y.; Sharma, P.R.; Jin, L.; Li, X.; Xu, B. 3D Printed Microheater Sensor-Integrated, Drug-Encapsulated Microneedle Patch System for Pain Management. *Adv. Healthc. Mater.* 2019, *8*, 1901170.

8. Sarabi, M.R.; Yetisen, A.K.; Tasoglu, S. Magnetic levitation for space exploration. *Trends Biotechnol.* 2022.

9. Tasoglu, S. Toilet-based continuous health monitoring using urine. *Nat. Rev. Urol.* 2022, *19*, 219–230.

10. Ronkainen, N.J.; Halsall, H.B.; Heineman, W.R. Electrochemical biosensors. *Chem Soc Rev* 2010, *39*, 1747–1763.

11. Akbari Nakhjavani, S.; Afsharan, H.; Khalilzadeh, B.; Ghahremani, M.H.; Carrara, S.; Omidi, Y. Gold and silver bio/nano-hybrids-based electrochemical immunosensor for ultrasensitive detection of carcinoembryonic antigen. *Biosens. Bioelectron.* 2019, *141*, 111439.

12. Nakhjavani, S.A.; Khalilzadeh, B.; Pakchin, P.S.; Saber, R.; Ghahremani, M.H.; Omidi, Y. A highly sensitive and reliable detection of CA15-3 in patient plasma with electrochemical biosensor labeled with magnetic beads. *Biosens. Bioelectron.* 2018, *122*, 8–15.

13. Zheng, H.; GhavamiNejad, A.; GhavamiNejad, P.; Samarikhalaj, M.; Giacca, A.; Poudineh, M. A Hydrogel Microneedle-assisted Assay Integrating Aptamer Probes and Fluorescence Detection for Reagentless Biomarker Quantification. *bioRxiv* 2021.

14. Wu, Y.; Tehrani, F.; Teymourian, H.; Mack, J.; Shaver, A.; Reynoso, M.; Kavner, J.; Huang, N.; Furmidge, A.; Duvvuri, A.; et al. Microneedle Aptamer-Based Sensors for Continuous, Real-Time Therapeutic Drug Monitoring. *Anal. Chem.* 2022, *94*, 8335–8345.

15. Tiangco, C.; Andar, A.; Sevilla, F., III; Rao, G.; Tolosa, L. Transdermal Glucose Monitoring Using Glucose Binding Protein-Based Fiber Optic Biosensor Coupled with Microneedles. *Sens. Transducers* 2018, *28*, 2–6.

16. Liu, Y.; Yu, Q.; Luo, X.; Yang, L.; Cui, Y. Continuous monitoring of diabetes with an integrated microneedle biosensing device through 3D printing. *Microsyst. Nanoeng.* 2021, *7*, 75.

17. Teymourian, H.; Moonla, C.; Tehrani, F.; Vargas, E.; Aghavali, R.; Barfidokht, A.; Tangkuaram, T.; Mercier, P.P.; Dassau, E.; Wang, J. Microneedle-Based Detection of Ketone Bodies along with Glucose and Lactate: Toward Real-Time Continuous Interstitial Fluid Monitoring of Diabetic Ketosis and Ketoacidosis. *Anal. Chem.* 2020, *92*, 2291–2300.

18. Ventrelli, L.; Marsilio Strambini, L.; Barillaro, G. Microneedles for Transdermal Biosensing: Current Picture and Future Direction. *Adv. Healthc. Mater.* 2015, *4*, 2606–2640.

19. Lee, K.J.; Jeong, S.S.; Roh, D.H.; Kim, D.Y.; Choi, H.-K.; Lee, E.H. A practical guide to the development of microneedle systems—In clinical trials or on the market. *Int. J. Pharm.* 2020, *573*, 118778.

20. Wang, M.; Hu, L.; Xu, C. Recent advances in the design of polymeric microneedles for transdermal drug delivery and biosensing. *Lab Chip* 2017, *17*, 1373–1387.

21. Bollella, P.; Sharma, S.; Cass, A.E.G.; Tasca, F.; Antiochia, R. Minimally Invasive Glucose Monitoring Using a Highly Porous Gold Microneedles-Based Biosensor: Characterization and Application in Artificial Interstitial Fluid. *Catalysts* 2019, *9*, 580.

22. Leng, F.; Zheng, M.; Xu, C. 3D-printed microneedles with open groove channels for liquid extraction. *Exploration* 2021, *1*, 20210109.

23. Mishra, R.K.; Vinu Mohan, A.M.; Soto, F.; Chrostowski, R.; Wang, J. A microneedle biosensor for minimally-invasive transdermal detection of nerve agents. *Analyst* 2017, *142*, 918–924.

24. Al Sulaiman, D.; Chang, J.Y.; Bennett, N.R.; Topouzi, H.; Higgins, C.A.; Irvine, D.J.; Ladame, S. Hydrogel-coated microneedle arrays for minimally invasive sampling and sensing of specific circulating nucleic acids from skin interstitial fluid. *ACS Nano* 2019, *13*, 9620–9628.

25. Linh, V.T.N.; Yim, S.-G.; Mun, C.; Yang, J.-Y.; Lee, S.; Yoo, Y.W.; Sung, D.K.; Lee, Y.-I.; Kim, D.-H.; Park, S.-G.; et al. Bioinspired plasmonic nanoflower-decorated microneedle for label-free intradermal sensing. *Appl. Surf. Sci.* 2021, *551*, 149411.

26. Park, J.E.; Yonet-Tanyeri, N.; Vander Ende, E.; Henry, A.-I.; Perez White, B.E.; Mrksich, M.; Van Duyne, R.P. Plasmonic Microneedle Arrays for in Situ Sensing with Surface-Enhanced Raman Spectroscopy (SERS). *Nano Lett.* 2019, *19*, 6862–6868.

27. Rezapour Sarabi, M.; Alseed, M.M.; Karagoz, A.A.; Tasoglu, S. Machine learning-enabled prediction of 3D-printed microneedle features. *Biosensors* 2022, *12*, 491.

28. Rezapour Sarabi, M.; Jiang, N.; Ozturk, E.; Yetisen, A.K.; Tasoglu, S. Biomedical optical fibers. *Lab Chip* 2021, *21*, 627–640.

29. Nair, S.S.; Mishra, S.K.; Kumar, D. Recent progress in conductive polymeric materials for biomedical applications. *Polym. Adv. Technol.* 2019, *30*, 2932–2953.

12 CHAPTER

A Laminar Flow-Based Microfluidic Tesla Pump via Lithography Enabled 3D Printing

Mohammed-Baker Habhab, Tania Ismail and Joe Fujiou Lo

Department of Mechanical Engineering, Bioengineering Program, University of Michigan at Dearborn, 2088 IAVS Building, 4901 Evergreen Rd., Dearborn, MI 48128, USA

ABSTRACT

Tesla turbine and its applications in power generation and fluid flow were demonstrated by Nicholas Tesla in 1913. However, its real-world implementations were limited by the difficulty to maintain laminar flow between rotor disks, transient efficiencies during rotor acceleration, and the lack of other applications that fully utilize the continuous flow outputs. All of the aforementioned limits of Tesla turbines can be addressed by scaling to the microfluidic flow regime. Demonstrated here is a microscale Tesla pump designed and fabricated using a Digital Light Processing (DLP) based 3D printer with 43 μm lateral and 30 μm thickness resolutions. The miniaturized pump is characterized by low Reynolds number of 1000 and a flow rate of up to 12.6 mL/min at 1200 rpm, unloaded. It is capable of driving a mixer network to generate microfluidic gradient. The continuous, laminar flow from Tesla turbines is well-suited to the needs of flow-sensitive microfluidics, where the integrated pump will enable numerous compact lab-on-a-chip applications.

Keywords: *Tesla turbine; Tesla valve; microfluidic; DLP; 3D printing*

1. INTRODUCTION

In the early 1900s, Nikola Tesla, a Serbian inventor, introduced a new turbine design that replaced the conventional impeller blades with smooth disks [1,2]. This design places multiple disks in parallel on a rotating shaft. Around the shaft are holes that allow fluids to flow orthogonal to the disks. Fluid is injected tangential to the disks and spirals across the disc surface towards the orthogonal holes at the center. Due to the viscous effects on

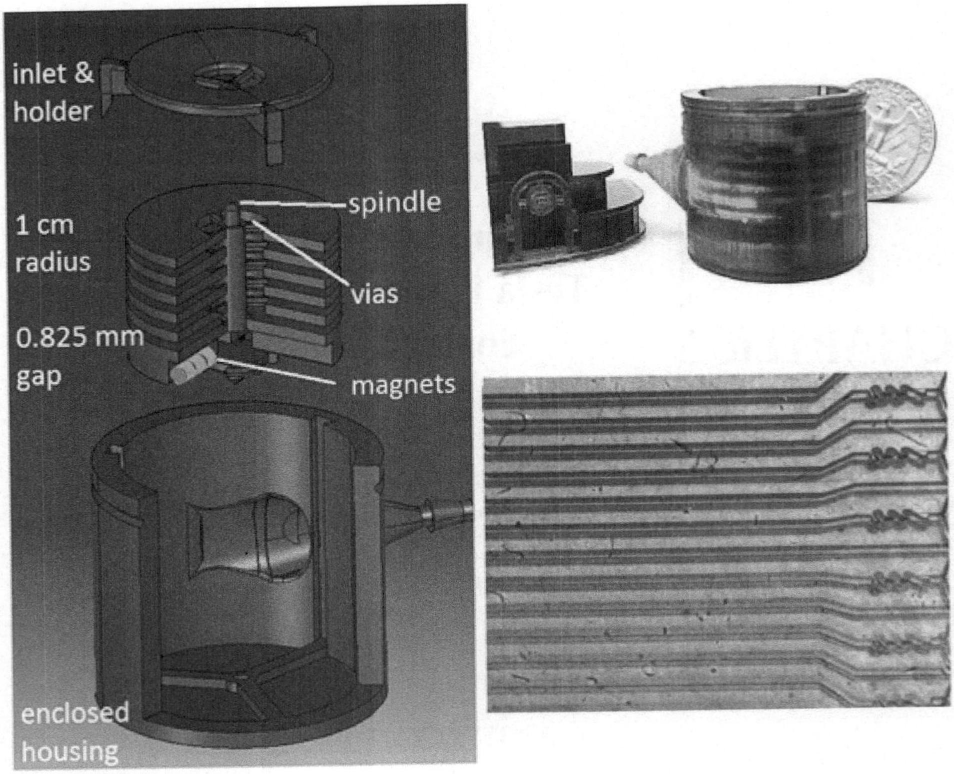

Graphical Abstract

the boundary surfaces of the parallel disks, the fluid then transfers its momentum towards the rotating disks continuously. As the discs begin to rotate, the path of the fluid begins to increase in length, which in turn increases the momentum transfer. In his 1913 patent for the "Tesla turbine", Tesla originally anticipated this design to be used for geothermal power generation with an efficiency over 80% [3,4,5]. However, the turbine can also be operated in reverse where rotation can be applied to drive a smooth tangential flow from the parallel disks. For example, pumping with Tesla turbine which is pulseless in nature has been demonstrated in a number of applications including the artificial heart [6,7]. Furthermore, the transfer of momentum is dependent on the boundary layer at the walls of the parallel disks, which can be improved by laminar flows within the low Reynold number regime. Thus the Tesla turbine is well suited to microfluidic flow. To achieve a miniature microfluidic Tesla (μTelsa) pump with complex disk and shaft geometries, a Digital Light Processing (DLP) based 3D printer was employed here (Projet 1200, 3DSystems, Rock Hill, SC, USA), enabling prototypes to be built with 43 μm × 43 μm × 30 μm voxel resolution. A multidisc rotor was completely enclosed in a 3D printed housing and magnetically coupled to a simple laboratory stir plate, reducing the complexity of mechanical drive and eliminating the need for fluidic sealing, Figure 1. Using this setup, the pump pressure, flow, and power output were characterized at multiple fluidic loads and visualized under a stereo microscope. Lastly, the pump was connected to a microfluidic mixer network with special Tesla valve geometries, demonstrating an all-Tesla-fluidics system for lab-on-a-chip applications.

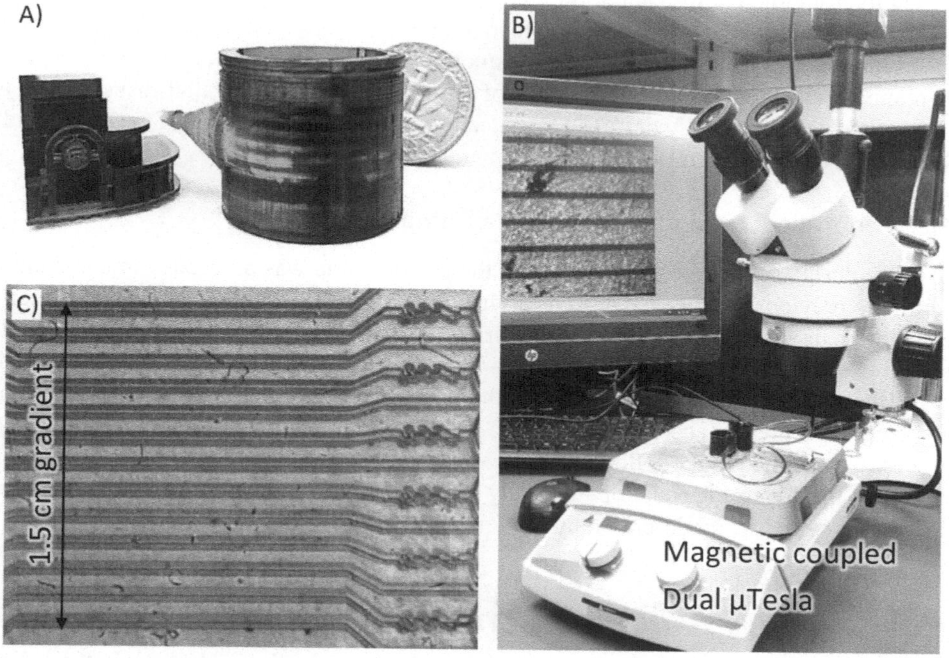

Figure 1. *3D printed μTesla pump. (A) The assembled pump is around an US quarter in diameter, printed in high resolution as demonstrated by the miniature engineering building of UM Dearborn; (B) Two μTesla pumps were coupled magnetically to the same stir plate to ensure exact pressure delivered to a microfluidic device, and visualized under a stereomicroscope; (C) Using this setup, gradients can be generated by μTesla pump to drive a microfluidic mixer network.*

1.1. Microfluidic Pumps

It can be asserted that a complete lab-on-a-chip device requires a source of fluidic power. Towards this, many efforts have been devoted to miniaturize pumps or create a driving force for flow in a microchannel. Typical external flow sources rely on screw-type syringe pumps, rotor-type peristaltic pumps, or press-driven pumps to control flow in the μL to nL per minute domain. Novel integrated microfluidic pumps include accoustofluidic pumps [8], the technology used in inkjet print heads [9]; electroosmotic pumps [10], which apply voltage across a diffusion membrane to control ionic flow; electrolysis gas pressure pumps [11]; capillary force siphoning [12], à la paper microfluidics [13,14]; and a number of valve-based peristaltic pumps [15,16] such as those applied in commercial miniaturized insulin pumps (STMicroelectronics, Santa Clara, NV, USA) [17]. Additionally, a number of aforementioned pumps, especially peristaltic ones, utilize passive and asymmetrical flow valves [18,19]. These special valves, which often coincidentally apply Tesla's other work in fluidics, the Tesla valve [20], in order to achieve low fluidic resistance in one flow direction, and much higher resistance in the opposite. However, most of the aforementioned microfluidic pumps cannot provide smooth laminar flow due to a pulsatile driving force like valves and acoustics, while others required special voltage that energized ionic membranes and produced flat-top electroosmotic flows. The benefit of a miniaturized Tesla pump leverages the fluidic mechanics of laminar flow within parallel disks, which is well-suited and specifically designed to match the low Reynolds number of the microfluidic channels that it is intended to drive.

1.2. Theoretical Design of the μ Tesla Pump

The design of the turbine geometry in the μTesla pump was proceeded in the manner similar to previously published designs of the Tesla turbine [21,22,23,24]. First, a cylindrical coordinate system was adapted to describe the spiraling flow path between the parallel turbine disks. Then, the flow conditions including Reynold number, etc. were defined. Next, the gap distance was derived from the flow conditions. And finally, the number of disks was specified in accordance with the flow rate requirements. Details of this design process is described in the following.

First, the governing equation for steady incompressible laminar flow between adjacent disks were cylindrically described by:

$$\frac{1}{r}\frac{\partial(rV_r)}{\partial r}+\frac{1}{r}\frac{\partial V_\theta}{\partial \theta}+\frac{\partial V_z}{\partial z}=0 \tag{1}$$

Vdenotes the velocity components where r,θ, are the radius and angle of the cylindrical coordinate system. The following assumptions were taken [25]: (1) the flow is two-dimensional and constant across the channel between the rotor plate surfaces and are mean velocities; (2) the flow field and velocity components are constant across the gaps, with body force represented by the wall shear at each (r,θ) location; (3) the flow field is radially symmetric and the inlet flow at rotor outer edge is uniform and θ derivatives of flow are therefore zero; (4) radial pressure gradients are negligible compared to angular momentum and frictional forces the wall. These assumptions reduce the continuity Equation (1) to:

$$\frac{1}{r}\frac{\partial(rV_r)}{\partial r}=0 \tag{2}$$

From the Navier-Stokes equation, the r-direction and θ-direction momentums reduce to:

$$V_r\frac{\partial V_r}{\partial r}-\frac{{V_\theta}^2}{r}=-\frac{1}{\rho}\left(\frac{\partial P}{\partial r}\right)+f_r \tag{3}$$

$$V_r\frac{\partial V_r}{\partial r}+\frac{V_r V_\theta}{r}=f_\theta \tag{4}$$

while the z-direction moment and pressure gradient are both zero.

Next, the following flow conditions were considered throughout the turbine design: In order to maintain laminar flows between the rotor disks, Reynolds number was designed at 1000, below the typically values associated with transition into the turbulent regime. The spin rate of the rotor ω was nominally set at 500 rpm (stir plate maximum of 1200 rpm). And the volumetric flow rate Q was designed to be 70 μL/min, an appropriate range for the microfluidic flows the pump is designed to drive. Thus, to calculate the gap distance, we start with the mass flow rate between the disk gaps:

$$VA\rho = 2\pi rbV_r\rho = \frac{\dot{m}_c}{n} \tag{5}$$

where m˙c is the total mass flow rate, b is the gap size between rotating plates, 2πrb2 is the circumferential disk gap area, and n is the number of disk gaps.

The gap thickness is related to the hydraulic diameter by:

$$D_h = 2b \tag{6}$$

and the Reynolds number *Re*, designed at 1000, is a function with hydraulic diameter, viscosity, and velocity of the fluid, here described by:

$$Re = \frac{\rho\Delta V_\theta D_h}{\mu} = \frac{\Delta V_\theta D_h}{\nu} \tag{7}$$

where ΔVθ is taken to be the tangential velocity at the disk edge, Dh is the hydraulic length, and ν is the kinetic viscosity, Dh can be specified at 1.65 mm giving a gap b of 0.826 mm.

Finally, the total volumetric flow rate is divided into individual flows between each pair of disks. And thus the total flow rate is described by:

$$N = n + 1, \text{ where } n = \dot{Q}/q \tag{8}$$

where Q˙ is the designed fluid flow rate, i.e., 70 μL/min, q is the flow rate between each pair of disks, n is the number of disk gaps, and N is thus the total number of disks.

On the other hand, the total mass flow rate m˙c is equal to the density times volumetric flow rate. Applying this to Equation (5) gives:

$$V_r 2\pi r_0 b\rho = \frac{\dot{m}_c}{n} = \frac{\rho \cdot \dot{Q}}{n} = \rho \cdot q, \text{ thus } q = V_r 2\pi r_0 b \tag{9}$$

Vr is taken to be the radial velocity of the fluid being pumped out of the disks, which is equal to the total design flow rate Q˙ divided by the cross sectional area of the gaps between the disks [26]. With the designed rotation speed, turbine disk radius of 1 cm, and bdefined at 0.826 mm, Vr was 2.57×10^{-6} m/s, q was 1.32×10^{-10} m^3/s, and *n* was ~5, giving us six disks. The μTesla design parameters are summarized in Table 1 and applied to the CAD design for 3D printing, Figure 2A.

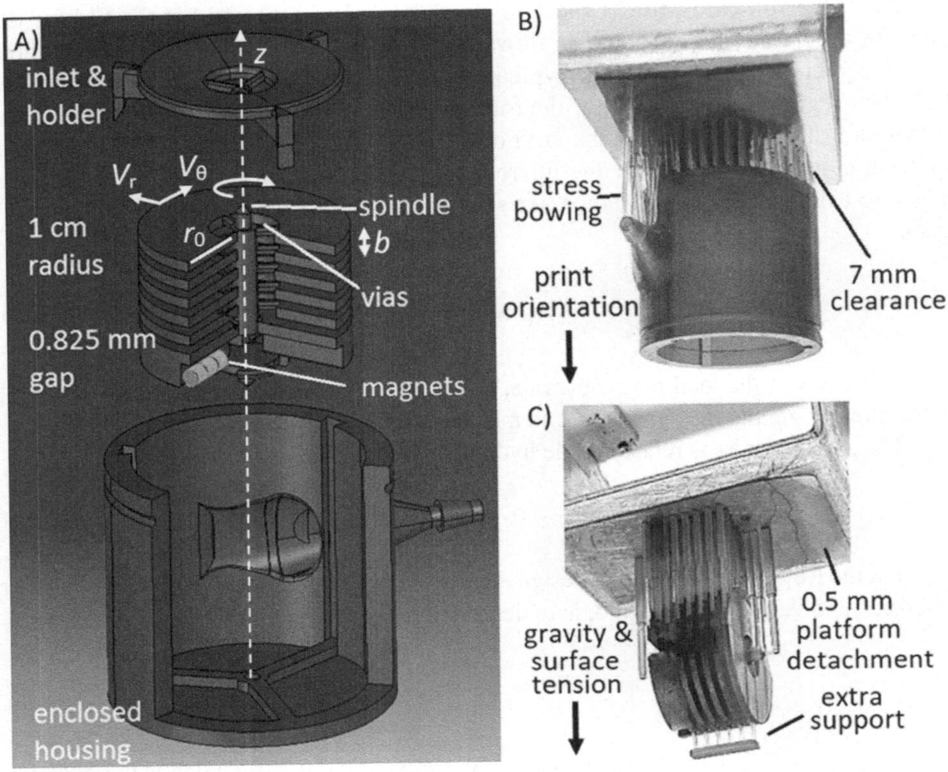

Figure 2. *Computer design and 3D printing of μTesla. (A) The μTesla pump was comprised of three parts: cap with pivot, the rotor with vias and pockets for magnets, and the enclosed housing with fluidic spout and bottom pivot. The computer design was then printed in a DLP 3D printer. The print was upside down as pictured in (B,C) and subjected to gravity pull; (B) To reduce surface tension pulling down large areas of the enclosed housing's bottom, substantial clearance was required along with numerous supporting posts to prevent excessive warping and detachment; (C) The rotor required extra support to keep the disks straight. Additional resin platform was added to ensure adhesion to the metal print stage.*

Table 1. μTesla Design Parameters.

Reynold Number	$Re_x = 1000$
Rotational speed	500 rpm → 0.5236 m/s
Disk radius	$r_0 = 1$ cm
Volumetric flow rate	$Q = 40$ μL/min
Disk Gap	b = 0.826 mm, as D_h = 1.65 mm
Number of disks	$N = 6$

2. RAPID PROTOTYPING USING DLP 3D PRINTER

Lithography based 3D printing requires the object to be built layer by layer from a bath of photo curable resin. This technique has a marvelous appearance as the object is literally "pulled out" of the resin bath. 3D printing has been used in the rapid prototyping of self-enclosed microfluidics [27,28], stereolithography-based microfluidics [29,30,31], miniaturized electronics [32], and microoptical systems [33,34,35], representing a *key* enabling technology to achieve intricate overhanging three-dimensional structures in microfabrication. This key technology was leveraged to shrink the complex μTesla geometries—with multiple disks, ports, vias—to a microscale dimension matching the microfluidic flow for which it is designed to drive.

Addressing DLP Printing Issues

The μTesla pump was printed in three parts—cap holder, rotor, and nozzled housing—and assembled to complete the device, Figure 2A. To translate the μTesla design into a 3D build, we must consider the DLP lithography issues in the following details. First, the minimum feature size differs from the actual DLP resolution (pixel pitch) or print thickness. The minimum wall thickness, for example, was found to be 0.5 mm to prevent collapse during printing. The minimum overhang and enclosed cross-sectional diameter was designed to be 1.5 mm, e.g., the nozzle lumen, as smaller dimensions would trap resin and the structure will not be resolved. However, supported, non-overhanging structures can be routinely achieved with 70–100 μm minimum feature sizes. Secondly, as each layer of the 3D build is exposed and pulled away from the print surface, the liquid-solid interface repeatedly exerts surface tension on the build structure. Hence adequate support structures for large lateral surfaces is required to counter this surface tension during printing. For example, the large base of the pump housing required numerous support beams with minimum of 7 mm clearance from the print stage, Figure 2B. If inadequate support leads to detachment or warpage, the resultant protruding material tends to splash unwanted print resin randomly, creating print defects. Thirdly, another technique to counteract surface tension was to include a bulk resin adhesion layer, which prevented detachment of supports near the edges of large structures, Figure 2C. In all final builds, this 0.5 mm thick support resin platform was included. Lastly, the lithographically printed structures often exhibited residual stress after the post-curing step. This was found to be partially reduced if the isopropanol washing step before post-curing was limited to two 60-s rinses, to limit solvent swelling of the polymer. Nevertheless, noticeable amount of residual stress remained, as seen in the bowed support beams, and were addressed by additional lateral support structures, Figure 2C.

3. RESULTS OF MTESLA PERFORMANCE

The μTesla was mounted to a standard stir plate with speed control graduated from 0 to 1200 rpm as shown in Figure 1. The μTesla rotor speed is coupled to the stir plate's rotation, which we have also verified via slow motion videos, Supplementary Video S1. Using this setup, the performances of the μTesla was quantified as follows.

3.1. µTesla Hydraulic Head and Stall Characteristics

The hydraulic pressure developed by the µTesla was characterized by measuring the height of the water column that was pumped up a connected Tygon tube with 1/16” inner diameter. Tygon tubes at varying lengths were also used for subsequent flow characterizations. First, the µTesla was primed with isopropanol to reduce the bubbles trapped between the rotor disks and tubing. Then, water was injected using a 5 mL syringe from the Tygon tubes. During the characterization, it was found that the pump rotor would stall at specific speeds if there was inadequate holding force from the magnets. Therefore, single, double, and triple 0.4 lb. magnets (2 mm $\phi \times 1.5$ mm, McMaster-Carr, Aurora, OH, USA) were placed at opposite ends of the rotor to test the stall characteristic when the µTesla was primed with water. The results showed that two sets of 3 magnets coupled end to end were required to eliminate stalling and provide speeds up to 1200 rpm, technically limited by our stirring plate, Figure 3. Furthermore, at 1200 rpm, the maximum hydraulic head achieved was 2.6 cm, equivalent to 253 Pa of developed pressure. In comparison, a microfluidic bladed turbine required almost 4000 rpm to produce 200 Pa of pressure [36].

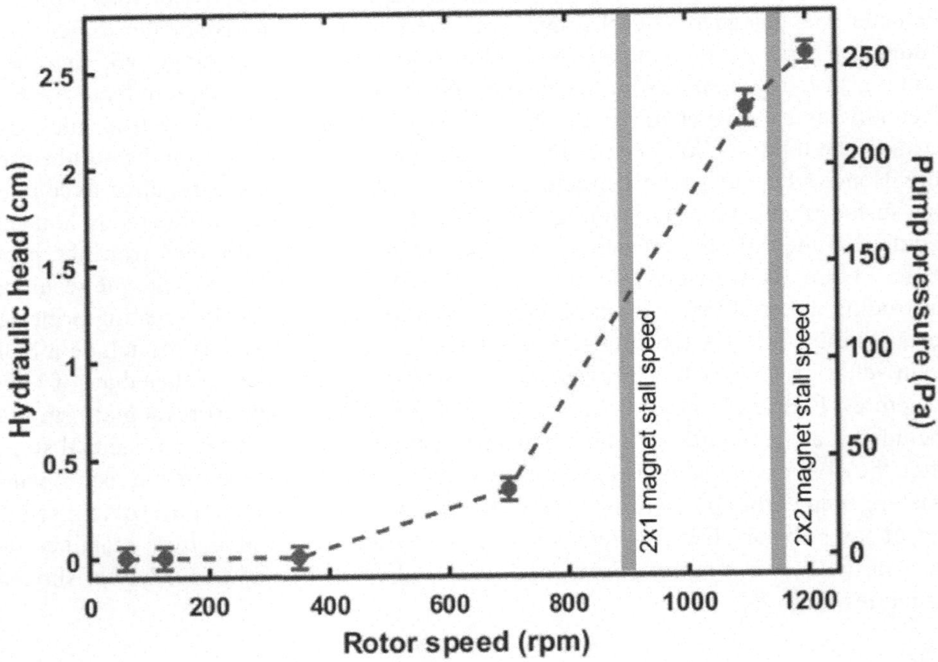

Figure 3. *Hydraulic head. The height above the µTesla pump to which the fluid was raised was characterized for rotor speeds from 60 to 1200 rpm (a 15 cm, 1/16 ID Tygon tubing was coupled for this measurement). With a pair of single magnet to couple the stir plate rotation, the rotor stalled around 900 rpm. With a pair of double magnets, stalling occurred around 1150 rpm. The total hydraulic head at 1200 rpm corresponds to 253 Pa of pressure. Error bar denotes standard deviation from 6 measurements.*

3.2. Flow Characteristics of μTesla at Different Resistances

Next, the μTesla pump was loaded with increasing fluidic resistances created by varying lengths of Tygon tubing from 15 to 45 cm, and connected to a microfluidic mixer (described in Section 4). The outlet tubing or microfluidic was placed at the same level as the pump spout to eliminate pressure differences due to height. The output volume from the tubing or microfluidics was then collected directly into a microcentrifuge tube for 30 s. This volume was then quantified via manual micro pipetting and calculated for the flow rate per minute. From Figure 4, it can be seen that the flow output increased with rotor speed, but began to taper close to 1200 rpm. The maximum flow output at 1200 rpm achieved with the 15 cm tubing was 12.6 mL/min. The flow across the mixer microfluidic achieved a significantly lower rate of 0–14.6 μL/min owing to its increased fluidic resistance. At a fixed rotor speed, flow output and resistance were inversely proportional, as expected from the constant pressure generated by the pump. Extrapolating the data from the 15–45 cm tubing, the mixer device has a fluidic resistance equivalent to 90 cm of tube length, with 1.04 MPa·s/m^3. This is comparable to the higher end of fluidic resistances typically found in microchannels [37].

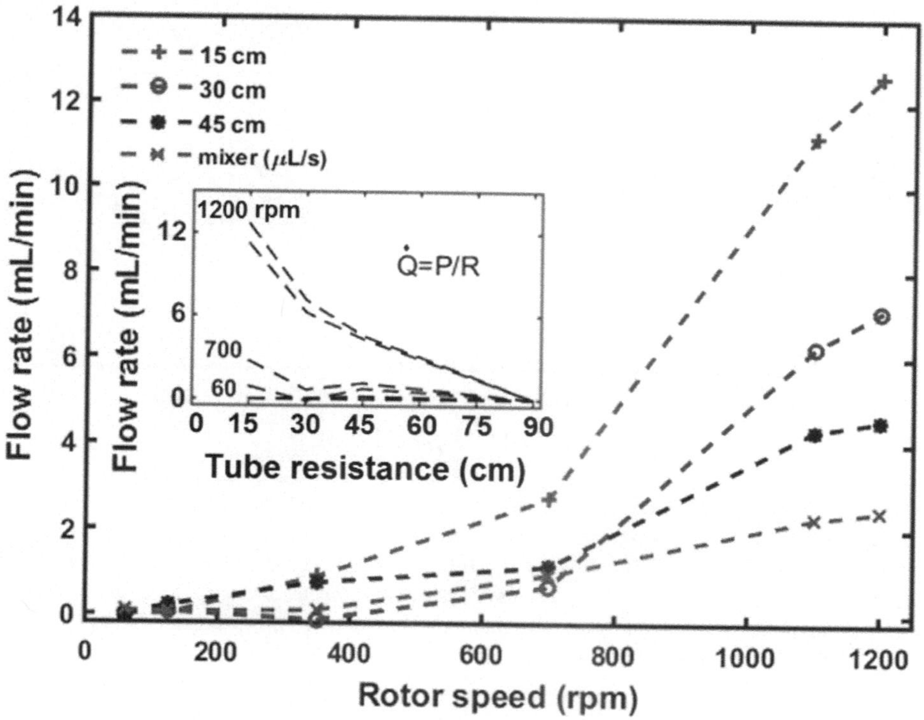

Figure 4. *Flow output at different loads. The flow characteristics of the μTesla pump under load was measured with different lengths of Tygon tubing (1/16" ID), and through a microfluidic mixer device. The flow rate across the mixer device is considerably lower at 0–14.6 μL/min, and plotted in a different unit. Inset: Flow and fluidic resistance is inversely proportional. By extrapolating the microfluidic flow rates, its resistance is equivalent to a tubing length of 90 cm, or 1.04 MPa·s/m3.*

3.3. μTesla Pump Power Calculations

Given the characteristic pressures and flow rates that were measured, the hydraulic power output of the μTesla pump was calculated using Power = $Q^{\cdot} \cdot \rho \cdot g \cdot \Delta h$ (where $Q^{\cdot}$ is flow in m^3/s and $\rho \cdot g \cdot \Delta h$ is from the hydraulic head) (Figure 5). The maximum power output of μTesla at 15 cm resistance (nearly unloaded) was 53 μW. The power after the mixer load was dropped to 37 nW, which meant that most of the output power was consumed by pumping the mixer device at 1200 rpm. Nevertheless, this power was capable of driving 0–14 μL/min of flow in the mixer device, adequate for general microfluidic applications.

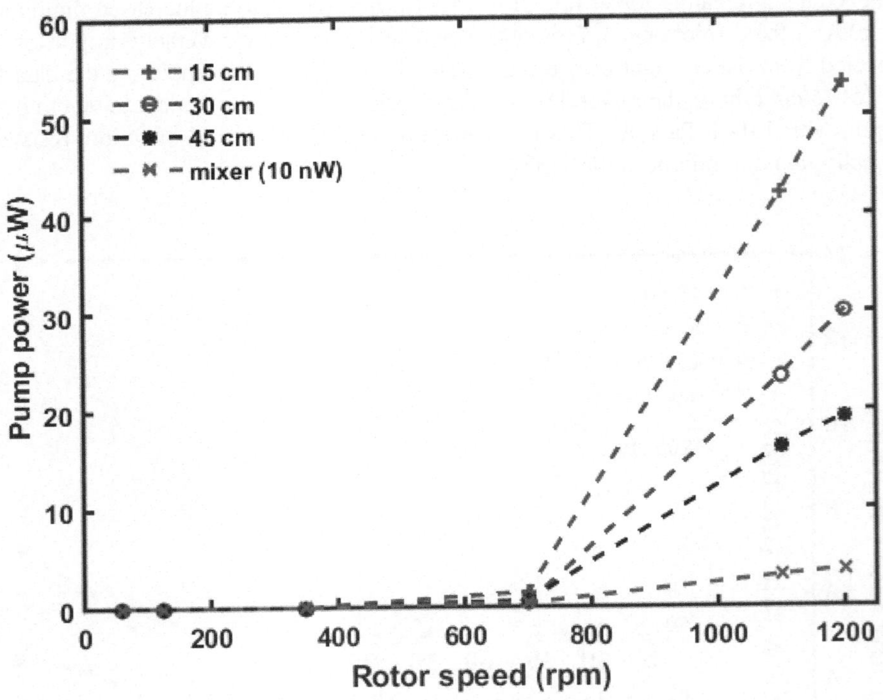

Figure 5. *Pump power at different loads. The hydraulic power can be calculated from the product of hydraulic head and flow rates. Power drops after different fluidic loads were also calculated, including load across a microfluidic mixer. The μTesla produced 53 μW of power at 1200 rpm across the lowest load of 15 cm. The power reduced to 37 nW across the mixer microfluidic load. The 53 μW of μTesla's nominal power output was more than adequate to drive the mixer device at flow rates of 0–14.6 μL/min, as shown in Figure 4.*

3.4. Laminar Flow Velocity Profile in Microchannels

The μTesla pump was connected to a microfluidic mixer and velocity profile inside the channel was quantified with nanoparticle velocimetry using dark field microscopy. 70 nm spherical gold nanoparticles were diluted in water and pipetted inside the microchannel before connecting the μTesla pump.

Flow across a channel of 500 μm by 500 μm cross section was then visualized under dark field to observe the nanoparticle scattering. The majority of the nanoparticles aggregated to form bright clusters easily visible even under short integration times. The pictures taken at 100 and 200 ms integration times, tailored to the speed of the flow, were analyzed in ImageJ. The calibrated distances of the particle streaks, Figure 6 inset, was divided by the integration time to yield particle velocity. As expected at the flow rates under 14 μL/min, velocity profile mimics the parabolic shapes in the laminar regime.

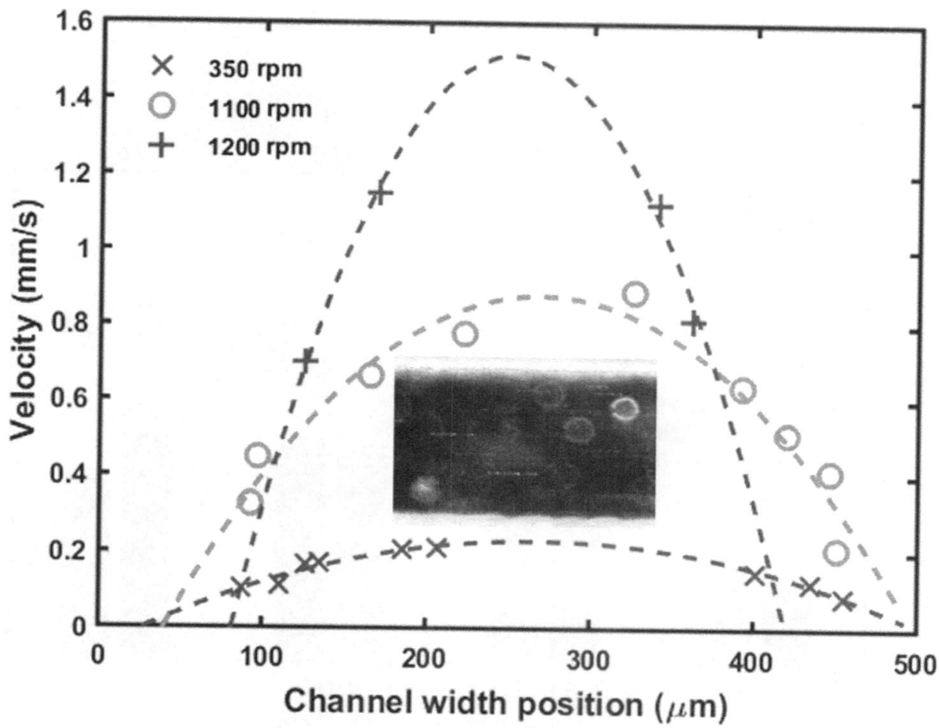

Figure 6. *Flow profile at different pump rates. To demonstrate that the pump is able to create laminar flow in the microfluidic mixer, a 500 μm wide section of a microchannel was imaged with gold nanoparticles in the flow stream (lines in insert). With the camera integration times fixed at 100–200 ms, the streak length was calibrated and calculated to yield the flow rate across the width of the channel. Parabolic flow profiles were observed at the three rotor speeds tested.*

4. RESULTS OF DRIVING THE TESLA INSPIRED MICROFLUIDIC MIXER

To demonstrate the usage, two μTesla pumps were coupled on the same laboratory stir plate, Figure 1, to guarantee the same flow rates that pumped red and blue food dye into the microfluidic mixer network (Figure 7 insert). This particular network incorporates mixer geometries inspired by the Tesla-valve, another fluidic work that Nicholas Tesla patented in 1920 [20]. The valves were typically used to create asymmetric flow resistance, where flow in one direction has much higher resistance than another. Here, it is used to create folding flows to provide convective mixing inside the gradient generator network. Instead of tuning the pump speed, additional fluidic resistance in Tygon tubing was added to the microfluidics to reduce the flow rates. The resultant color mixing was captured under the stereomicroscope and analyzed by ImageJ. The blue and red channels were analyzed to find their 20% to 80% intensities, and the width between these were plotted against the fluidic resistances applied. As expected, we observed higher mixing at higher resistance, or low flow rates, Figure 7. The maximum gradient width achieved at 3 nL/min flow rate was 4 mm wide. This gradient was a demonstration of an all-Tesla-fluidics based system for microfluidic applications.

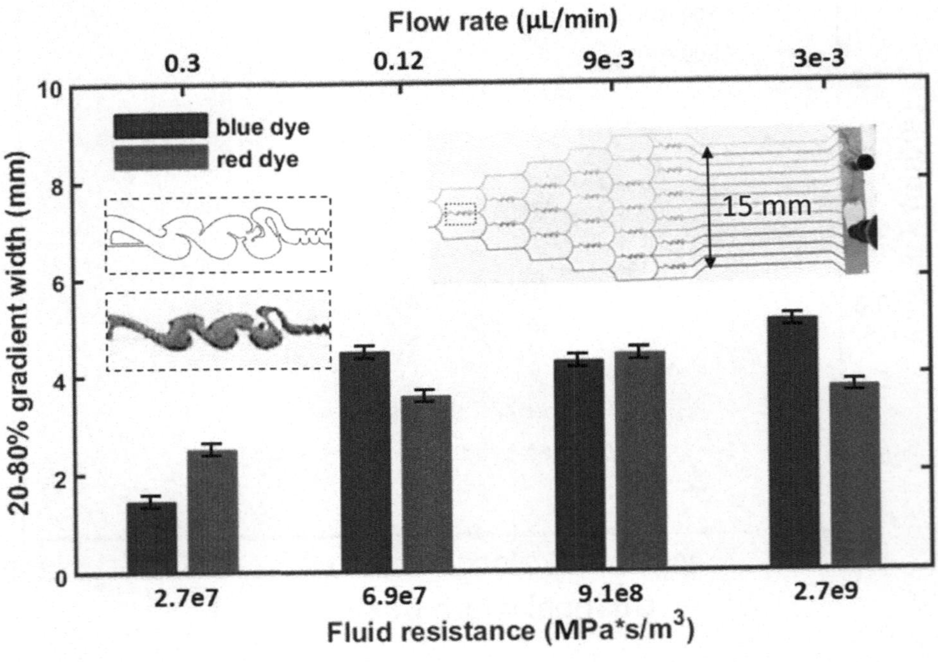

Figure 7. *µTesla driven, Tesla-inspired mixer microfluidics. The all-Tesla-fluidic system generated gradients using blue and red dyes at flow rates as low as 3 nL/min. The mixer incorporates flow folding structures similar to Tesla valves (CAD and enlarged insert on left). The widest gradient achieved spanned 4 mm across (20%–80% intensity), as analyzed from their RGB channels in ImageJ. Errors bars denote maximum and minimum values from image analysis.*

5. FUTURE WORKS

Further miniaturization and integration into a single microfluidic substrate can open the doors to a number of µTesla based lab-on-a-chip applications. In order to achieve this, three future improvements should be made: (1) The print resolution can be improved, either through higher print resolution or lower surface tension platforms used in competing DLP systems; (2) the fluid path along the axial holes and the spout apertures needs to be optimized to account for differential pressure drops across the disks [38]; (3) electromagnetic integration would eliminate the stir plate and provide a self-powered device that can be visualized under both upright and inverted microscopes. Future studies on the temporal responses of the μTesla's flow, both through simulation and flow visualization between the disks (a challenging task at the microscale level) can also reveal additional performance and efficiency characteristics of the μTesla pump. These works represent many exciting research directions this current work may lead to in the future.

6. CONCLUSIONS

A miniaturized Tesla turbine fluidic pump was designed and fabricated using DLP based 3D printing. The turbine geometries were designed with low Reynolds number of 1000. The DLP lithography 3D printing required the optimization of feature sizes, support structures, and gap to platform distances to minimize surface tension and residual stress that may warp the printed parts. The resultant μTesla pumps were coupled magnetically to a lab stirring plate to provide rotor actuation up to 1200 rpm, where it achieved a pump pressure of 253 Pa and maximum flow greater than 12 mL/min. Coupling this μTesla pump to a microchannel mixer with Tesla-valve inspired geometries generated gradients using color dyes. In this manner, a complete pump-and-microfluidic system was realized based on Tesla's fluidic theories. Future miniaturization by optimizing the print resolution and integrating electronics has the potential to create a low cost, low complexity, self-powered, unified lab-on-a-chip device for bioassay applications.

Acknowledgments

This work was supported by the University of Michigan Dearborn Office of Research and Sponsored Projects (ORSP) undergraduate research funds and the Mechanical Engineering Department senior design funds.

Author Contributions

The concept was conceived by Joe Fujiou Lo, Mohammed-Baker Habhab, and Tania Ismail. The design of the μTesla dimensions were done by Mohammed-Baker Habhab. The design and fabrication of the Tesla inspired microfluidics were done by Mohammed-Baker Habhab and Tania Ismail. The μTesla pump characterizations were done by Mohammed-Baker Habhab and Joe Fujiou Lo. The microfluidic gradient generations were done by Mohammed-Baker Habhab and Joe Fujiou Lo. The data analysis was done by Mohammed-Baker Habhab.

REFERENCES

1. Tesla, N. Fluid Propulsion. U.S. Patent 1061142 A, 6 May 1913.
2. Tesla, N. Turbine. U.S. Patent 1061206 A, 6 May 1913.
3. Steidel, R.; Weiss, H. *Performance Test of a Bladeless Turbine for Geothermal Applications*; Report No. UCID-17068; Lawrence Livermore Laboratory: Livermore, CA, USA, 1974.

4. Ho-Yan, B.P. Tesla turbine for pico hydro applications. *Guelph Eng. J.* 2011, *4*, 1–8.
5. Hoya, G.P.; Guha, A. The design of a test rig and study of the performance and efficiency of a Tesla disc turbine. *J. Power Energy* 2009, *223*, 451–465.
6. Miller, G.E.; Sidhu, A.; Fink, R.; Etter, B.D. Evaluation of a multiple disk centrifugal pump as an artificial ventricle. *Artif. Organs* 1993, *17*, 590–592.
7. Miller, G.E.; Madigan, M.; Fink, R. A preliminary flow visualization study in a multiple disk centrifugal artificial ventricle. *Artif. Organs* 1995, *19*, 680–684.
8. Du, X.Y.; Fu, Y.Q.; Luo, J.K.; Flewitt, A.J.; Milne, W.I. Microfluidic pumps employing surface acoustic waves generated in ZnO thin films. *J. Appl. Phys.* 2009, *105*, 024508.
9. Gallas, Q.; Holman, R.; Nishida, T.; Carroll, B.; Sheplak, M.; Cattafesta, L. Lumped element modeling of piezoelectric-driven synthetic jet actuators. *AIAA J.* 2003, *41*, 240–247.
10. Urbanski, J.P.; Levitan, J.A.; Burch, D.N.; Thorsen, T.; Bazant, M.Z. The effect of step height on the performance of three-dimensional ac electro-osmotic microfluidic pumps. *J. Colloid Interface Sci.* 2007, *309*, 332–341.
11. Munyan, J.W.; Fuentes, H.V.; Draper, M.; Kelly, R.T.; Woolley, A.T. Electrically actuated, pressure-driven microfluidic pumps. *Lab Chip* 2003, *3*, 217–220.
12. Steigert, J.; Brenner, T.; Grumann, M.; Riegger, L.; Lutz, S.; Zengerle, R.; Ducrée, J. Integrated siphon-based metering and sedimentation of whole blood on a hydrophilic lab-on-a-disk. *Biomed. Microdevices* 2007, *9*, 675–679.
13. Martinez, A.W.; Phillips, S.T.; Butte, M.J.; Whitesides, G.M. Patterned paper as a platform for inexpensive, low-volume, portable bioassays. *Angew. Chem. Int. Ed.* 2007, *46*, 1318–1320.
14. Osborn, J.L.; Lutz, B.; Fu, E.; Kauffman, P.; Stevens, D.Y.; Yager, P. Microfluidics without pumps: Reinventing the T-sensor and H-filter in paper networks. *Lab Chip* 2010, *10*, 2659–2665.
15. Unger, M.A.; Chou, H.P.; Thorsen, T.; Scherer, A.; Quake, S.R. Monolithic microfabricated valves and pumps by multilayer soft lithography. *Science* 2000, *288*, 113–116.
16. Grover, W.H.; Skelley, A.M.; Liu, C.N.; Lagally, E.T.; Mathies, R.A. Monolithic membrane valves and diaphragm pumps for practical large-scale integration into glass microfluidic devices. *Sens. Actuators B Chem.* 2003, *89*, 315–323.
17. Anhalt, H.; Bohannon, N.J. Insulin patch pumps: Their development and future in closed-loop systems. *Diabetes Technol. Ther.* 2010, *12*, S51–S58.
18. Nguyen, N.T.; Truong, T.Q.; Wong, K.K.; Ho, S.S.; Low, C.L.N. Micro check valves for integration into polymeric microfluidic devices. *J. Micromech. Microeng.* 2003, *14*, 69.
19. Hong, C.C.; Choi, J.W.; Ahn, C.H. A novel in-plane passive microfluidic mixer with modified Tesla structures. *Lab Chip* 2004, *4*, 109–113.
20. Tesla, N. Valvular Conduit. U.S. Patent 1329559 A, 3 February 1920.
21. Ladino, A.F.R. Numerical Simulation of the Flow Field in a Friction-Type Turbine (Tesla Turbine). Ph.D. Thesis, National University of Colombia, Bogotá, Colombia, 2004.
22. Rice, W. An analytical and experimental investigation of multiple-disk turbines. *J. Eng. Power* 1965, *87*, 29–36.
23. Rice, W. Tesla turbomachinery. In *Handbook of Turbomachinery*; Logan, E., Ed.; Marcel Dekker: New York, NY, USA, 2003; pp. 861–874.
24. Lawn, M.J., Jr.; Rice, W. Calculated design data for the multiple-disk turbine using incompressible fluid. *J. Fluids Eng.* 1974, *96*, 252–258.

25. Carey, V.P. Assessment of Tesla turbine performance for small scale Rankine combined heat and power systems. *J. Eng. Gas Turbines Power* 2010, *132*, 122301.
26. Sengupta, S.; Guha, A. A theory of Tesla disc turbine. *J. Power Energy* 2012, *226*, 650–663.
27. Sochol, R.D.; Sweet, E.; Glick, C.C.; Venkatesh, S.; Avetisyan, A.; Ekman, K.F.; Raulinaitis, A.; Tsai, A.; Wienkers, A.; Korner, K.; et al. 3D printed microfluidic circuitry via multi-jet-based additive manufacturing. *Lab Chip* 2016, *16*, 668–678.
28. Kitson, P.J.; Rosnes, M.H.; Sans, V.; Dragone, V.; Cronin, L. Configurable 3D-Printed millifluidic and microfluidic 'lab on a chip' reaction ware devices. *Lab Chip* 2012, *12*, 3267–3271.
29. Au, A.K.; Bhattacharjee, N.; Horowitz, L.F.; Chang, T.C.; Folch, A. 3D-printed microfluidic automation. *Lab Chip* 2015, *15*, 1934–1941.
30. Bishop, G.W.; Satterwhite, J.E.; Bhakta, S.; Kadimisetty, K.; Gillette, K.M.; Chen, E.; Rusling, J.F. 3D-Printed Fluidic Devices for Nanoparticle Preparation and Flow-Injection Amperometry Using Integrated Prussian Blue Nanoparticle-Modified Electrodes. *Anal. Chem.* 2015, *87*, 5437–5443.
31. Zhu, F.; Skommer, J.; Macdonald, M.P.; Friedrich, T.; Kaslin, J.; Wlodkowic, D. Three-dimensional printed millifluidic devices for zebrafish embryo tests. *Biomicrofluidics* 2015, *9*, 046502.
32. Wu, S.Y.; Yang, C.; Hsu, W.; Lin, L. 3D-printed microelectronics for integrated circuitry and passive wireless sensors. *Microsyst. Nanoeng.* 2015.
33. Zhang, C.; Anzalone, N.C.; Faria, R.P.; Pearce, J.M. Open-source 3D-printable optics equipment. *PLoS ONE* 2013, *8*, e59840.
34. Willis, K.; Brockmeyer, E.; Hudson, S.; Poupyrev, I. Printed optics: 3D printing of embedded optical elements for interactive devices. In Proceedings of the 25th Annual ACM Symposium on User Interface Software and Technology, Cambridge, MA, USA, 7–10 October 2012; pp. 589–598.
35. Zou, L.; Koslakiewicz, R.; Mahmoud, M.; Fahs, M.; Liu, R.; Lo, J.F. 3D Printed miniaturized spectral system for collagen fluorescence lifetime measurements. *J. Biomed. Opt.* 2016, *21*, 075001.
36. Lam, R.H.; Li, W.J. A digitally controllable polymer-based microfluidic mixing module array. *Micromachines* 2012, *3*, 279–294.
37. Godwin, L.A.; Deal, K.S.; Hoepfner, L.D.; Jackson, L.A.; Easley, C.J. Measurement of microchannel fluidic resistance with a standard voltage meter. *Anal. Chim. Acta* 2013, *758*, 101–107.
38. Guha, A.; Smiley, B. Experiment and analysis for an improved design of the inlet and nozzle in Tesla disc turbines. *J. Power Energy* 2010, *224*, 261–277.

13 CHAPTER

Arbitrarily Accessible 3D Microfluidic Device for Combinatorial High-Throughput Drug Screening

Zhuofa Chen [1], Weizhi Li [2], Gihoon Choi [1], Xiaonan Yang [1], Jun Miao [2], Liwang Cui [2] and Weihua Guan [1,3]

[1] *Department of Electrical Engineering, Pennsylvania State University, University Park, PA 16802, USA*

[2] *Department of Entomology, Pennsylvania State University, University Park, PA 16802, USA*

[3] *Department of Biomedical Engineering, Pennsylvania State University, University Park, PA 16802, USA*

ABSTRACT

Microfluidics-based drug-screening systems have enabled efficient and high-throughput drug screening, but their routine uses in ordinary labs are limited due to the complexity involved in device fabrication and system setup. In this work, we report an easy-to-use and low-cost arbitrarily accessible 3D microfluidic device that can be easily adopted by various labs to perform combinatorial assays for high-throughput drug screening. The device is capable of precisely performing automatic and simultaneous reagent loading and aliquoting tasks and performing multistep assays with arbitrary sequences. The device is not intended to compete with other microfluidic technologies regarding ultra-low reaction volume. Instead, its freedom from tubing or pumping systems and easy operation makes it an ideal platform for routine high-throughput drug screening outside traditional microfluidic labs. The functionality and quantitative reliability of the 3D microfluidic device were demonstrated with a histone acetyltransferase-based drug-screening assay using the recombinant Plasmodium falciparum GCN5 enzyme, benchmarked with a traditional microtiter plate-based method. This arbitrarily accessible, multistep capable, low-cost, and easy-to-use device can be widely adopted in various combinatorial assays beyond high-throughput drug screening.

Keywords: *microfluidic; high-throughput; combinatorial; multiplex; drug screening*

1. INTRODUCTION

High-throughput techniques are critically needed for efficient screening of pharmaceutically valuable lead compounds. The rapid progress of high-throughput screening (HTS) has enabled parallel analysis of thousands of reactions in order to identify effective compounds for a particular biological process. Most of the current HTS technologies involve robotics for automatic liquid and plate handling (96-, 384-, and 1536-well) [1,2,3,4,5]). Although the throughput has been increased tremendously by robotic technology as compared to manual operations [6], the high cost associated with the instrument has restricted many researchers from performing HTS independently. In addition, the volume required in the microtiter plate (100 μL and above) translates into a high cost for biological samples and drug libraries. As a result, there is a strong desire to develop low-cost technologies that require less sample and reagent consumption. To that end, the microfluidics-based drug-screening platform has been proposed due to its reduced reagent consumption, low cost, and high throughput [7,8,9,10].

In the past few years, significant progress has been achieved in developing microfluidics-based drug-screening components and systems. For example, the valve-based high-content microfluidic chamber arrays for a cell-based drug screening assay have been demonstrated [11,12,13,14,15]. The flow diffusive mixing-based systems provided a versatile method to generate on-chip concentration gradients [16,17,18,19,20,21]. The droplet-based microfluidics has shown nanoliter to picoliter compartmentalization in a high-throughput manner [22,23,24,25,26]. In addition, a low-cost slip chip provided a facile and economic method for performing microfluidic reactions without pumps or valves [10]. Recently, low-cost and easy-to-use 3D microfluidic chips have been introduced for high-throughput screening [27,28]. Despite impressive progress in microfluidics-based drug screening, challenges still remain. First, performing a multistep (>2 steps) combinatorial assay is a big challenge for most existing systems due to the difficulty in fluidic control. Second, microfluidic systems that require peripheral tubing and pumping system are too complex to operate by non-microfluidic researchers. Third, though there are commercial microwell plates in the volume range of a few microliters (e.g., Corning, Labcyte etc.), the reagent loading and aliquoting schemes are still labor-intensive due to its serial nature. As a result, there is a strong desire to develop an easy-to-use, low-cost, and multistep-assay-compatible microfluidic system that can be used by ordinary labs to perform HTS.

In this study, we describe an arbitrarily accessible, tubing-and-pumping-system-free, and multistep-assay-compatible 3D microfluidic device for HTS that can be routinely adopted by ordinary labs. Our microfluidic devices use a few hundred nanoliters to microliter to avoid the evaporation challenges common in nanoliter- to picoliter-scale multistep assays [29,30]. The device consists of sample-loading chips and auto-aliquoting chips for rapid, precise, automatic, and simultaneous heterogeneous/homogeneous reagent array formation. As a proof of concept, the device was validated with a recombinant *Plasmodium falciparum* GCN5 (*Pf*GCN5) enzyme-based drug-screening assay. The *Pf*GCN5 enzyme plays an important role in the regulation of chromatin structure and thus gene transcription in malaria parasites [31,32]; inhibition of *Pf*GCN5 activity is believed to be a potential target for novel chemotherapies of malaria [33,34,35]. The excellent quantitative agreement between the arbitrarily accessible 3D microfluidic device and the gold-standard 96-well microtiter plate demonstrates the reliability, robustness, and accuracy of the 3D microfluidic device. As the tubing-and-pumping-system-free 3D microfluidic device capable of performing multi-step combinatorial assays with arbitrary sequences, we believe it offers unprecedented opportunities for low-cost and high-throughput drug screening in various laboratory settings.

2. MATERIALS AND METHODS

2.1. Materials and Chemicals

Polymethylmethacrylate (PMMA) sheets were from ePlastics. Food dye was from AmeriColor. HAT Activity Fluorometric Assay Kit was obtained from Biovision (Biovision Incorporated, Milpitas, CA, USA). Calcein and histone acetyltransferase (HAT) inhibitors (curcumin, cyclopentylidene-[4-(4′-chlorophenyl) thiazol-2-yl] hydrazine (CPTH-2), and (2R, 3S)-rel-4-methylene-5-oxo-2-propyltetrahydrofuran-3-carboxylic acid (MB-3)) were obtained from Sigma-Aldrich (St. Louis, MO, USA).

2.2. Device Design and Fabrication

The 3D microfluidic device consists of (1) a sample-loading chip; and (2) an auto-aliquoting chip (Figure S1). Both chip patterns were designed in AutoCAD 2015 (Autodesk Inc., San Rafael, CA, USA). The chips were fabricated with PMMA, which was patterned by the CO_2 laser cutting machine (Epilog Laser Inc., Golden, CO, USA) with a power of 100%, a speed of 25% (for sample-loading chip of 1.6 mm thick), 30% (for auto-aliquoting chip of 1 mm thick), 60% (for spacer of 0.8 mm thick), and a frequency of 5000 Hz. After laser cutting, both chips were thoroughly cleaned in distilled water and isopropyl alcohol for 10 min. Both chips were then sandwiched between two metal plates and annealed in the oven for 24 h at 90 °C for stress relief. The sample-loading chip was then used as it is, whereas the auto-aliquoting chip was surface-coated with a thin layer of hydrophobic wax by laminating with the wax paper at 120 °C (Apache AL13P). The fabricated chips were kept in a sealed package before use. More detailed fabrication information can be found in Supplementary Materials.

2.3. Device Operation

The arbitrarily accessible 3D microfluidic device starts in a high-throughput manner to form heterogeneous or homogeneous reagent arrays. (1) For heterogeneous array formation (Figure 1a), reagents with various concentrations were prepared and diluted in a 96-well plate and then dispensed onto the sample-loading chip with an HTS-compatible pipetting system (left illustration in Figure 1a). Afterward, the auto-aliquoting chip was simply placed on top of the sample-loading chip, with a spacer of 0.8 mm height (middle illustration in Figure 1a). The auto-aliquoting chip was then separated from the sample-loading chip. Due to the capillary force, the auto-aliquoting chip efficiently formed a uniform and heterogeneous reagent array, each of a volume of ~2 μL (right illustration in Figure 1a); (2) For homogeneous array formation (Figure 1b), a facile scraping process is sufficient to form a uniform array with the help of the capillary force. After forming various desired heterogeneous and homogeneous reagents arrays, a combinatorial assay can be performed by aligning and arbitrarily combining these microarrays (Figure 1c). Figure 1d shows the photograph of the device with the auto-aliquoting chip placed on top of the sample-loading chip. We did not find any liquid evaporation issues during the time course of the experiment.

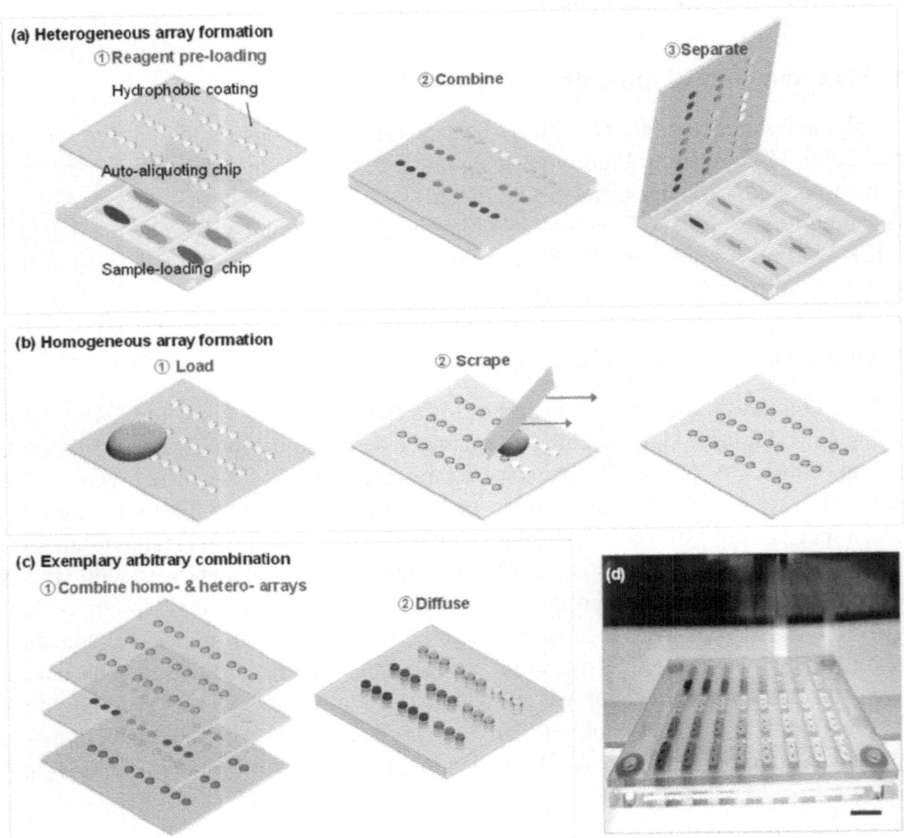

Figure 1. *Schematic demonstration and photograph of the 3D microfluidic device. (a) Heterogeneous array formation. Heterogeneous reagents were loaded by high-throughput pipetting system onto the sample-loading chip. The pinning effect holds the liquid on each sample-loading island. By combining and separating the auto-aliquoting chip with the sample-loading chip (with a 0.8 mm high spacer), a heterogeneous reagent array is automatically formed; (b) Homogeneous array formation. Homogeneous reagent (e.g., enzyme solution or histone acetyltransferase (HAT) reaction mixture solution) was simply scraped over the wax-treated auto-aliquoting chip; (c) Exemplary arbitrary combination of three auto-aliquoting chips. The mixing of various reagents is achieved by the diffusion process; (d) Photograph of the device with the auto-aliquoting chip placed on top of the sample-loading chip. Note the schematic is not drawn to scale. The scale bar in the photograph is 1 cm.*

2.4. Device Validation

To evaluate the uniformity and reaction performance of the 3D microfluidic device, we carried out validation experiments using the plain food dyes as well as the fluorescent calcein dye. For the food dye experiment, different colored food dyes were loaded and aliquoted to form a heterogeneous array and a homogenous array, using the methods mentioned above. For the fluorescent calcein dye experiment, calcein with concentrations ranging from 0.02 μM to 2.5 μM was used to form a heterogeneous array. A homogeneous array was formed with 1.25 μM calcein using the scraping method. The homogeneous and heterogeneous chips were then aligned and combined to evaluate the uniformity and accuracy of the device after mixing.

2.5. PfGCN5 Enzyme Purification

To purify the *Pf*GCN5 HAT from the malaria parasite *Plasmodium falciparum*, a PTP (ProtC-TEV-ProtA) tag was added to the C-terminus of the endogenous *Pf*GCN5 gene in the malaria parasite as described before [36,37]. Briefly, ~1 kb fragment from the C-terminus end (without the stop codon) of *Pf*GCN5 was amplified and fused to the PTP tag and 3′ UTR region of the *P. berghei dhfr-ts* gene and cloned into the transfection vector pHD22Y with the human DHFR resistance cassette. Parasite transfection, drug selection, and cloning were performed as described [36,38]. Positive clones were verified by integration-specific PCR. The malaria parasites with PTP-tagged *Pf*GCN5 were cultured using a standard procedure [39] and the *Pf*GCN5 was purified by IgG beads under the native condition as described previously [36]. The purified protein was stored at −80 °C until enzymatic analysis.

2.6. HAT Assay for PfGCN5-Based Drug Screening

The 3D microfluidic device was applied in the HAT-based drug-screening assay to evaluate various drugs (inhibitors) against the purified *Pf*GCN5. In this assay, *Pf*GCN5 catalyzes the transfer of the acetyl group from acetyl–CoA to an H3 histone peptide, generating acetylated peptide and CoA–SH. The CoA–SH reacts with the developer to produce signaling fluorophore that is detected at Ex/Em = 535/587 nm. We first performed the reference experiment to investigate the compatibility of the 3D microfluidic device to the HAT-based assay and to establish the standard curve for the end-product (CoA–SH) according to the manual of HAT Activity Fluorometric Assay Kit. In addition, a pilot HAT-based drug-screening assay using positive controls (*Pf*GCN5 enzyme and HAT reaction mixture) and negative controls (buffer and HAT reaction mixture) was carried out on the 3D microfluidic device to determine the Z′-factor and thus the HTS potential of the assay [40]. Finally, several established inhibitors against the *Pf*GCN5 enzyme, including MB-3, CPTH-2, and curcumin [34,41,42], were used as reference drugs on the 3D microfluidic device and compared with the gold-standard 96-well microtiter plate method. The *Pf*GCN5 inhibitors were loaded onto the auto-aliquoting chip by the heterogeneous array formation process as described above, whereas *Pf*GCN5 and the HAT reaction mixture solution were loaded by the homogeneous array formation process. Detailed information of the experiment and the determination of the Z′-factor and the half maximal inhibitory concentration (IC_{50}) are provided in the Supplementary Materials. The inhibitor chip, *Pf*GCN5 chip, and the HAT reaction mixture solution chip were then aligned and combined with an adaptor and incubated in a custom-built box at room temperature for 40 min in the dark. The combined chips were put on the homemade adapter and analyzed by the Typhoon scanner (GE 9410). With the adapter, the combined plates are 2 mm away from the surface of the scanner to prevent contaminating the scanner. We choose the 3 mm height mode for scanning with Ex/Em = 532/580 nm. The images were analyzed by ImageJ (NIH).

3. RESULTS AND DISCUSSION

3.1. Device Design Theoretical Considerations

The sample-loading chip relies on the pinning effect to operate. For a liquid of an equilibrium contact angle θ on a solid surface moving towards a three-phase (liquid/vapor/solid) edge, the droplet will be pinned at the edge due to the increased liquid/vapor interface area and the increased activation barrier to move. The contact angle on the edge will increase from θ to θ + α and the droplet cannot move over the edge until the contact angle exceeds θ + α, as shown in Figure 2a, where α is the surface bending angle [43,44,45]. Our sample-loading chip was fabricated by cutting through rectangle-shaped windows with α = 90°, thus the maximum contact angle of a droplet on the sample-loading island is θ + 90°. As shown in Figure 2b,c, the as-fabricated PMMA chip has a contact angle of 60° and the measured contact angle on the edge of the sample-loading island is 150°, corresponding well to the theoretical prediction. This tremendously increased the contact angle and activation energy and helped prevent reagent leaking and mixing during the operation of the sample-loading chip.

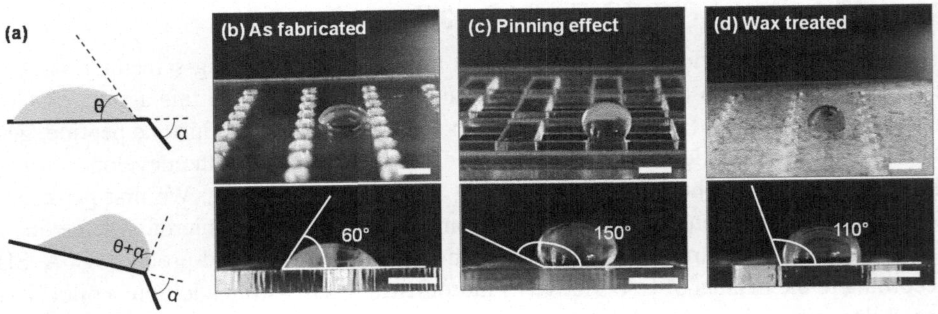

Figure 2. *Pinning effect and chip surface treatment. (a) A droplet on a solid surface with a contact angle of θ, which will be increased up to θ + α when moving towards a three-phase edge; (b) The as-fabricated PMMA chip shows a contact angle of 60°; (c) The droplet pinned on the sample-loading chip shows a 150° contact angle; (d) The wax treated auto-aliquoting chip shows a contact angle of 110°. The scale bars in the images are 3 mm.*

Another important aspect of the device operation is the surface wettability. For an as-fabricated PMMA chip without hydrophobic treatment, the observable liquid trace would be left on the chip surface. As a result, a hydrophobic surface is desirable. Accordingly, we performed wax treatment on the auto-aliquoting chip to obtain a hydrophobic surface. As shown in Figure 2d, the contact angle on the wax-treated auto-aliquoting chip increased to 110°. The wax on the device surface reduces the surface free energy and thus increases the liquid contact angle, consistent with the previous report [46]. The hydrophobic property of the auto-aliquoting chip made the subsequent operations of the 3D device more reliable. First, it minimized the liquid trace left on the chip surface during the array formation processes. Second, it prevented the liquid from leaking out when several auto-aliquoting chips were combined for reagent mixing and reaction.

The heterogeneous array formation process is achieved by the capillary force. Once the auto-aliquoting chip is in contact with the sample-loading chip, the capillary force automatically drives the liquid into the auto-aliquoting chip array. At equilibrium, the upward force Fup=2πaγLVcosθ is in balance with the downward gravity force Fdown=ρghπa2where γLV is the surface tension between liquid/vapor, a and hh are the radius and the depth of the well, respectively, θ is the contact angle of liquid on the solid surface, ρ is the liquid density, and g is the acceleration of gravity [47]. For effective liquid loading, the capillary force needs to overcome the gravity force, which requires a≤2γLVcosθ/ρgh. Each well in the auto-aliquoting chip used in the experiment has a radius of 750 μm and height of 1000 μm. This results in a capillary force of 171 μN, much larger than the gravity barrier of 17.3 μN, and thus an efficient sample auto-aliquoting process is guaranteed (Figure 1a).

3.2. Quantitative Performances of the 3D Device

To establish the proof of concept for the proposed device, we tested the 3D microfluidic device with plain food dyes and a fluorescent calcein dye. We assessed the 3D microfluidic device by evaluating the loading efficiency, uniformity, and accuracy. Figure 3a–c shows the testing results with the food dyes. A heterogeneous array was generated with food dyes of different colors by a heterogeneous array formation process (Figure 3a). No observable dye trace was left over the auto-aliquoting chip surface after combining/separating with the sample-loading chip, indicating a highly efficient and effective array formation process. The homogeneous array with a single dye was formed by a homogeneous array formation process (Figure 3b). Due to the strong hydrophobicity of the chip surface, the scraping process only filled the wells without leaving a liquid trace on the chip surface. By combining these two arrays, a mixing process occurs through the diffusion process (Figure 3c). The color gradients indicated an effective mixing process. Leaking was not observed between the two arrays, indicating no cross-contamination during the combining process. The result with the food dyes validated the operating principle of the 3D microfluidic device.

To further characterize the quantitative performance of the 3D microfluidic device, we carried out the test with the fluorescent calcein dye. A heterogeneous array was formed with different concentrations of calcein dye (0.02, 0.04, 0.08, 0.16, 0.31, 0.63, 1.25, and 2.50 μM) using the heterogeneous array formation process, while a homogeneous array was formed with 1.25 μM of calcein using the homogeneous array formation process (Figure 3d,e). Again, no dye trace was detected on the array surfaces. Figure 3f shows the image of the mixed chips after combining the heterogeneous and homogeneous arrays. The quantitative performance can be evaluated from images of Figure 3d–f. First, Figure

3g shows the measurement of the fluorescence intensities of the heterogeneous array. The excellent fit (R^2 = 0.98) in the linear range and small variations among replicates demonstrate that the heterogeneous array formation process is successful and reliable. Second, Figure 3h shows the measurement of the fluorescence intensities of the homogeneous array formed by scraping method. The fluorescence intensity variation of each array well was less than 2.3%, indicating a uniform homogeneous array formation process. Figure 3i shows the measured and the expected fluorescence intensity for the combined chip from Figure 3d,e. The excellent quantitative agreement confirms a thorough reaction mixing was achieved.

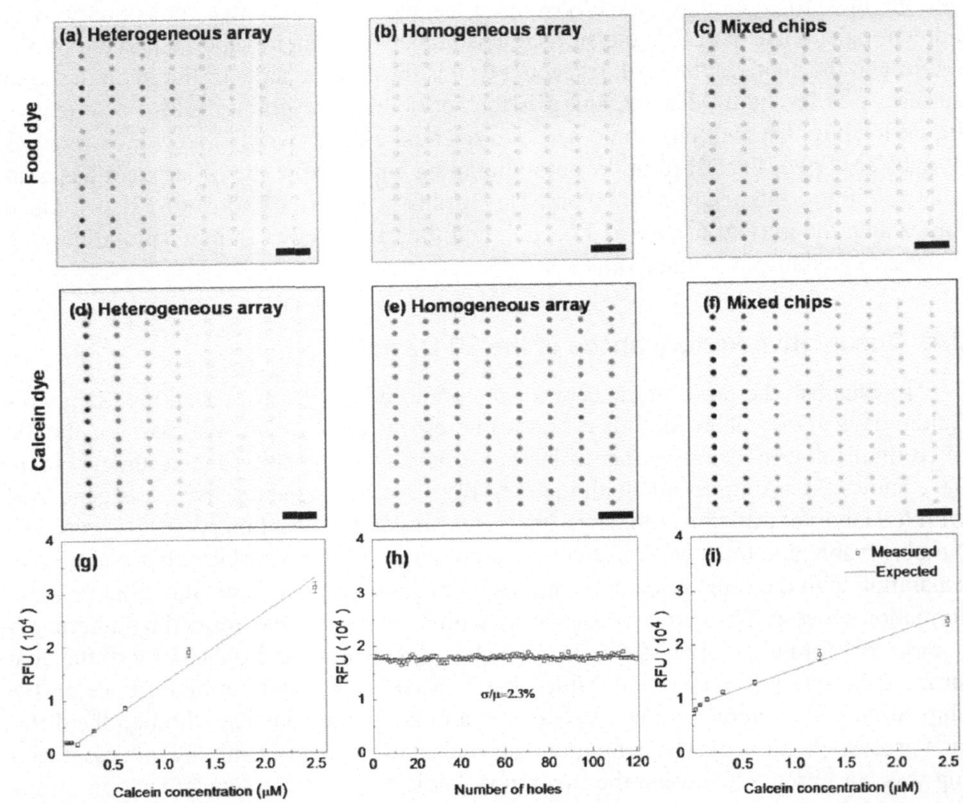

Figure 3. *Validation of the 3D microfluidic device with plain food dyes and fluorescent calcein dye. (a–c) Food dye validation. The image shows the heterogeneous array chip consisting of different colors (a); homogeneous array chip of a single color (b); and the mixed chips (c); (d–f) Fluorescent calcein dye validation. The scanning images show the heterogeneous array chip (d); homogeneous array chip (e); and the mixed chips (f); (g–i) Measured fluorescence intensity for the heterogeneous array (g); homogeneous array (h); and the mixed chip (i); corresponding to images in (d–f); respectively. The fluorescence variation (σ/μ) within a homogeneous array is ~2.3% for 120 representative wells. The excellent agreement between the measured and the expected fluorescence intensity in the mixed chip shows that the mixing is reliable and thorough. Error bars correspond to 15 replicates. The scale bars in the images are 10 mm.*

3.3. PfGCN5-Based Malaria Drug Screening

After validation of the device performance using dyes, we performed drug screening using the *Pf*GCN5 HAT on the 3D microfluidic device. We first performed a control experiment to investigate the suitability of the 3D microfluidic device for the HAT-based assay and to establish a standard curve for the end-product (CoA–SH) according to the manual of HAT Activity Fluorometric Assay Kit. The result was compared with that of the 96-well microtiter plate method as the gold standard. Figure 4a illustrates the reaction associated with the HAT-based assay [48]. The *Pf*GCN5 enzyme catalyzes the acetylation reaction that generates CoA–SH. The CoA–SH reacts with the developer to produce a signaling fluorophore. Figure 4b,c show the fluorescence intensity as a function of CoA–SH

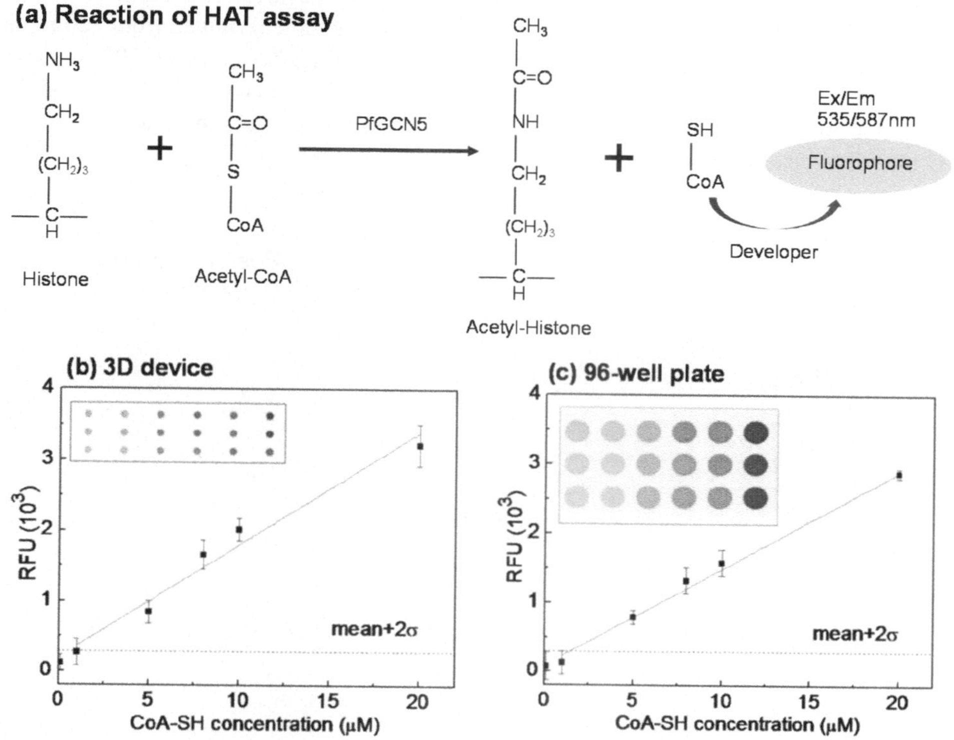

Figure 4. *(a) Schematic illustration of HAT-based assay reaction. The PfGCN5 enzyme catalyzes the acetylation and generates CoA–SH. The CoA–SH then reacts with the developer to produce signaling fluorophore; (b,c) the fluorescence intensity as a function of CoA–SH concentration on the 3D microfluidic device (b) and the 96-well microtiter plate (c). The insets show the scanning images corresponding to the plots in (b,c). The reaction volume in the 3D microfluidic device is 2 μL, 50 times less than the 100 μL in the 96-well microtiter plate. The red line is fitting for the linear range, which shows a good fit of R2 = 0.93 and 0.98 for the 3D microfluidic device and the 96-well microtiter plate, respectively. The dashed line indicated the optical scanner's limit of detection, which is determined as two standard deviations above the background from the negative control (buffer and HAT reaction mixture). Error bars correspond to three replicates.*

concentration reacting with the developer in the 3D microfluidic device and the 96-well microtiter plate, respectively. The plots show the standard curve of CoA–SH with a good fit for the 3D microfluidic device ($R^2 = 0.93$) and the 96-well microtiter plate ($R^2 = 0.98$), respectively. While the 3D microfluidic device consumes only 2 μL of reagents per reaction (a 50 times reduction as compared to the volume of 100 μL in a 96-well microtiter plate), the quantitative performance of the 3D microfluidic device and the 96-well microtiter plate remained the same (Figure 4b,c). This confirmed the compatibility of the 3D microfluidic device for the HAT-based drug-screening assay.

Pilot HAT-based drug-screening assays using positive and negative controls were carried out on the 3D microfluidic device to determine the Z′-factor, an indicator of the HTS potential of the assay [40]. Figure 5 shows the fluorescence intensity of the positive and negative controls of the HAT-based drug-screening assay. The positive controls have a mean value of 35,576 and a standard deviation of 613. The negative controls have a mean value of 28,149 and a standard deviation of 608. The Z′-factor determined from 120 replicates in three independent experiments on the 3D microfluidic device was 0.507, suggesting that it is suitable for HTS [40]. The small variation of positive controls and negative controls from different experiments also indicates the stability and reproducibility of the 3D microfluidic device.

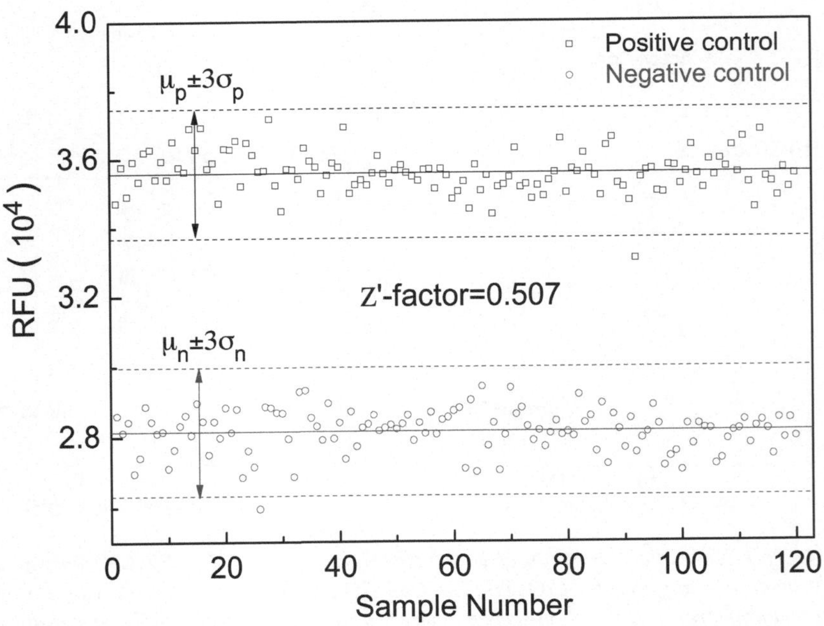

Figure 5. *Z′-factor determination of the HAT-based drug-screening assay on the 3D microfluidic device. The solid horizontal lines show the mean values of the positive control and negative control. Black and red dash lines display 3 standard deviations (σ) from the mean value of positive control and negative control data set. The Z′-factor is determined to be 0.507 according to* $Z'=1-(3\sigma p+3\sigma n)/|\mu p-\mu n|$. *A Z′-factor in the range between 0.5 and 1 indicates that the assay is suitable for HTS.*

After confirming the compatibility of the HAT-based assay with the 3D microfluidic device, MB-3, CPTH-2, and curcumin were used as reference compounds on the 3D microfluidic device and compared with the 96-well microtiter plate method. Figure 6 showed that MB-3, CPTH-2, and curcumin are strong inhibitors of the *Pf*GCN5 HAT, consistent with previous reports [34,41,42]. Higher concentrations of these compounds inhibited more *Pf*GCN5 enzyme and resulted in less CoA–SH (Figure 4a), leading to lower fluorescence intensity. The IC_{50} values determined on the 3D microfluidic device and in the 96-well microtiter plate were comparable, with a variation smaller than 11% (120 and 134 μM for MB-3, 127 and 116 μM for CPTH-2, 3.85 and 3.38 μM for curcumin using the 3D microfluidic device and the 96-well microtiter plate, respectively). The excellent agreement between the two results demonstrates the reliability, robustness, and accuracy of the 3D microfluidic device.

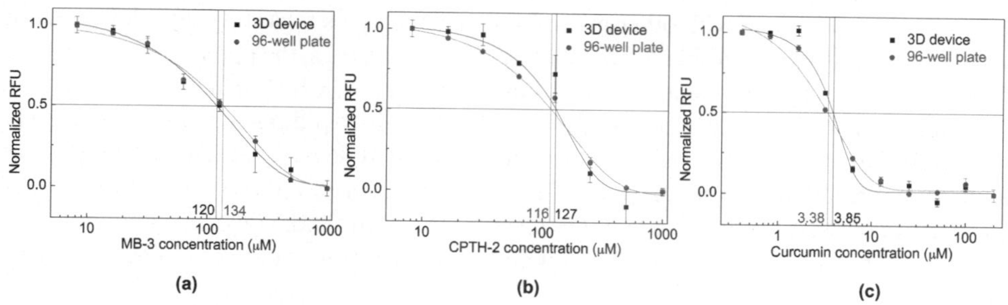

Figure 6. *Drug response curves of the PfGCN5 enzyme to (a) MB-3; (b) CPTH-2 and (c) curcumin on the 3D microfluidic device (black) and 96-well microtiter plate (red). PfGCN5 was tested under serial concentrations of drugs. The plots show the normalized relative fluorescence unit as a function of drug concentration, where IC50 is the drug concentration at which normalized relative fluorescence units (RFU) is 0.5, indicating 50% inhibition. The extracted IC50 are indicated on the plot. Error bars correspond to three replicates.*

3.4. Versatility and Adaptability of the 3D Microfluidic Device

It is worth noting that we do not envision the 3D microfluidic device to compete with picoliter to nanoliter technologies (e.g., droplet microfluidics [22,25]). The advantages of our 3D microfluidic device are the following: (1) High-throughput. By adopting the heterogeneous and homogeneous array formation processes, the 3D microfluidic device allows parallel reagent loading and aliquoting in a much faster method (a few seconds) than the serial loading process using a pipetting system (minutes to hours); (2) Multiplexity. With the automatic heterogeneous array formation process from a library of reagents, various reagents can be simultaneously arrayed using the 3D microfluidic device. It is noteworthy that reagents from different auto-aliquoting chips can be mixed in parallel by the combinatorial process, ensuring simultaneous incubation for each individual reactions; (3) Versatility. The combinatorial nature of the 3D microfluidic device allows for any multistep assay with arbitrary incubation sequence. A device capable of performing the multistep assay is highly desirable since a variety of biological assays require a series of combinatorial reactions.

Besides, the 3D microfluidic device is versatile with various readout methods, as the detection can be easily extended to various imaging setups; (4) Low-Cost. Both the device materials (PMMA, 0.04 dollars per chip) and the fabrication process (laser cutting) is cost-effective. The device operation requires no tubing and pumping instruments and thus can be easily adapted to other labs without microfluidics experience and microfluidics infrastructures.

4. CONCLUSIONS

We report an easy-to-use, scalable, and cost-effective 3D arbitrary microfluidic device for an HTS purpose. Using an HAT-based malaria drug-screening assay as a model system, we evaluated the drug-screening performance and benchmarked it with the gold-standard 96-well microtiter plate method. The demonstrated 3D arbitrary microfluidic device is capable of easily performing automatic and simultaneous reagent aliquoting due to its parallel nature, as well as performing an arbitrary sequence of assay steps. The device is not intended to compete with other microfluidic technologies regarding ultra-low reaction volumes (e.g., pL–nL in droplet microfluidics [22,25]). Instead, the operation of the device does not require any peripheral tubing or pumping systems, which makes it an ideal platform for ordinary labs to perform high-throughout, multistep drug-screening assays. The current device can be easily scaled up to an array size of 1056 with reagent consumption of ~100 nL per reaction. A thorough screening of a small chemical library against *Pf*GCN5 is currently underway. We believe that the 3D arbitrary microfluidic device is not only valuable as a general platform for high-throughput drug-screening, but could also be widely extended to various other bioassays.

Acknowledgments

This project was partly supported by the Penn State CTSI Grant (UL Tr000127) from the National Center for Advancing Translational Sciences, National Institutes of Health. The content is solely the responsibility of the authors and does not necessarily represent the official views of the NIH. This project was also supported by Penn State Startup Fund (Guan).

Author Contributions

Zhuofa Chen conceived and performed the research; Weihua Guan supervised the study; Zhuofa Chen and Weizhi Li performed the drug-screening assay; Jun Miao extracted the *Pf*GCN5; Zhuofa Chen, Xiaonan Yang, and Gihoon Choi analyzed the data; Zhuofa Chen wrote the manuscript and discussed with Weihua Guan, Liwang Cui, and Jun Mao.

REFERENCES

1. Shi, X.X.; Sha, S.S.; Liu, L.K.; Li, X.; Ma, Y.F. A 96-well microtiter plate assay for high-throughput screening of Mycobacterium tuberculosis dTDP-d-glucose 4,6-dehydratase inhibitors. *Anal. Biochem.* 2016, *498*, 53–58.
2. Semedo, M.C.; Karmali, A.; Fonseca, L. A novel colorimetric assay of beta-d-glucans in basidiomycete strains by alcian blue dye in a 96-well microtiter plate. *Biotechnol. Prog.* 2015, *31*, 1526–1535.

3. Booij, P.; Vethaak, A.D.; Leonards, P.E.G.; Sjollema, S.B.; Kool, J.; de Voogt, P.; Lamoree, M.H. Identification of photosynthesis inhibitors of pelagic marine algae using 96-well plate microfractionation for enhanced throughput in effect-directed analysis. *Environ. Sci. Technol.* 2014, *48*, 8003–8011.
4. Judson, R.; Kavlock, R.; Martin, M.; Reif, D.; Houck, K.; Knudsen, T.; Richard, A.; Tice, R.R.; Whelan, M.; Xia, M.H.; et al. Perspectives on validation of high-throughput assays supporting 21st century toxicity testing. *ALTEX* 2013, *30*, 51–66.
5. Attene-Ramos, M.S.; Miller, N.; Huang, R.L.; Michael, S.; Itkin, M.; Kavlock, R.J.; Austin, C.P.; Shinn, P.; Simeonov, A.; Tice, R.R.; et al. The Tox21 robotic platform for the assessment of environmental chemicals—From vision to reality. *Drug Discov. Today* 2013, *18*, 716–723.
6. Conway, M.K.; Gerger, M.J.; Balay, E.E.; O'Connell, R.; Hanson, S.; Daily, N.J.; Wakatsuki, T. Scalable 96-well plate based iPSC culture and production using a robotic liquid handling system. *J. Vis. Exp.* 2015, *99*, e52755.
7. Du, G.; Fang, Q.; den Toonder, J.M. Microfluidics for cell-based high throughput screening platforms—A review. *Anal. Chim. Acta* 2016, *903*, 36–50.
8. Li, X.; Zhang, X.; Zhao, S.; Wang, J.; Liu, G.; Du, Y. Micro-scaffold array chip for upgrading cell-based high-throughput drug testing to 3D using benchtop equipment. *Lab Chip* 2014, *14*, 471–481.
9. Neuzi, P.; Giselbrecht, S.; Lange, K.; Huang, T.J.; Manz, A. Revisiting lab-on-a-chip technology for drug discovery. *Nat. Rev. Drug Discov.* 2012, *11*, 620–632.
10. Du, W.; Li, L.; Nichols, K.P.; Ismagilov, R.F. SlipChip. *Lab Chip* 2009, *9*, 2286–2292.
11. Kim, J.; Taylor, D.; Agrawal, N.; Wang, H.; Kim, H.; Han, A.; Rege, K.; Jayaraman, A. A programmable microfluidic cell array for combinatorial drug screening. *Lab Chip* 2012, *12*, 1813–1822.
12. Park, E.S.; Brown, A.C.; DiFeo, M.A.; Barker, T.H.; Lu, H. Continuously perfused, non-cross-contaminating microfluidic chamber array for studying cellular responses to orthogonal combinations of matrix and soluble signals. *Lab Chip* 2010, *10*, 571–580.
13. Wang, Z.H.; Kim, M.C.; Marquez, M.; Thorsen, T. High-density microfluidic arrays for cell cytotoxicity analysis. *Lab Chip* 2007, *7*, 740–745.
14. Kane, B.J.; Zinner, M.J.; Yarmush, M.L.; Toner, M. Liver-specific functional studies in a microfluidic array of primary mammalian hepatocytes. *Anal. Chem.* 2006, *78*, 4291–4298.
15. Unger, M.A.; Chou, H.P.; Thorsen, T.; Scherer, A.; Quake, S.R. Monolithic microfabricated valves and pumps by multilayer soft lithography. *Science* 2000, *288*, 113–116.
16. Yang, C.G.; Wu, Y.F.; Xu, Z.R.; Wang, J.H. A radial microfluidic concentration gradient generator with high-density channels for cell apoptosis assay. *Lab Chip* 2011, *11*, 3305–3312.
17. Selimovic, S.; Sim, W.Y.; Kirn, S.B.; Jang, Y.H.; Lee, W.G.; Khabiry, M.; Bae, H.; Jambovane, S.; Hong, J.W.; Khademhosseini, A. Generating nonlinear concentration gradients in microfluidic devices for cell studies. *Anal. Chem.* 2011, *83*, 2020–2028.
18. Sugiura, S.; Hattori, K.; Kanamori, T. Microfluidic serial dilution cell-based assay for analyzing drug dose response over a wide concentration range. *Anal. Chem.* 2010, *82*, 8278–8282.
19. Ye, N.N.; Qin, J.H.; Shi, W.W.; Liu, X.; Lin, B.C. Cell-based high content screening using an integrated microfluidic device. *Lab Chip* 2007, *7*, 1696–1704.
20. Qin, J.H.; Ye, N.N.; Liu, X.; Lin, B.C. Microfluidic devices for the analysis of apoptosis. *Electrophoresis* 2005, *26*, 3780–3788.
21. Jeon, N.L.; Dertinger, S.K.W.; Chiu, D.T.; Choi, I.S.; Stroock, A.D.; Whitesides, G.M. Generation of solution and surface gradients using microfluidic systems. *Langmuir* 2000, *16*, 8311–8316.

22. Guo, M.T.; Rotem, A.; Heyman, J.A.; Weitz, D.A. Droplet microfluidics for high-throughput biological assays. *Lab Chip* 2012, *12*, 2146–2155.

23. Baraban, L.; Bertholle, F.; Salverda, M.L.M.; Bremond, N.; Panizza, P.; Baudry, J.; de Visser, J.A.G.M.; Bibette, J. Millifluidic droplet analyser for microbiology. *Lab Chip* 2011, *11*, 4057–4062.

24. Trivedi, V.; Doshi, A.; Kurup, G.K.; Ereifej, E.; Vandevord, P.J.; Basu, A.S. A modular approach for the generation, storage, mixing, and detection of droplet libraries for high throughput screening. *Lab Chip* 2010, *10*, 2433–2442.

25. Brouzes, E.; Medkova, M.; Savenelli, N.; Marran, D.; Twardowski, M.; Hutchison, J.B.; Rothberg, J.M.; Link, D.R.; Perrimon, N.; Samuels, M.L. Droplet microfluidic technology for single-cell high-throughput screening. *Proc. Natl. Acad. Sci. USA* 2009, *106*, 14195–14200.

26. Clausell-Tormos, J.; Lieber, D.; Baret, J.C.; El-Harrak, A.; Miller, O.J.; Frenz, L.; Blouwolff, J.; Humphry, K.J.; Koster, S.; Duan, H.; et al. Droplet-based microfluidic platforms for the encapsulation and screening of mammalian cells and multicellular organisms. *Chem. Biol.* 2008, *15*, 875.

27. Yan, X.; Wang, J.; Zhu, L.; Lowrey, J.J.; Zhang, Y.; Hou, W.; Dong, J.; Du, Y. A ready-to-use, versatile, multiplex-able three-dimensional scaffold-based immunoassay chip for high throughput hepatotoxicity evaluation. *Lab Chip* 2015, *15*, 2634–2646.

28. Gong, X.Q.; Yi, X.; Xiao, K.; Li, S.; Kodzius, R.; Qin, J.H.; Wen, W.J. Wax-bonding 3D microfluidic chips. *Lab Chip* 2010, *10*, 2622–2627.

29. Sun, Y.N.; Chen, X.D.; Zhou, X.G.; Zhu, J.B.; Yu, Y.D. Droplet-in-oil array for picoliter-scale analysis based on sequential inkjet printing. *Lab Chip* 2015, *15*, 2429–2436.

30. Neugebauer, S.; Evans, S.R.; Aguilar, Z.P.; Mosbach, M.; Fritsch, I.; Schuhmann, W. Analysis in ultrasmall volumes: Microdispensing of picoliter droplets and analysis without protection from evaporation. *Anal. Chem.* 2004, *76*, 458–463.

31. Cui, L.W.; Miao, J. Chromatin-mediated epigenetic regulation in the malaria parasite plasmodium falciparum. *Eukaryot. Cell* 2010, *9*, 1138–1149.

32. LaCount, D.J.; Vignali, M.; Chettier, R.; Phansalkar, A.; Bell, R.; Hesselberth, J.R.; Schoenfeld, L.W.; Ota, I.; Sahasrabudhe, S.; Kurschner, C.; et al. A protein interaction network of the malaria parasite *Plasmodium falciparum*. *Nature* 2005, *438*, 103–107.

33. Dekker, F.J.; van den Bosch, T.; Martin, N.I. Small molecule inhibitors of histone acetyltransferases and deacetylases are potential drugs for inflammatory diseases. *Drug Discov. Today* 2014, *19*, 654–660.

34. Chimenti, F.; Bizzarri, B.; Maccioni, E.; Secci, D.; Bolasco, A.; Chimenti, P.; Fioravanti, R.; Granese, A.; Carradori, S.; Tosi, F.; et al. A novel histone acetyltransferase inhibitor modulating Gcn5 Network: Cyclopentylidene-[4-(4′-chlorophenyl)thiazol-2-yl)hydrazone. *J. Med. Chem.* 2009, *52*, 530–536.

35. Cui, L.; Miao, J.; Furuya, T.; Fan, Q.; Li, X.; Rathod, P.K.; Su, X.Z.; Cui, L. Histone acetyltransferase inhibitor anacardic acid causes changes in global gene expression during in vitro *Plasmodium falciparum* development. *Eukaryot. Cell* 2008, *7*, 1200–1210.

36. Miao, J.; Fan, Q.; Cui, L.; Li, X.L.; Wang, H.Y.; Ning, G.; Reese, J.C.; Cui, L.W. The MYST family histone acetyltransferase regulates gene expression and cell cycle in malaria parasite *Plasmodium falciparum*. *Mol. Microbiol.* 2010, *78*, 883–902.

37. Fidock, D.A.; Wellems, T.E. Transformation with human dihydrofolate reductase renders malaria parasites insensitive to WR99210 but does not affect the intrinsic activity of proguanil. *Proc. Natl. Acad. Sci. USA* 1997, *94*, 10931–10936.

38. Miao, J.; Li, J.F.; Fan, Q.; Li, X.L.; Li, X.Y.; Cui, L.W. The Puf-family RNA-binding protein PfPuf2 regulates sexual development and sex differentiation in the malaria parasite *Plasmodium falciparum*. *J. Cell. Sci.* 2010, *123*, 1039–1049.

39. Trager, W.; Jensen, J.B. Human malaria parasites in continuous culture. *Science* 1976, *193*, 673–675.

40. Zhang, J.H.; Chung, T.D.Y.; Oldenburg, K.R. A simple statistical parameter for use in evaluation and validation of high throughput screening assays. *J. Biomol. Screen.* 1999, *4*, 67–73.

41. Secci, D.; Carradori, S.; Bizzarri, B.; Bolasco, A.; Ballario, P.; Patramani, Z.; Fragapane, P.; Vernarecci, S.; Canzonetta, C.; Filetici, P. Synthesis of a novel series of thiazole-based histone acetyltransferase inhibitors. *Bioorgan. Med. Chem.* 2014, *22*, 1680–1689.

42. Shukla, A.; Singh, A.; Singh, A.; Pathak, L.P.; Shrivastava, N.; Tripathi, P.K.; Singh, M.P.; Singh, K. Inhibition of *P. falciparum* PFATP6 by curcumin and its derivatives: A bioinformatic study. *Cell Mol. Biol.* 2012, *58*, 182–186.

43. Kurogi, K.; Yan, H.; Tsujii, K. Importance of pinning effect of wetting in super water-repellent surfaces. *Colloid Surface A* 2008, *317*, 592–597.

44. Gao, L.C.; McCarthy, T.J. Contact angle hysteresis explained. *Langmuir* 2006, *22*, 6234–6237.

45. Oliver, J.F.; Huh, C.; Mason, S.G. Resistance to spreading of liquids by sharp edges. *J. Colloid Interface Sci.* 1977, *59*, 568–581.

46. Fowkes, F.M.; Harkins, W.D. The state of monolayers adsorbed at the interface solid-aqueous solution. *J. Am. Chem. Soc.* 1940, *62*, 3377–3386.

47. Rodriguez-Valverde, M.A.; Tirado Miranda, M. Derivation of Jurin's law revisited. *Eur. J. Phys.* 2011, *32*, 49–54.

48. You, L.Y.; Nie, J.Y.; Sun, W.J.; Zheng, Z.Q.; Yang, X.J. Lysine acetylation: Enzymes, bromodomains and links to different diseases. *Essays Biochem.* 2012, *52*, 1–12.

14 CHAPTER

Reconfigurable Microfluidic Magnetic Valve Arrays: Towards a Radiotherapy-Compatible Spheroid Culture Platform for the Combinatorial Screening of Cancer Therapies

Alexandre R. Brunet [1,2,3], **Frédérique Labelle** [4], **Philip Wong** [2,3,5] **and Thomas Gervais** [1,2,3,4]

[1] *Biomedical Engineering Institute, École Polytechnique de Montréal, Montreal, H3T 1J4, Canada*

[2] *Institut du Cancer de Montréal, Université de Montréal, Montréal, H3T 1J4, Canada*

[3] *Centre de Recherche du Centre Hospitalier de l'Université de Montréal, H2X 0A9, Montréal, Canada*

[4] *Department of Engineering Physics, École Polytechnique de Montréal, Montreal, H3T 1J4, Canada*

[5] *Department of Radiation Oncology, Centre Hospitalier de l'Université de Montréal, Montréal, H2X 0A9, Canada*

ABSTRACT

We introduce here a microfluidic cell culture platform or spheroid culture chamber array (SCCA) that can synthesize, culture, and enable fluorescence imaging of 3D cell aggregates (typically spheroids) directly on-chip while specifying the flow of reagents in each chamber via the use of an array of passive magnetic valves. The SCCA valves demonstrated sufficient resistance to burst (above 100 mBar), including after receiving radiotherapy (RT) doses of up to 8 Gy combined with standard 37 °C incubation for up to 7 days, enabling the simultaneous synthesis of multiple spheroids from different cell lines on the same array. Our results suggest that SCCA would be an asset in drug discovery processes, seeking to identify combinatorial treatments.

Keywords: *microfluidics; radiotherapy; chemotherapy; spheroids; drug discovery; valves*

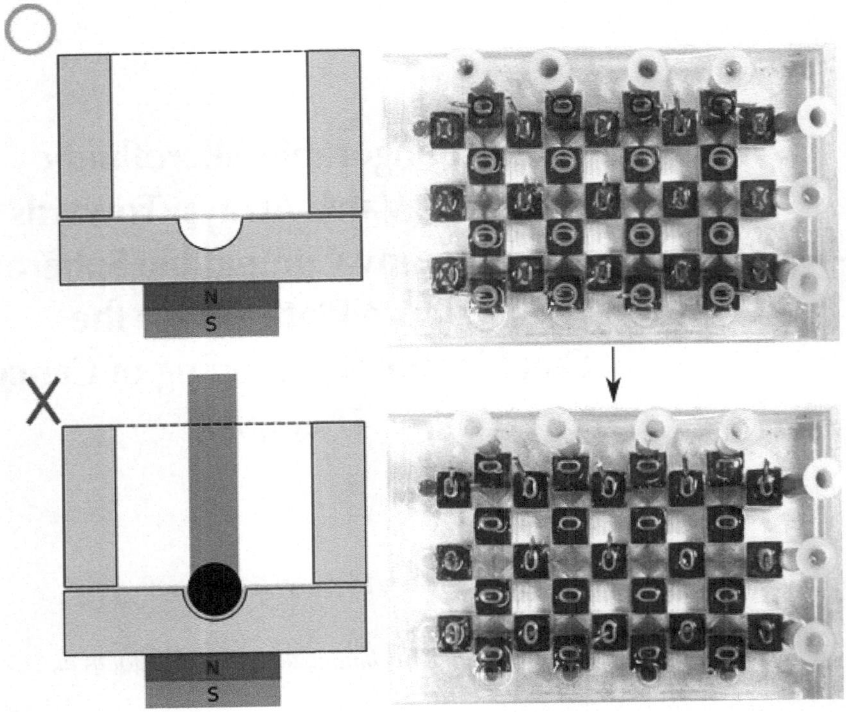

Graphical Abstract

1. INTRODUCTION

Cancer treatment has evolved with an increasing number of molecular targeted agents being approved for different diseases [1]. Molecular targeted agents are often developed as monotherapies first, then explored in combination with standard cytotoxic chemotherapies (CT), but seldom with radiotherapy (RT). Nevertheless, there have been examples where combinations of drug/drug [2] or drug/RT [3,4,5,6] during a treatment led to a synergistic effect to either reduce drug toxicity [3] and/or induce a stronger response against cancerous cells than when treatments were used independently. Such results show that some drugs, whose development was halted because a lack of efficacy as monotherapy, could still hold the potential to treat cancer when combined with other drugs or RT. Furthermore, it is well established that cancer cell in monolayer culture tend to be significantly more sensitive to RT than typical in vivo tissue [7,8,9,10]. Therefore, tridimensional (3D) models, such as spheroids, are expected to better assess treatment efficacy as they present more cell-to-cell interactions [8,9,10,11] and could thus hold the key to improved drug screening.

In addition, in the context of microfluidics, spheroids of various sizes can easily be obtained in large numbers [10,12,13] at relatively low costs. In devices made exclusively of polydimethylsiloxane (PDMS), the natural permeability of the material to gas (O_2 and

CO_2) permits culture medium replenishment every 24 h or more instead of requiring a constant perfusion [13,14].

In order to study treatment synergy in a high throughput format, spheroid forming microfluidic chips must also enable combinatorial testing of drug interactions. Microfluidics enable the use of valves that can separate culture chambers, creating an M × N array A for which each element amn corresponds to a different culture condition [15,16]. An example of how CT and RT can be performed on a spheroid culture chamber array (SCCA) is shown in Figure 1a.

Microfluidic valves can be divided into actively and passively actuated, the former requiring an external power source to be actuated. Active valves can generally be remotely controlled and can be actuated via pressure [15,16,17,18], heat [18,19], or magnetism using solenoids [18,20,21]. Many active valves have already been integrated into arrays and multiplexers for monolayer cultures [15,16]. However, since they need external power sources to be actuated, they are more expensive and it can be challenging to have them operated by non-specialists. On the other hand, passive microfluidic valves do not need external power sources to be operated, making them easier to handle. They are typically actuated mechanically, chemically or magnetically. Mechanic valves are typically closed by a user that presses on or screws into the actuator [19]. Chemically actuated passive valves respond to a change in concentration of an effector (typically pH) in the surrounding fluid [22]. Both types are slower to operate than magnetic valves, which are usually more compact, resist to higher pressures and are easily toggled between opened/closed states [23,24,25].

In this study, we designed and fabricated a microfluidic SCCA entirely of PDMS (Figure 1b) combined with a passive magnetic valve array that enables 3D cell cultures. Each culture chamber can grow up to 25 spheroids under the same condition. We demonstrate a simple valve design that enables easy manipulation and reversible separation of culture chambers within the device at low cost, which could lead to study of up to 12 treatment conditions on the same SCCA.

To ensure reliability of the valves, their resistance to pressure was tested under various conditions relevant to cell culture (different incubation times and RT doses). The diffusion of small particles across the valves was also studied in order to ensure complete separation of samples during a study. Finally, flow routing and spheroid culture experiments were performed in the device to demonstrate its potential.

2. MATERIALS AND METHODS

2.1. Description of the Design

2.1.1. Valve Design

The designed passive magnetic valve was inspired by Gaspar et al. [25] and uses a permanent magnet and a metal rod for actuation, as shown in the side view (Figure 1c). The actuation of the valve is triggered by inserting the metal rod into an opening on top of the channel (Figure 1d) and a magnet underneath. The attraction between the two deforms a membrane, closing the valve. This design was selected for its low cost and complexity level, compactness and high resistance to pressure. Valves are also individually

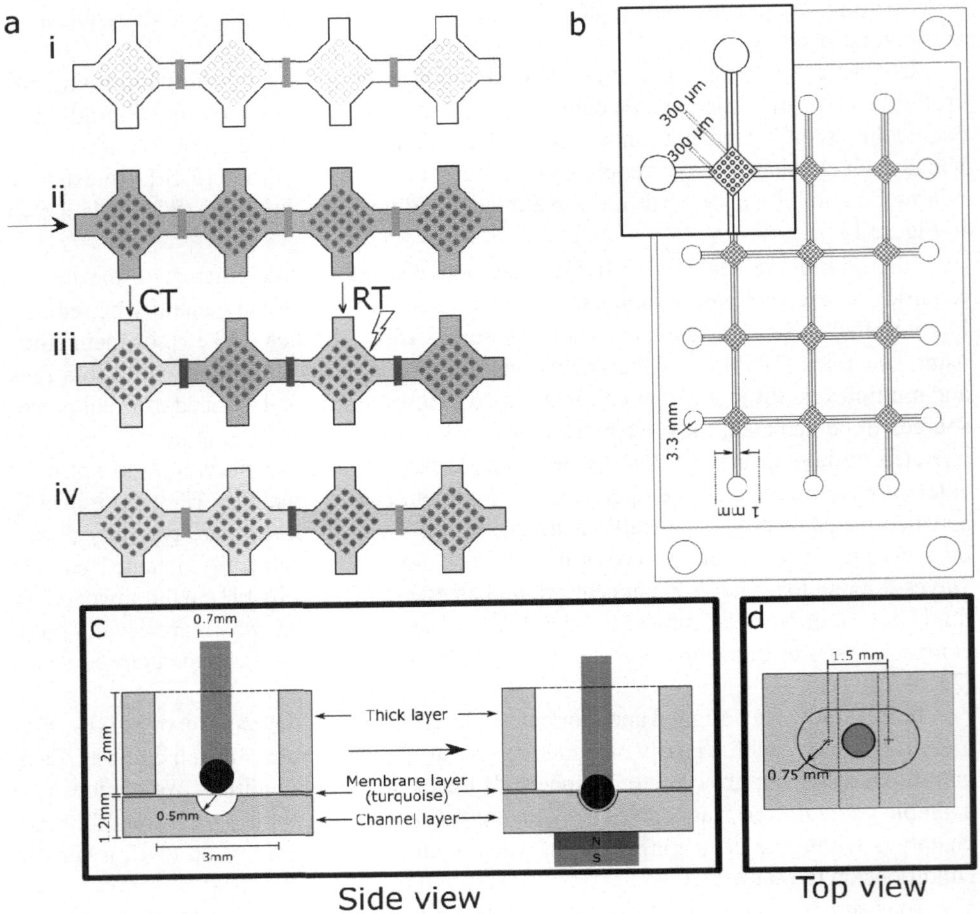

Figure 1. *Elements involved in a spheroid culture chamber array (SCCA) design. (a) Working principle of a microfluidic SCCA: (i) all valves are open (shown in green between each culture chamber containing 5 × 5 wells) to fill the device with a cell suspension; (ii) cells are left to sediment in each well to form spheroids; (iii) some valves are closed (depicted in red), so that different treatments (chemotherapy (CT) or radiotherapy (RT)) can be applied to different columns of culture chambers independently; (iv) each valve can be opened independently to create a specific flow path for a reagent through the device; (b) Mold used for SCCA's channel layer; (c) Side view of the microfluidic valves used: when a magnet is inserted under the channel, the magnetic polydimethylsiloxane (mPDMS)-tipped (black) metal rod gets pulled down, deforming the membrane and closing the channel; (d) Top view of microfluidic valves used, the ovaloid opening represents the hole through which the metal rod (grey) is inserted.*

addressable to increase the versatility of the system. The valve consists of a first layer of 1.2 mm of PDMS (channel layer) that contains a semi-cylindrical microchannel (radius of 0.5 mm) of at least 2 mm, topped with a 100 μm-thick PDMS membrane (turquoise line in Figure 1d) and another layer of 2 mm of PDMS (thick layer) that contains a 1.5

mm × 3 mm ovaloid opening. To create a conformal seal of the channel by the membrane, the tip of the metal rod was rounded by adding a drop of magnetic PDMS (mPDMS, see Section 2.2.4).

2.1.2. Spheroid Culture Chamber Arrays (SCCA)

We developed a 5.06 × 7.62 cm^2 (2 × 3 in^2) SCCA that contained four vertical and three horizontal semi-cylindrical channels (valves), each interconnected and a few millimeters long (Figure 1b). At the end of each channel, there were 3.3 mm inlets and at the junction of each channel were diamond-shaped culture chambers. Cells were cultured into spheroids in each chamber, which consisted of 25 (5 × 5) cubic square wells of 300 μm on a side separated by 300 μm. The size of the wells was chosen as small as possible to match the resolution of the printer and to be able to keep the channel layer as thin as possible to maximize the magnetic force. Between each chamber was a valve to isolate culture media, resulting in a flexible flow control in the device. The chambers were separated in order to be compatible with the spatial resolution of RT devices and to facilitate manipulation of the valves.

2.2. Fabrication and Assembly

2.2.1. Molds and Substrates

SCCAs were fabricated out of PDMS (Sylgard 184 Silicone Elastomer Kit, Dow Corning, Auburn, MI, USA). First, the molds were designed on a CAD software (Catia V5R20, Dassault Systèmes, Vélizy-Villacoublay, France) and the STL files were then processed using the Asiga software (Asiga Composer 1.2.3, Alexandria, Australia). A layer thickness of 50 μm was chosen and PLASClear v2.0 was selected as the resin. The parts were then 3D printed individually by using an Asiga Pro2 for the SCCA and an Asiga Freeform Pico 39 for the individual test valves (ITV) to evaluate resistance to burst. Both were post-processed as recommended by Comina et al. [26]. Briefly, the device was dehydrated in two consecutive 5 min baths in isopropanol, followed by a 5 min curing step under UV light, a 5 min incubation in a heated 70% ethanol ultrasonic bath, a quick dry using a nitrogen gun and an application of a thin layer of paint (Aztek opaque black, Testors, Vernon Hills, IL, USA) using a generic dual-action gravity feed airbrush. This post-processing was necessary in order to harden the molds and allow PDMS to cure properly during the molding step.

2.2.2. Device Molding and Assembly

The thick layer and channel layer were constructed by pouring PDMS mixed at a 10:1 ratio into the 3D printed molds and allowing it to cure at 80 °C for 1 h (Figure 2a). For the SCCA, the molds were screwed to a polymethylmetacrylate (PMMA) plate to prevent the PLASClear from curling up when heated and minimize variation in the thickness of the layers. The substrate used for membrane fabrication was a 4-inch glass wafer that was silanized for 24 h by using trichloro(octadecyl)silane under vacuum. PDMS mixed at a 20:1 ratio was poured over the silanized glass wafer to create the membrane layer (Figure 2b). Afterwards, the wafer was spin-coated using a Cee 200CBX (Brewer Science, Rolla, MO, USA) at 300 rpm for 60 s and was cured at room temperature for 72 h.

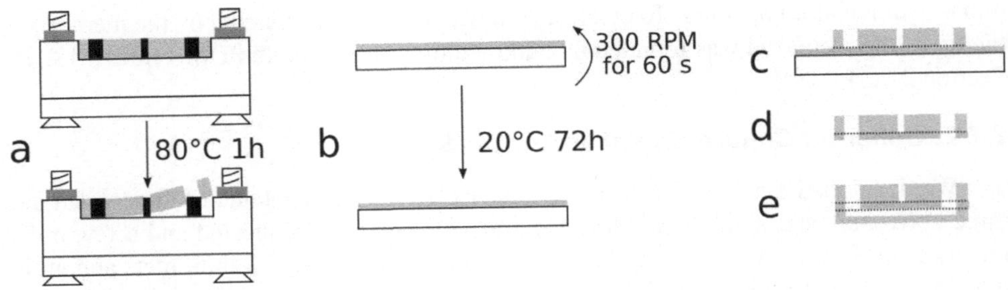

Figure 2. *Fabrication and assembly of the spheroid culture chamber array (SCCA): (a) The 3D printed molds of the thick and channel layers were screwed onto polymethylmetacrylate (PMMA) plates, polydimethylsiloxane (PDMS) at 10:1 was poured inside the molds, cured at 80 °C for 1 h and removed from the molds; (b) For the membrane layer, PDMS at 20:1 was poured over a silanized glass wafer before being rotated at 300 rpm for 1 min and cured at room temperature for 72 h; (c) The thick layer and the membrane layer (still mounted on its substrate) were plasma bonded together; (d) Holes corresponding to the inlets/outlets were punched into the membrane layer using the thick layer as a guide; (e) The other side of the membrane layer and the channel layer were plasma bonded together.*

2.2.3. Device Assembly

Both the thick layer and the channel layer were removed from their respective molds using tweezers. The layers were bonded by following the sequences showed in Figure 2c–e and by using an atmospheric plasma (Dyne-A-Mite, Enercon, Menomonee Falls, MI, USA). The surfaces to be bonded were first exposed to the plasma for 60 s for SCCA and 30 s for the ITV and then pressed together. First, the thick layer was bonded on the membrane (still attached to its substrate; Figure 2c), then the assembly was detached from the substrate using a scalpel, and holes were punched into the membrane at the spots corresponding to the inlets/outlets (Figure 2d). Afterwards, the membrane/thick layer were manually aligned and bonded to the channel layer using the atmospheric plasma (Figure 2e).

2.2.4. Valves Reusable Parts

The metal rods were created from sections of paper clips tipped with drops of mPDMS. First, a section of 2 cm of the paper clip was cut using metal scissors, producing a metal rod with a flat and a sharp end. The flat end of the metal rod was dipped in uncured mPDMS, obtained by mixing 10:1 PDMS with magnetite (Fe_3O_4, Alpha Chemicals, Cape Girardeau, MO, USA) at a 50/50 weight ratio [20,27]. The metal rod was then put in an oven at 80 °C for 1 h.

The magnets used for this project were 6.35 mm (1/4 in) square neodymium magnets (B444-N52, K&J Magnetics, Pipersville, PA, USA) and chosen because of their small size and high pull force. In order to place the magnets directly under the corresponding valves, a magnet-holder of PMMA was designed. The file for the magnet-holder was generated in Inkscape (Version 0.91, Inkscape: Open Source Scalable Vector Graphics Editor) before being uploaded to a Trotec Speedy 300 CO_2 laser cutter (Trotec, Marchtrenk, Austria), which was used to cut the part from a 6.35 mm PMMA plate. Once the plate was made, the magnets were inserted manually to ensure that each magnet's polarity would be the opposite of the adjacent magnets. The metal rods, neodymium magnets and magnet holder were all reusable and could be sterilized using conventional methods, even though this was not tested in the present study, since these parts were never in direct contact with fluids or biologic material.

2.3. Experimental Procedure

2.3.1. Experimental Protocol for Valve Resistance

The ITVs (Figure 3b) were used to evaluate the valves' resistance to applied pressures and diffusion of small molecules. Through the course of the experiments, the procedure described in Table 1 was followed, with the valves being left open and the devices left untouched during incubation, unless otherwise specified.

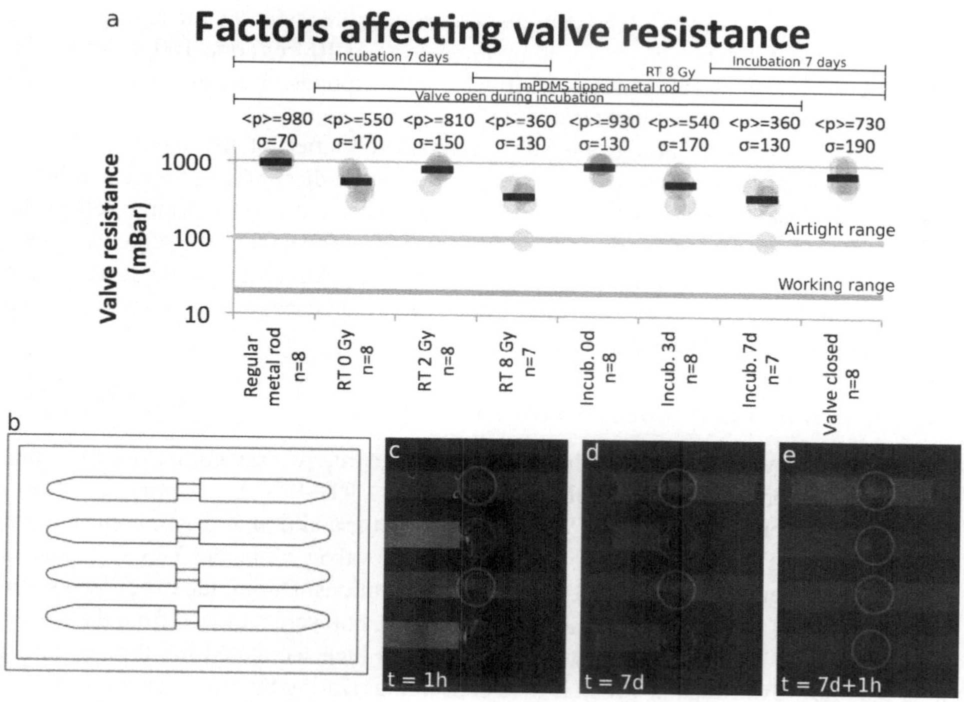

Figure 3. *Characterization of valve resistance to flow and diffusive mass transfer (all data shown) (a) Effect of various parameters (type of metal rod used, radiotherapy (RT) dose, incubation time and state of valve in the incubator) over valve resistance in mBar (1 mBar = 1 cm of H_2O); (b) Schematics of device channel layer mold used for individual test valves (ITV); (c–e) Small molecule (333 Da) diffusion across the closed valve. The valves in green represent open valves, while the ones in red are closed.*

Table 1. *Device preparation and experimental procedure.*

Day 0	Assembly of the device Air bubbles removal through manual filling with 70% ethanol and isopropanol Injection of pluronic solution (10 mg/mL) Incubation for 24 h at room temperature
Day 1	Flush with water to remove excess pluronic Fill with cell medium (DMEM/F12) RT treatment of 0 to 8 Gy (8 Gy considered default)
Day 1–X	Incubation at 37 °C for 0 to 6 days (6 days considered default)

RT was performed using a Gammacell 3000 (Best Theratronics, Kanata, ON, Canada). Valve resistance was tested on day 7 for each device by using an AF1 dual unit pressure pump (Elveflow, Paris, France) and visualized in bright field using a stereoscope (Olympus SZX16, Olympus, Tokyo, Japan). To observe valve resistance, the pressure pump was connected to one end of the channel and liquid was allowed to fill the channel until it crossed the valve. The pump was stopped while closing the valve and the applied pressure increased until a movement of the meniscus was perceptible under the microscope. Every 20 s, the applied pressure was increased by increments of 10 from 0 to 100 mBar and by increments of 100 from 100 to 1000 mBar, which corresponds to the maximum pressure attainable with the pump that was used.

To observe diffusion through the valve, immediately after RT treatment, half of the valves (two out of four) in each device were closed and a drop of a fluorescein solution (M422-05, JT Baker, Center Valley, PA, USA) was deposited into each inlet (left of the device). The systems were then imaged in green fluorescence using the aforementioned stereoscope, and put into the incubator at 37 °C, 5% CO_2. After 7 days, the devices were removed from the incubator and imaged with the same stereoscope. The valves were then opened and the systems were imaged one hour later.

2.3.2. Experimental Protocol for SCCA

For the SCCA validation, experiments were divided into two categories: one for actual cell culture and one for flow control (using food dyes). The SCCA was prepared by following the same procedure as the ITVs (Table 1), with the addition of plastic connectors at each inlet/outlet. For the flow control experiments, the valves of the device were adjusted in different configurations and filled with food dye solutions. During the experiments, the plastic connectors were left in the inlets (top and right), but were removed from the outlets (bottom and left) to induce a variation of pressure, which facilitated the flow of liquid rather than flow by diffusion, which is time consuming. During the whole experiment, the outlets were frequently emptied using a pipette in order to maintain fluid flow.

For the cell culture experiments, after RT treatment, the SCCA was seeded with untreated cell suspensions of the ovarian cell lines OV1946 [28] at 2.5 × 10^6 cells/mL and OV90 [29] at 5 × 10^6 cells/mL. These cell lines were genetically engineered to express enhanced green fluorescent protein (OV1946 and OV90) and mCardinal (OV90). The device was incubated at 37 °C, 5% CO_2 for 24 h with all valves within horizontal channels closed. Subsequently, the medium was replaced and the devices were imaged using a

Nikon ECLIPSE TE300 microscope (Nikon, Tokyo, Japan) in bright field and fluorescence using a Nikon super high pressure mercury lamp. Cell medium was replaced every 24 h, before imaging.

3. RESULTS AND DISCUSSION

3.1. Valve Resistance

From the valve resistance results, we determined two limits that described valve behavior under pressure. The *working limit* is defined as a resistance of 20 mBar. We set this limit since a regular device should not exceed 10 cm in length and the pressure that would be generated by tilting the device at 90° should be of approximately 10 mBar (1 mBar being equivalent to 1 cm pressure head in H_2O). By adding a security factor of two, we obtain the said limit of 20 mBar. The *airtight limit* represents a valve resistance of 100 mBar and was defined as the pressure that could be applied by softly pressing the surface of the system as to remove air bubbles. These limits provided restrictions and guidelines on device handling.

To validate the valve design and its function, we compiled all data from valve resistance (Figure 3). Mainly four factors can influence valve resistance: the metal rod used, the RT dose, the incubation time and the valve state (opened/closed) in the incubator. Each of these aspects was tested in ITVs made of four separate channels shut by a 2-mm valve placed at mid-point in each channel (Figure 3b). PDMS delamination at the edges of the main channel occurred at pressures nearing 500 mBar when the bonding between the PDMS layers failed before valve burst (around 5% of cases). In such cases, the valve resistance was defined as the highest pressure attained at a safe level before device rupture. In 50% of the results for pressures of 900 mBar and above, the valves appeared to resist the maximum pressure (1000 mBar) for a few seconds, but then leaked quickly. This could be attributed to a delamination of the PDMS around the valve. Although the valve itself would not directly cause this, valve resistance was again defined, using a worst case scenario approach, as the pressure level before the leak was detected (900 mBar).

3.1.1. Pressure Resistance

As shown in the two first columns of Figure 3a, the use of a regular metal rod resulted in a higher valve resistance than the ones tipped with mPDMS. This was to be expected, since a higher iron density corresponds to a higher pull force. However, this increase in resistance was offset by a greater tendency for the valves to tear along the rod's edges. The tearing was observed when the ITVs were exposed to longer incubation times or higher temperatures (data not shown). Since valve robustness and reusability is a more stringent design constraint than burst pressure, which is easily achieved experimentally (Figure 3a), soft, round mPDMS tipped rods offer a significant improvement in device performance.

Figure 3a demonstrates the relationship between valve resistance, RT dose and incubation time. As the dose increased from 2 to 8 Gy, the valve resistance decreased. This effect is in contradiction with what was observed by Briganti et al. [30], where a slight and non-significant difference between irradiated and untreated PDMS was observed We hypothesize that these changes could be attributable to an increase in polymer cross-linking in the bulk of the PDMS leading to an increase in brittleness, as observed elsewhere [31]. At RT doses of up to 8 Gy, the device was systematically capable of withstanding pressures over 100 mBar, which represented the airtight limit.

We also tested the effect of incubation length at 37 °C on the valve resistance. There appeared to be a direct correlation between time spent in the incubator and valve resistance. This relationship was expected as PDMS hardens over time and the curing process is greatly sped up when the temperature increases. Figure 3a also shows that the device resistance stayed over 100 mBar, the airtight limit, for a period of up to 7 days in the incubator, which was compatible with typical cell culture conditions. These results demonstrated that the valve resistance of our device achieved airtight limits even after being exposed to RT doses of 8 Gy and 7 days incubation at 37 °C.

The last aspect to consider when evaluating valve resistance was the state of the valve between uses. The two last columns of Figure 3a show that if a valve was kept in closed state during incubation, the resistance was greater than if kept in open state. When kept in closed state for a long period of time, the PDMS membrane took and held the shape of the channel, causing it to become more resistant in future operations. In order to ensure that the channels kept in closed position were still usable, we filled them with an aqueous solution and did not notice any impact on the flow when compared to the ones that were kept opened. This result showed that, whenever possible, valves should be kept in closed state during incubation, which is normally desired to ensure reagent segregation inside the chambers.

3.1.2. *Diffusion Resistance*

Figure 3c–e shows the difference in diffusion-based transport of a fluorescein solution between opened and closed valves. A fluorescein solution that was loaded into the inlets (left) of each channel while the valves were left opened (first and third channels) or closed (second and fourth channels). As shown in Figure 3c, one hour after the injection, fluorescein had diffused across the whole channels when valves were kept open, but could not reach the far side of the closed ones. One week later (Figure 3d), the far side still remained untouched by fluorescein, but air (black meniscus) was present in the channel. After opening the valves, diffusion occurred in the whole channels (Figure 3e), showing that the channel could still be opened. The test conclusively demonstrated that the valves can prevent the diffusion of molecules of at least 333 Da (molecular mass of fluorescein) [32] and, consequently, could likely resist the diffusion of most biological signaling factors, such as cytokines (8–50 kDa), implicated in a variety of cellular pathways and signaling. This experiment also highlighted the occurrence of evaporation in the device, which could be explained by the breathability of PDMS. This side effect was compensated in cell culture experiments by changing the medium in the system every 24 h.

3.2. Fluid Routing Using Valve Arrays in the SCCA

After characterizing valve performance on ITVs, we proceeded to determine their collective behavior to reroute flows inside the SCCA. We applied a sequence of changes in the path conformation of channels of a SCCA to reroute aqueous solutions. Figure 4 shows the selection of flow paths realized during the course of a given experiment where four different reagents (illustrated by the dyes) were used to combinatorially generate four different reagent conditions in individual chambers (see Supplementary Video S1). It was possible to fill the entire device from one inlet (Figure 4b), the vertical inlets (Figure 3c) and the horizontal inlets (Figure 3d). No reagents leaked or followed a path other than the desired one. It would also be possible to do RT directly on chip with the use of a linear accelerator that can target individual compartments within the SCCA (not shown). In theory, up to M × N different reagents, cell lines, and RT doses combinations could be created using this matrix array.

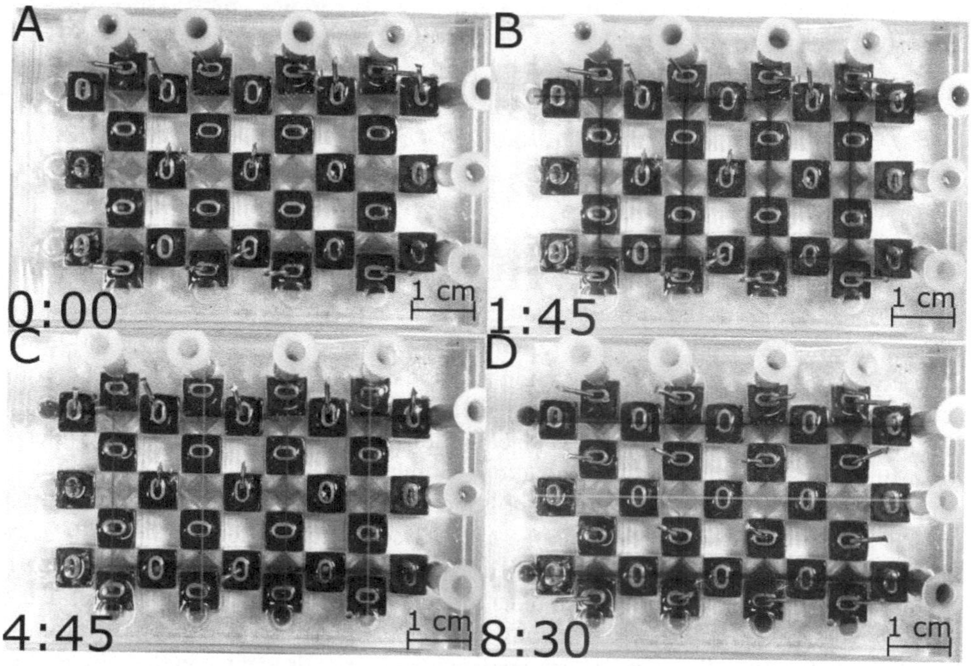

Figure 4. *Screen captures from a video (see Supplementary Video S1) illustrating a typical flow rerouting experiment inside a SCCA (A) Device before loading at t = 0 min; (B) Loading of the device with blue dye through one inlet (top right) and one outlet (top left) at t = 1.75 min; (C) Vertical loading (top to bottom) of the device with red and green dyes at t= 4.75 min; (D) Horizontal loading (right to left) of the device with blue and yellow dyes at t = 8.5 min. The arrows indicate the direction of the flow in the channel and the color of the aqueous solution used.*

The SCCA enabled the culture and bright light and fluorescent imaging of spheroids from different cell lines on one device. As illustrated in Figure 5, spheroids of OV1946 and OV90 were cultured and imaged directly on the same chip after 24 h, after change of cell medium. Both cell suspensions formed spheroids in the SCCA within 24 h. The cell density of the suspension injected in the system controlled the size of the spheroids and the dimensions of the well-defined maximum size attainable. Around 25% of the wells were empty by the time of the imaging, which could be caused by insufficient trapping, due to the size of the spheroids formed. In order to grow bigger spheroids, the size of the wells and the suspension density should be increased, since optimal trapping is achieved for spheroids whose diameter is at most half the size of the wells [14]. However, such a change affects the distance between the metal bar and the magnet and would require a slight increase in the magnetic field to maintain similar valve resistance and reliability.

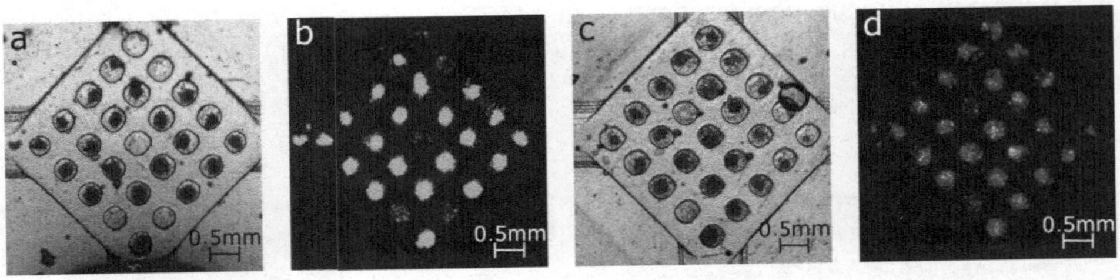

Figure 5. *Cell culture in chamber array after 24 h incubation and after change of medium: OV1946 seeded at 2.5 × 10⁶ cells/mL in (a) bright field and (b) fluorescence; OV90 seeded at 5 × 10⁶ cells/mL in (c) bright field and (d) fluorescence.*

4. DISCUSSION AND CONCLUSIONS

We have designed a microfluidic cell culture chamber array with passive magnetic valves that can be manipulated individually and resist pressures of at least 100 mBar after exposure to RT doses of 8 Gy and incubation at 37 °C for up to a week. We evaluated our device for the growth of spheroids and assessment of RT treatments directly on-chip. Our device was compatible with conditions that permit cell culture experiments for testing different treatment combinations simultaneously. It holds the potential to simultaneously study any M × N number of treatment conditions on the same chip, even though a 3 × 4 array was defined as the optimal configuration for a 5.06 × 7.62 cm^2 footprint. The current bottleneck in chip fabrication is the membrane curing time (72 h), yet these dead times could be avoided in large scale manufacturing, as membranes could be pre-synthesized. Thus, the SCCA using permanent magnetic valve array is a promising platform for drug discovery or personalized medicine for identifying combinatorial cancer treatments on spheroid forming cells, which may find applications in other fields of microfluidics where magnetic valving needs to be combined with a see-through configuration for chip imaging.

Acknowledgments

We thank Anne-Marie Mes-Masson and Adriana Orimoto for supplying the ovarian cell lines used. We acknowledge the Cancer Research Society for an operating grant (#20276). A.R.B. acknowledges funding from a Canderel Scholarship. P.W. acknowledges funding from the Fonds de Recherche Québec Santé. T.G. acknowledges funding from the Natural Sciences and Engineering Council of Canada (NSERC).

Author Contributions

A.R.B. designed the system, performed the valve resistance and flow control experiments and wrote the paper; F.L. performed cell culture experiments; P.W and T.G designed the research and wrote the paper.

REFERENCES

1. Mosquera, C.; Maglic, D.; Zervos, E.E. Molecular targeted therapy for pancreatic adenocarcinoma: A review of completed and ongoing late phase clinical trials. *Cancer Genet.* 2016, *209*, 567–581.
2. Reardon, D.A.; Gilbert, M.R.; Wick, W.; Liau, L. Immunotherapy for neuro-oncology: The critical rationale for combinatorial therapy. *Neuro Oncol.* 2015, *17*, vii32–vii40.
3. Stupp, R.; Mason, W.P.; Van Den Bent, M.J.; Weller, M.; Fisher, B.; Taphoorn, M.J.B.; Belanger, K.; Brandes, A.A.; Marosi, C.; Bogdahn, U.; et al. Radiotherapy plus Concomitant and Adjuvant Temozolomide for Glioblastoma. *N. Engl. J. Med.* 2005, *35210*, 987–996.
4. Bonner, J.A.; Harari, P.M.; Giralt, J.; Azarnia, N.; Shin, D.M.; Cohen, R.B.; Jones, C.U.; Sur, R.; Raben, D.; Jassem, J.; et al. Radiotherapy plus Cetuximab for Squamous-Cell Carcinoma of the Head and Neck. *N. Engl. J. Med.* 2006, *354*, 567–578.
5. Wong, P.; Houghton, P.; Kirsch, D.G.; Finkelstein, S.E.; Monjazeb, A.M.; Xu-Welliver, M.; Dicker, A.P.; Ahmed, M.; Vikram, B.; Teicher, B.A.; et al. Combining targeted agents with modern radiotherapy in soft tissue sarcomas. *J. Natl. Cancer Inst.* 2014, *106*, 16–18.
6. Hutchison, S.; Rae, C.; Tesson, M.; Babich, J.W.; Boyd, M.; Mairs, R.J. Evaluation of Melanin-Targeted Radiotherapy in Combination with Radiosensitizing Drugs for the Treatment of Melanoma. *Cancer Oncol. Res.* 2014, *2*, 58–67.
7. Falkenberg, N.; Höfig, I.; Rosemann, M.; Szumielewski, J.; Richter, S.; Schorpp, K.; Hadian, K.; Aubele, M.; Atkinson, M.J.; Anastasov, N. Three-dimensional microtissues essentially contribute to preclinical validations of therapeutic targets in breast cancer. *Cancer Med.* 2016, *5*, 703–710.
8. Morgan, M.M.; Johnson, B.P.; Livingston, M.K.; Schuler, L.A.; Alarid, E.T.; Sung, K.E.; Beebe, D.J. Personalized in vitro cancer models to predict therapeutic response: Challenges and a framework for improvement. *Pharmacol. Ther.* 2016, *165*, 79–92.
9. Kadletz, L.; Heiduschka, G.; Domayer, J.; Schmid, R.; Enzenhofer, E.; Thurnher, D. Evaluation of spheroid head and neck squamous cell carcinoma cell models in comparison to monolayer cultures. *Oncol. Lett.* 2015, 1281–1286.
10. Weiswald, L.-B.; Bellet, D.; Dangles-Marie, V. Spherical Cancer Models in Tumor Biology. *Neoplasia* 2015, *17*, 1–15.
11. Choi, J.; Lee, E.K.; Choo, J.; Yuh, J.; Hong, J.W. Micro 3D cell culture systems for cellular behavior studies: Culture matrices, devices, substrates, and in-situ sensing methods. *Biotechnol. J.* 2015, *10*, 1682–1688.
12. Achilli, T.-M.; Meyer, J.; Morgan, J.R. Advances in the formation, use and understanding of multi-cellular spheroids. *Expert Opin. Biol. Ther.* 2012, *12*, 1347–1360.
13. McMillan, K.S.; McCluskey, A.G.; Sorensen, A.; Boyd, M.; Zagnoni, M. Emulsion technologies for multicellular tumour spheroid radiation assays. *Analyst* 2016, *141*, 100–110.
14. Rousset, N.; Monet, F.; Gervais, T. Simulation-assisted design of microfluidic sample traps for optimal trapping and culture of non-adherent single cells, tissues, and spheroids. *Sci. Rep.* 2017, *7*, 245.
15. Thorsen, T.; Maerkl, S.J.; Quake, S.R. Microfluidic large-scale integration. *Science* 2002, *298*, 580–584.
16. Park, E.S.; Brown, A.C.; DiFeo, M.A.; Barker, T.H.; Lu, H. Continuously perfused, non-cross-contaminating microfluidic chamber array for studying cellular responses to orthogonal combinations of matrix and soluble signals. *Lab Chip* 2010, *10*, 571–580.

17. Unger, M.A.; Chou, H.-P.; Thorsen, T.; Scherer, A.; Quake, S.R. Monolithic Microfabricated Valves and Pumps by Multilayer Soft Lithography. *Science* 2000, *288*, 113–116.

18. Folch, A. *Introduction to BioMEMS*; CRC Press: Boca Raton, FL, USA, 2012; pp. 145–167.

19. Li, A.; Khosla, A.; Drewbrook, C.; Gray, B.L. Fabrication and testing of thermally responsive hydrogel-based actuators using polymer heater elements for flexible microvalves. In Proceedings of the SPIE, San Francisco, CA, USA, 22–27 January 2011; Volume 7929.

20. Jackson, W.C.; Tran, H.D.; O'Brien, M.J.; Rabinovich, E.; Lopez, G.P. Rapid prototyping of active microfluidic components based on magnetically modified elastomeric materials. *J. Vac. Sci. Technol. B Microelectron. Nanom. Struct.* 2001, *19*, 596.

21. Pamme, N. Magnetism and microfluidics. *Lab Chip* 2006, *6*, 24–38.

22. Beebe, D.J.; Moore, J.S.; Bauer, J.M.; Yu, Q.; Liu, R.H.; Devadoss, C.; Jo, B.-H. Functional hydrogel structures for autonomous flow control inside microfluidic channels. *Nature* 2000, *404*, 588–590.

23. Rahbar, M.; Seyfollahi, S.; Khosla, A.; Gray, B.L.; Shannon, L. Fabrication Process for Electromagnetic Actuators Compatible with Polymer Based Microfluidic Devices. *ECS Trans.* 2012, *41*, 7–17.

24. Pirmoradi, F.; Cheng, L.; Chiao, M. A magnetic poly(dimethylesiloxane) composite membrane incorporated with uniformly dispersed, coated iron oxide nanoparticles. *J. Micromech. Microeng.* 2010, *20*, 15032–15037.

25. Gaspar, A.; Piyasena, M.E.; Daroczi, L.; Gomez, F.A. Magnetically controlled flexible valve for flow manipulation in polymer microfluidic devices. *Microfluid. Nanofluid.* 2008, *4*, 525–531.

26. Comina, G.; Suska, A.; Filippini, D. PDMS lab-on-a-chip fabrication using 3D printed templates. *Lab Chip* 2014, *14*, 424–430.

27. Yamanishi, Y.; Lin, Y.-C.; Arai, F. Magnetically modified PDMS microtools for micro particle manipulation. In Proceedings of the 2007 IEEE/RSJ International Conference on Intelligent Robots and Systems, San Diego, CA, USA, 29 October–2 November 2007; pp. 753–758.

28. Ouellet, V.; Zietarska, M.; Portelance, L.; Lafontaine, J.; Madore, J.; Puiffe, M.-L.; Arcand, S.L.; Shen, Z.; Hébert, J.; Tonin, P.N.; et al. Characterization of three new serous epithelial ovarian cancer cell lines. *BMC Cancer* 2008, *8*, 152.

29. Lounis, H.; Mes-Masson, A.M.; Dion, F.; Bradley, W.E.; Seymour, R.J.; Provencher, D.; Tonin, P.N. Mapping of chromosome 3p deletions in human epithelial ovarian tumors. *Oncogene* 1998, *17*, 2359–2365.

30. Briganti, E.; Kayal, T.A.; Kull, S.; Losi, P.; Spiller, D.; Tonlorenzi, S.; Berti, D.; Soldani, G. The effect of gamma irradiation on physical-mechanical properties and cytotoxicity of polyurethane-polydimethylsiloxane microfibrillar vascular grafts. *J. Mater. Sci. Mater. Med.* 2010.

31. Palsule, A.S.; Clarson, S.J.; Widenhouse, C.W. Gamma Irradiation of Silicones. *J. Inorg. Organomet. Polym. Mater.* 2008, *18*, 207–221.

32. Baker, J.T. Fluorescein Bar 100G M422-05. Available online: http://www.labplanet.com/jt-baker-fluorescein-bar-100g-m422-05.html (accessed on 11 July 2017).

INDEX

R

S

T

U

V

W

X

#-1